T0132739

The Equitable Forest

The Equitable Forest

Diversity, Community, and Resource Management

EDITED BY

Carol J. Pierce Colfer

RESOURCES FOR THE FUTURE
Washington, DC, USA

CENTER FOR INTERNATIONAL
FORESTRY RESEARCH
Bogor, Indonesia

Copyright © 2005 by Resources for the Future. All rights reserved.

Printed in the United States of America

No part of this publication may be reproduced by any means, whether electronic or mechanical, without written permission. Requests to photocopy items for classroom or other educational use should be sent to the Copyright Clearance Center, Inc., Suite 910, 222 Rosewood Drive, Danvers, MA 01923, USA (fax +1 978 646 8600; www.copyright.com). All other permissions requests should be sent directly to the publisher at the address below.

An RFF Press book
Published by Resources for the Future
1616 P Street NW
Washington, DC 20036–1400
USA
www.rffpress.org

A copublication of Resources for the Future (www.rff.org) and the Center for International Forestry Research (www.cifor.org)

Library of Congress Cataloging-in-Publication Data

The equitable forest : diversity, community, and resource management / edited by
    Carol J. Pierce Colfer.
        p. cm.
    "An RFF Press book"—T.p. verso.
    ISBN 1-891853-77-5 (cloth : alk. paper) — ISBN 1-891853-78-3 (pbk. : alk. paper)
    1. Forest management—Tropics—Citizen participation. 2. Sustainable forestry
    —Tropics—Citizen participation.   I. Colfer, Carol J. Pierce.
    SD247.E58 2004
    634.9'2'0913—dc22   2004014209

The paper in this book meets the guidelines for permanence and durability of the Committee on Production Guidelines for Book Longevity of the Council on Library Resources. This book was designed and typeset in Bembo and Gill Sans by Chrysalis Editorial. It was copyedited by Carol Rosen. The cover was designed by Devin Keithley. Cover inset photo by Edmond Dounias titled "The converging fate of generations. An old man and a child in a Punan village of the Upper Tubu watershed, East Kalimantan, Indonesia."

The findings, interpretations, and conclusions offered in this publication are those of the contributors. They do not necessarily represent the views of Resources for the Future, its directors, or officers, and they do not necessarily represent the views of any other organization that has participated in the development of this publication.

The geographical boundaries and titles depicted in this publication, whether in maps, other illustrations, or text, do not imply any judgment or opinion about the legal status of a territory on the part of Resources for the Future, the Center for International Forestry Research, or any other organization that has participated in the preparation of this publication.

ISBN 1-891853-77-5 (cloth)   ISBN 1-891853-78-3 (paper)

About Resources for the Future *and* RFF Press

RESOURCES FOR THE FUTURE (RFF) improves environmental and natural resource policymaking worldwide through independent social science research of the highest caliber. Founded in 1952, RFF pioneered the application of economics as a tool for developing more effective policy about the use and conservation of natural resources. Its scholars continue to employ social science methods to analyze critical issues concerning pollution control, energy policy, land and water use, hazardous waste, climate change, biodiversity, and the environmental challenges of developing countries.

RFF PRESS supports the mission of RFF by publishing book-length works that present a broad range of approaches to the study of natural resources and the environment. Its authors and editors include RFF staff, researchers from the larger academic and policy communities, and journalists. Audiences for publications by RFF Press include all of the participants in the policymaking process—scholars, the media, advocacy groups, NGOs, professionals in business and government, and the public.

About the Center for International Forestry Research

THE CENTER FOR INTERNATIONAL FORESTRY RESEARCH (CIFOR) was established in 1993 as part of the Consultative Group on International Agricultural Research (CGIAR) in response to global concerns about the social, environmental, and economic consequences of forest loss and degradation. CIFOR research produces knowledge and methods needed to improve the well-being of forest-dependent people and to help tropical countries manage their forests wisely for sustained benefits. This research is done in more than two dozen countries, in cooperation with numerous partners. Since it was founded, CIFOR has also played a central role in influencing global and national forestry policies.

Resources for the Future

Directors

Catherine G. Abbott	E. Linn Draper Jr.	Michael A. Mantell
Vicky A. Bailey	Mohamed El-Ashry	James F. O'Grady Jr.
Joan Z. Bernstein	Dod A. Fraser	Steven W. Percy
Julia Carabias Lillo	Kathryn S. Fuller	Mark A. Pisano
Norman L. Christensen Jr.	Mary A. Gade	Robert F.X. Sillerman
Maureen L. Cropper	David G. Hawkins	Robert N. Stavins
W. Bowman Cutter	Lawrence H. Linden	Joseph E. Stiglitz
John M. Deutch	Lawrence U. Luchini	Edward L. Strohbehn Jr.
	Jim Maddy	

Officers

Robert E. Grady, *Chair*
Frank E. Loy, *Vice Chair*
Paul R. Portney, *President*
Edward F. Hand, *Vice President–Finance and Administration*
Lesli A. Creedon, *Vice President–External Affairs*

Editorial Advisers for RFF Press

Walter A. Rosenbaum, *University of Florida*
Jeffrey K. Stine, *Smithsonian Institution*

Dedication

To my brother, David Brian Pierce, who, amidst his recurrent goading on gender issues, has steadfastly loved and supported me. I hope he reads this book.

To my two sons-in-law, Stephen Steinert and Stephen Sweitzer, who have brought my daughters, Megan and Amy, such happiness, in their own so very different ways. May they continue to do so!

Contents

Foreword

*T*HE SUBJECT OF EQUITY IS considerably underrepresented, locally and globally, in the discourse and practice relating to development. Yet, it should be at the very center of the development process, whether as ultimate outcome of or as precondition for development. While references to equity—and more often to its absence—can be found at various levels of rhetoric about the state of our civilization, it is astounding to note that the development process often leads to more inequity, not less. Political and policy attention that focuses on this phenomenon, and puts in place appropriate mechanisms to assess and redress such problems, is not readily evident. This book, which explores the concept of equity and related issues within the domain of forests, is therefore a welcome addition to the literature and to the advocacy for equity in analysis, planning, and management of forests.

The concept of equity is as complex as the requirements for its realization. In relation to forests alone, equity encompasses elements of security of tenure and access to resources; stakeholder participation in decisionmaking and management; definition and defense of intellectual property rights; gender-related roles and rewards; and sharing of benefits, including intergenerational ethics. Advancing on any of these fronts requires good empirical analysis of the incidence and determinants of inequity. The cross-regional case studies presented in *The Equitable Forest* from Asia, Africa, and Latin America help to illuminate this understanding—as a prelude to more effective action.

The sharing of power may be regarded as a sine qua non for the sharing of benefits. This opinion recognizes that those who already enjoy economic and political influence are able to secure, amplify, and perpetuate their own individual or group interest relative to those who do not. These processes form a feedback loop that reinforces existing inequities. Carol Colfer (in the Introduction) concludes, on the basis of the cases presented, that "what is clear is the fact that certain segments of populations do not have the straightforward access to either resources or voice that is easily accessible to other parties." No wonder the equity gap is widening! But how can those who begin by being at a disadvantage in the development process break into this closed circle that maintains and strengthens the current conditions? How can development policy and practice more consciously help them to do so? This book explores, in relation to forests, a variety of enabling factors that might allow disadvantaged groups to sustain their livelihoods and their well-being. Thus, case examples presented reveal ways in which conventional approaches, like training villagers and technicians, technical assistance on income generation or marketing, capacity and confidence-building among various stakeholders, governance reforms at the village level, and attention to gender and ethnic differences in natural resource management, *when combined with responsive and flexible facilitation*, can contribute to devolved decisionmaking and management by local communities. The approaches described in this book—emphasizing external facilitation and local collective action—help to reduce the difficult, conceptual discrepancy between the global concerns to enhance equity and self-determination, on the one hand, and the fact that many local systems are inherently inequitable, on the other. These authors assume the importance of both equity and self-determination and the necessity for all the people in communities to fashion their own evolving system together. This book focuses on efforts to move toward the notion of an equitable forest, at a pace and in a direction that is acceptable to local people.

Within such interventions, specific attention needs to be paid to the situation of women. *The Equitable Forest* delivers on this: it reveals the disproportionate responsibility and involvement of women in managing forests, while tracing their lack of involvement in decisionmaking that has denied them the ability to be involved and share in power; it presents cases where this imbalance and inequity are being addressed; it makes powerful advocacy on behalf of historically disadvantaged groups in relation to the sharing of benefits from forests; and it signals the potential for moving along a path of sustainable and equitable forest management if the contribution of such marginalized groups is maximized and appropriately acknowledged.

No automatic solutions to inequitable arrangements are offered by the authors, as it is recognized that sustainable and equitable forest management cannot be divorced from the economic, political, social, and cultural parameters that surround forests. This is amply demonstrated by the particularities of local places in the three continents represented in the book. This heterogeneity calls forth exposition of a variety of theories that seek to explain the political relationships within and among groups, which have implications for realizing equity in forest management; however, no single theory is advocated. There is a suitable emphasis on finding the solution within the context of the local reality.

The contributions remind us that neither forests nor communities are static and that resilience in both natural and human systems is desirable. Thus, the merits of participatory research and adaptive management, in which attention to political as well as technical matters can be addressed, are underscored.

In the final analysis, however, research such as this will be assessed by the extent to which it makes a difference to the hundreds of millions of people who directly depend on forests for their lives and livelihood. It is therefore salutary to note that the authors crystallize promising approaches that would enable forest managers to contribute to that outcome.

ANGELA CROPPER
Chair of the Board of Trustees of the Center for International
Forestry Research and President of the Cropper Foundation

About the Contributors

George Akwah, a Cameroonian anthropologist who was part of the ACM–Cameroon team, is now assistant coordinator for Innovative Resources Management and works in Cameroon and the Democratic Republic of Congo.

Njau Anau is an Indonesian working as a consultant for CIFOR in Bulungan Research Forest, East Kalimantan, Indonesia.

Omaira Bolaños is a Colombian anthropologist from the University of Florida in Gainesville, Florida.

Constance Campbell is a social science and biodiversity advisor at the U.S. Agency for International Development.

Guilhermina Cayres is a Brazilian specialist in sustainable development working as a consultant. She was part of the eastern Amazon ACM team.

Avecita Chicchón is a Peruvian anthropologist working for the John D. and Catherine T. MacArthur Foundation in Chicago, Illinois.

Carol J. Pierce Colfer is an American anthropologist and public health specialist working for CIFOR.

Peter Cronkleton is an American anthropologist who coordinated CIFOR's ACM work in Bolivia and is now working on grassroots networking in South America.

Sushma Dangol is a Nepalese forester working for NORMS, a Nepali NGO, in partnership with CIFOR.

Mariteuw Chimère Diaw is a Senegalese anthropologist working for CIFOR and leading the ACM teams in Cameroon and Ghana.

Stepi Hakim, an Indonesian wildlife biologist and natural resource manager, coordinated CIFOR's research in Pasir, East Kalimantan and is now working as a forestry specialist with the European Union-Ministry of Forestry/Indonesia Forest Liaison Bureau in Jakarta.

Miriam van Heist is an ecologist working as a consultant at CIFOR in Bogor, Indonesia.

Ramses Iwan is a forest specialist for CIFOR in Bulungan Research Forest, East Kalimantan, Indonesia.

Trikurnianti Kusumanto is an agronomist who has been coordinating CIFOR's ACM field research in Jambi, Sumatra since 2000.

Godwin Limberg is an agronomist and a consultant working as a field team leader for CIFOR in Bulungan Research Forest, East Kalimantan, Indonesia.

Frank Matose, a Zimbabwean anthropologist and former CIFOR team leader in Harare, is senior researcher and program manager of the Center for Applied Social Studies, University of Zimbabwe and the Program for Land and Agrarian Studies at the University of the Western Cape program on People-Centered Approaches to Natural Resource Management.

Tendayi Mutimukuru is a Zimbabwean specialist in social learning. She worked as a Participatory Action Researcher with CIFOR in Mafungautsi State Forest in Gokwe District in central Zimbabwe.

Nontokozo Nemarundwe is a Zimbabwean anthropologist who now coordinates CIFOR's ACM work in Zimbabwe.

Joachim Nguiébouri is a field researcher working primarily in the Campo Ma'an National Park as part of the ACM-Cameroon team.

Westphalen Nuñes is a community forester and coordinator for Amazonian issues for the National Fund for the Environment of the Brazilian Ministry for the Environment and part of the eastern Amazon ACM team.

Richard Nyirenda, a Zimbabwean forester who worked as a member of the Zimbabwe ACM team, is now pursuing his M.Sc. at the University of Wales, Swansea.

Phil René Oyono is a Cameroonian sociologist and member of the ACM–Cameroon team.

Richard Piland is an American anthropologist specializing in tropical agriculture and now working as a natural sciences teacher in Chicago, Illinois.

Benno Pokorny, a German forester who coordinated CIFOR's ACM research in Brazil, is now in charge of tropical forestry at the Institute for Forest Management and Silviculture at the Forest Faculty of the University of Freiburg.

Noemi Miyasaki Porro is a Brazilian anthropologist who worked with the ACM team in western Brazil, and is now working as a consultant out of Belem, Brazil.

Marianne Schmink is an American anthropologist and professor at the University of Florida who linked CIFOR scientists with the ACM field team in Acre, Brazil.

Bevlyne Sithole is a Zimbabwean anthropologist who worked as a consultant at CIFOR in Bogor.

Samantha Stone is an American anthropologist from the University of Florida in Gainesville, Florida who conducted her doctoral research in the ACM site. She is now working in Brazil as a consultant on grassroots networking.

Made Sudana is an Indonesian forester working as a consultant for CIFOR in Bulungan Research Forest, East Kalimantan, Indonesia.

Anne Marie Tiani, a Cameroonian ecologist who worked on the ACM–Cameroon team, works for Innovative Resource Management in Yaoundé, Cameroon.

Eva Wollenberg is an American scientist at CIFOR who specializes in community forests, livelihoods, and devolution.

Acknowledgments

*T*HIS VOLUME HAS BEEN PRODUCED with the financial assistance of the following donors, whose contributions we gratefully acknowledge: the Asian Development Bank (ADB, RETA 5812: Planning for Sustainability of Forests through Adaptive Comanagement); the UK Department for International Development (DFID); the European Union (EU, B7-6201 Tropical Forestry Budget Line); the International Tropical Timber Organization (ITTO project pd 12/97 rev. 1(f): Forest, Science, and Sustainability: Bulungan Model Forest); the International Fund for Agricultural Development (IFAD, Making Space for Local Forest Management in Asia Project); the United States Agency for International Development (USAID); the German Agency for Technical Cooperation (GTZ); and, of course, the Center for International Forestry Research (CIFOR). The views expressed herein are those of the authors and can in no way be taken to reflect the official opinion of the donors.

The contributors and I are also grateful to our home institutions. They have all graciously accepted our continued involvement in preparing these analyses and have supported our efforts in many other day-to-day ways.

We would also like to thank the parade of editors and others who have helped us in the production of this book. These include Olivia Vent at Cornell University and the staff of RFF Press: Don Reisman, Jessica Palmer, Vanessa Mallory, Carol Rosen, and John Deever. At the Center for International Forestry Research, we were consistently aided by Rahayu Koesnadi, Atie Puntodewo, Gideon Suharyanto, and Yani Saloh. Cornell University provided an excellent atmosphere in which to pull together these chapters, particularly the Department of Natural Resources, which hosted Colfer during her 2002–2003 sabbatical, and Mann Library.

We would also like to thank, most fundamentally, the people in the organizations and communities with whom we have come in contact in the course of our research. Participatory action research is a joint effort by facilitators and the others involved. To an even greater extent than usual, we are indebted to our collaborators, in all their diversity, for their acceptance of our intrusions and for their ultimate enthusiasm for what we have all been trying to do. We hope and expect they will carry on.

INTRODUCTION

The Struggle for Equity in Forest Management

Carol J. Pierce Colfer

The growing debate on gender and the environment is developing partly in response to the need to move beyond an undifferentiated "community" in thinking about rural environmental change. Importantly, it asks whether and how women's relationship with the environment and its resources is distinct from men's; and consequently, about women's roles and interests in environmental protection (Leach 1994, 23).

*E*VIDENCE OF DIFFERENCES between women's relations with their environment and men's has been available and widely known for several decades. That understanding, however, has neither been widely acknowledged by forest managers nor reflected in formal forest management practice. Acceptance of this difference, and its incorporation into research and practice, has been much greater in the field of agriculture, where research on "women in development" and later on "gender and diversity" has been common. In the last decade, since Leach made this statement, some progress has been made in forestry—though nowhere near enough.

This book recounts a variety of efforts that are currently under way "to move beyond an undifferentiated 'community'" in forest management. Our experience also draws out the important parallels between the problems that have been identified in women's situations compared with men's and in the relations between various kinds of marginalized and elite groups. These range from community members interacting with outside change agents in Brazil, to interactions between Punan hunter gatherers and more "socially acceptable" Dayak groups in Indonesia, to relations among caste and ethnic groups in Nepal. The important roles of those still more powerful—timber companies, local governments, and conservation agencies—will not be missed in this collection. All of these kinds of differences have emerged as significant in attempts to manage forests and other natural resources better. Although these issues are by no means limited to tropical forest conditions, our primary interest has been in that context, and this book reflects that preoccupation.

This introductory chapter sets the stage for the analyses that follow. I first discuss briefly the global conditions that have prompted the research reported here. This is followed by a description of the research program that served as an umbrella for most of the analyses, with reference to pertinent theory. Eight promising techniques used in the research—selected based on local contexts—are then presented. I then introduce the authors, along with a brief synopsis of each chapter. Finally, I discuss some of my own personal perspectives, in light of the call for more self-disclosure from authors writing about gender, diversity, and other social issues.

The Dispossession of Forest Dwellers

Forest communities are often dispossessed, comparatively powerless to deal with intrusions related to government, to conservation, to plantation development and to logging (cf. Colfer and Byron 2001). In our research, we have found widespread confusion in forest areas about access issues such as tenure and use rights, on the one hand, and about "voice" or participation issues, on the other; specifically, who has the rights and responsibilities to manage forests? Some stakeholders—defined as individuals or groups with an interest in a particular forest—have more access and stronger voices than others. We have argued elsewhere that local communities need special attention in the sustainable management of forests (Colfer et al. 1999a).

However, the powerlessness of members of these communities is relative. Communities tend not to be homogeneous, and different segments of these communities have differential access to power in their relations with each other and with outsiders (Antona and Babin 2001; Baviskar 2001; Engel et al. 2001; Leach and Fairhead 2001). Braidotti and others (1994, *113*), for instance, argue that community forestry efforts "[use] the rhetoric of people's participation without differentiating between the often conflicting interests of the peasantry along the axes of class, caste, and gender" (see Rocheleau and Slocum 1995; Shiva 1989; Venkateswaran 1995, for similar observations). Cooke (2002, *115*) notes the same heterogeneity and argues that such in-group/out-group differences suggest a real vulnerability on the part of the out-groups who will need careful attention if participatory research is not to do more harm than good. Our collection is concerned with how such differences affect forest communities' success in dealing with both their local problems and with the various outsiders knocking at their doors. We have a special concern with how these differences affect the process of developing forest management systems that are more adaptive and more collaborative—something that most of the authors in this book were trying to catalyze in their respective locales. We are equally concerned with how the process affects those involved in it.

This volume begins by placing the research effort in a historical and theoretical context. We then cluster the cases geographically, with Part I on Asia (Indonesia and Nepal), Part II on Africa (Cameroon and Zimbabwe), and Part III on South America (Bolivia, Brazil, and Peru). The parallels between gender and diversity issues will be clear in the contributions.

The book concludes with some generalizations drawn from these cases and some suggestions on their relevance for would-be forest managers. The intent is to provide some significant lessons learned.

CIFOR'S ACM Research in Context

In early 1998, two projects at the Center for International Forestry Research (CIFOR) were, unbeknownst to each other, heading in a similar direction.[1] The scientists working on one project, on criteria and indicators (C&I) for sustainable forest management,[2] had come to the reluctant, but perhaps obvious, conclusion that developing good criteria and indicators was insufficient to address the problems confronting us in tropical forests. Something needed to be done to ensure that the *conditions* identified in the C&I *actually existed* in forests and communities. At the same time, members of another CIFOR project, focusing on devolution[3] and livelihoods, were coming to similar conclusions. The focus of that project was on the naturally occurring adaptiveness of traditional or local systems. Forest conditions are continually changing, communities themselves are continually changing, and local communities are adapting to all these changes. We concluded there was scope for collaboration, and we put our respective strengths together in a new and larger program called Local People, Devolution, and Adaptive Collaborative Management of Forests[4]—out of which most of the research reported here has come.

The combined program, which has involved 30 research sites in 11 countries (see brief descriptions of relevant sites at the openings to Parts I, II, and III), involved the use of participatory action research (PAR) as a central feature.[5] ACM facilitators worked with local communities and other stakeholders to catalyze an adaptive and collaborative process designed to improve local human and natural well-being. The first step in our process was to conduct a series of context studies (on stakeholder identification, historical trends, policy, C&I assessment, and an assessment of existing levels of adaptiveness and collaborativeness), followed shortly by initiation of PAR (see Prabhu 2000, for the evolving ACM Handbook; or Hartanto et al. 2003). A central feature of PAR is the direct and proactive involvement of local communities in problem identification, planning, monitoring, and re-planning, in a cyclical loop over time (cf. Allen 2001; Fisher 1995; Greenwood and Levin 1998).

Throughout the research, we first facilitated interaction and collaboration among stakeholders in and around each research site (a horizontal function). A second focus was on strengthening two way communication links between local communities and relevant hierarchies, whether government or business (a vertical function)—recently also called for by Mohan (2002, *164*). Thirdly, our research has emphasized the importance of social learning in management (a diachronic function; cf. Abbott and Guijt 1998; Malla et al. 2001; and collections by Guijt et al. 2000; Wollenberg et al. 2001). Most fundamentally the program has been designed to assess whether or not we can catalyze a process that will empower local communities in their interactions with others, in such a way that both communities and the environment can benefit. We have used a variety of tools (e.g., future scenarios, C&I monitoring, CORMAS, CIMAT, CO-Learn, CO-View, focus groups, pebble sorting, transects, and others)[6]—selected to be compatible with local human conditions—in efforts to bring about the required collective action.

We tried to "practice what we preach," in using adaptive, social learning approaches to planning and decisionmaking within our own program (also called for recently by Taylor, who commented on "the irony of organizations that profess to empower communities but have no equivalent mechanisms for empowering their own staff" (2002, *126*)). We also monitored our own activities to allow for a thorough evaluation of what worked and did not work in the process.

This collection brings together the work of a group of researchers, guided by the broad umbrella of CIFOR's ACM program. Although this program specified a series of steps to be undertaken on the ACM Main Sites (described above), these were not explicitly focused on gender or equity issues. Our efforts in this realm were guided by the central question or concern: How can we facilitate processes that integrate input from the various segments of forest communities in a just or fair way? The diversity that exists in forest communities, and between them and other stakeholders, precludes standardized answers to that question. Our hope, however, has been that by both encouraging researchers to address this issue, and by facilitating maximal freedom to pursue the subject in locally appropriate ways, we might gain insights that could be used more widely. The result has been the potpourri of findings and methods described in this volume; they provide unusually rich insights for forest managers in other contexts.

Although our research was not framed within any one theoretical framework, we were collectively influenced by a number of theoretical perspectives, which I briefly mention here. The recent literature on gender is more abundant than that on the diversity issues addressed in this book. Vandana Shiva (1989) has been a prominent and controversial proponent of the view that women have something special to contribute in natural resource management, because of their unique and inherently close relationship with nature. Sachs (1997, 5) talks about how feminists have moved beyond the familiar liberal, radical and Marxist categories. She talks of women's ways of knowing (after Harding 1986), situated knowledge (after Haraway 1991), and the importance of recognizing the differences in women's experience based on race, class, ethnicity, sexuality and nationality (Sachs 1997, 19–25); (see also Venkateswaran 1957; Guijt and Shah 1998). Rocheleau et al. (1996, 3) summarize the major schools of feminist scholarship and activism relating to the environment as ecofeminists, feminist environmentalists, socialist feminists, feminist poststructuralists, and environmentalists. These editors go on to propose yet another school of thought, feminist political ecology, which draws on ideas from feminist cultural ecology (e.g., Fortmann and Bruce 1988; Leach 1994), political ecology (e.g., Schmink and Wood 1992), feminist geography (e.g., Momsen and Townsend 1987; Nightingale 2003), and feminist political economy (Stamp 1989). We have been particularly influenced by collections relating to women and natural resources (Momsen and Townsend 1987; Fortmann and Bruce 1988; Diamond and Orenstein 1990; Croll and Parkin 1992; Sigot et al. 1995; Thomas-Slayter et al. 1995; Townsend et al. 1995; Rocheleau et al. 1996; Sachs 1997), as well as insightful analyses like those of Braidotti et al. 1994; Leach 1994; Reardon 1995; Venkateswaran 1995; Kurian 2000; Marchbank 2000; and others.

As we examined the relations between various segments within communities, and between community groups and outsiders, we found many parallels to gender relations. Although we found fewer writers on more general equity issues in forests, several have had a serious impact on our thinking. Wollenberg and others (2001) nicely summarize many of the issues that pertain to diversity per se. They identify the central issue as being "how the interests of the least powerful groups can be accommodated." Van den Berg and Biesbrouck (2000) present a nice example of this interest, in looking at the divergences between the needs and interests of pygmies and Bantu forest dwellers in Cameroon (cf. also Chapter 5).

The discussion in Wollenberg et al. 2001 of how to deal with the divergent interests found in forest settings notes two strategies: (a) concentrating on action points that enable multiple positions to coexist and increase opportunities for "win–

win" situations, and (b) focusing on managing conflict through recognizing and coordinating these differences. They say

> In this second strategy, facilitating interaction, communication, learning and the application of a common process for making decisions in ways that are fair to all groups—especially disadvantaged groups—is where the obstacles to accommodation are highest and the need for new insights about effective interventions are greatest (2001, *204*).

The Equitable Forest includes contributions of relevance to both of these strategies. We have collectively struggled with questions relating to the meaning and distribution of power. Should we talk about people having power as one has brown hair (probably not) or a particular skill (more likely), or is it more meaningful to think of power as a current that flows and can thus be switched on or increased? Is power part of a zero-sum game? Nor is it clear what the implications of one interpretation or another are for development practice and specifically for the concept of empowerment (discussed further in the concluding chapter). What *is* clear is the fact that certain segments of populations do not have the straightforward access to either resources or voice that is easily accessible to other parties. Our central concern has been how to render such circumstances—so common in forested areas—more equitable.

We have been suspicious of "formulaic participative technologies" (Hailey 2002) from the start and have embraced the kind of pluralistic perspective encouraged by the writers in the Wollenberg et al. collection (2001). Having been, like many, somewhat put off by the excessive jargon in much postmodern literature, I was surprised to read Rudel and Gerson's (1999) clear summary of the main ideas of postmodernism. According to them, postmodernism includes

- A rejection of comprehensive explanations and grand theories
- Social fluidity, in which ever-changing social conditions require flexibility and adaptability in people, who respond by changing themselves
- In rejecting universal truth, an emphasis on the local and particular
- Polyvocality, or the existence of multiple groups with different concerns and voices
- Attention to the interpretation of signs and texts in which meaning is contingent on social relations.

These ideas are indeed central to our own perspectives, though we are firmly grounded in practical concerns. Pertinent theoretical discussions are provided in a number of the contributions (see particularly Chapters 1 and 4). Subsequent chapters further address central concepts like PAR (Chapter 2), community forestry (Chapter 3), and C&I (Chapter 11).

Eight Promising Approaches for Forest Managers

Here I highlight eight of our promising approaches to strengthening equity in forest management that come out of the research reported here (though there are numerous small ways in which the teams attempt to address the equity issue on a routine basis). The most fundamental point is that each method was selected to respond to an issue in the locale. The teams examined their contexts, identified issues needing attention, and selected appropriate methods to address those issues. These are some of the approaches the team members took:

1. Training in Transformative Learning

The Zimbabwe ACM team working in and around Mafungautsi Forest Reserve[7] decided early on that the community's approach to change was far too passive for the kind of participatory action research we envisioned. After considerable discussion, they opted to bring in a group specializing in Training for Transformation, building on Paulo Freire's approach to learning (Freire 1970; Freire and Freire 1994). This approach maintains that education can either domesticate or liberate; and its proponents work toward inculcating an interest and ability in critically analyzing one's own conditions. The team found that after the training was complete, community members were much more able and willing to contribute their thoughts and ideas about improving forest management. Most specifically, they found dramatic increases in women's involvement in decisionmaking fora (Chapter 9).

2. Devolution of Community-Level Authority to Smaller Groups

In Nepal, the ACM teams[8] found fairly consistent domination of forest management in the community forests by local elites, often in cooperation with the District Forest Officers (consistent with the observations of others, such as Malla 2001 and Malla et al. 2001). These communities were among the most diverse in the whole ACM sample, with up to 18 different ethnic and caste groups, for instance; and their "Great Tradition" is explicitly hierarchical, making the involvement of marginalized groups unusually difficult. After a series of workshops involving local communities and other stakeholders, focusing on criteria and indicators for sustainable forest management (defined to include human well being), the community groups decided to devolve some authority from the central forest user group committees to the hamlet (*tole*) level. These hamlets were typically composed of a single ethnic group and in any event were much more homogeneous in composition than the entire communities. The new approach involved people meeting in their own hamlet to discuss the important issues relating to the annual operating plan for the community forest, for instance. An elected representative from each hamlet then brought these ideas to the forest user group committee. The field teams saw a real increase in involvement by the marginalized groups, including women and the stigmatized caste and ethnic groups (Chapter 3).

3. Use of Multi-Stakeholder Workshops

A process of official decentralization began in Indonesia on January 1, 2001. Our field team[9] in Pasir, East Kalimantan, began their research just as this process was getting under way, which provided them with both problems and opportunities. In response to both the lack of power traditionally available to local communities, and the uncertainty about how to proceed under the new guidelines on the part of government officials, the team was able to convene several multi-stakeholder workshops, specifically focused on planning. At the first such meeting, they worked toward sharing a common vision for the future and planning joint activities designed to contribute to more sustainable use of forests in the area. Several such meetings were held, interspersed with single-stakeholder meetings, with increasing competence on the part of community members and increasing receptivity on the part of the government officials. Particularly thorny issues were avoided at first, but by June 2002, it was even possible to address issues relating to the conflicts between local communities and logging companies (Chapter 2).

4. Use of Multi-Stakeholder Visioning Exercises

Future scenario approaches were used in several countries (Bolivia, Indonesia, Malawi, Zimbabwe) to catalyze a sharing of perspectives among actors. Here we focus on the Bolivian experience, where there was community confusion about the implications of, and the distribution of benefits from, a timber management project. The method described by Wollenberg et al. (2000) was used, in which participants are asked to imagine an ideal future. The community was broken into subgroups of older males, women, and young people. Participants were asked to imagine (and draw) a point, five years in the future, when their timber management plan was working perfectly. They were asked to describe their households, their communities, and the forests, indicating what changed and what stayed the same, what benefits were produced. The groups then joined together and compared and discussed their respective ideal futures, focusing specifically on benefits. Women were found to be interested in using community benefits for household improvements; the youth expressed unusual interest in public meeting places; men focused on community, household and forests. The interests and concerns of community subgroups could be clarified and easily shared using this method (Chapter 13).

5. Collaboration with Nongovernmental Organizations

In Romwe, Zimbabwe, Nemarundwe (Chapter 7) found women to be concerned about their access to land and to decisionmaking. She found that the supporting role of an NGO had been critical in making land formally available to women and increasing their voice in decisionmaking about garden lands. With the support of CARE International, a subset of women was able to get legal certificates to land, in contradistinction to traditional norms relating to land tenure. The women were also able to assemble as a group, centered on the garden, to confront an inequitable decision by the traditional leader (*sabuku*) (cf. also the section on Maranhão in Chapter 12).

6. Use of Criteria and Indicators

In Cameroon, the mutual ignorance among stakeholders about each other's perspectives was identified as a problem early on. A series of multi-stakeholder workshops was held, during which the concept of criteria and indicators for sustainable forest management was explained. The workshop participants divided into smaller groups to identify and select criteria and indicators that they considered important in assessing the sustainability of forest management (including human well-being). In plenary, the discussion was carefully facilitated, so the marginalized groups, such as villagers, had an opportunity to express their views. Participants were almost unanimously enthusiastic about this approach, as a means of sharing and learning about different perspectives—a step considered essential for progress in future collaborative management (Chapters 4 and 6). Similar workshops were held in the Philippines (Hartanto et al. 2002) and Nepal (New Era Team 2002).

7. Boundary Demarcation to Facilitate Coordination among Stakeholders

In Malinau, East Kalimantan, Indonesia, boundaries between communities were not clearly defined, representative of widespread confusion in the country about access to land and resources. The ACM group[10] undertook a complex, three-year process of trying to map the territories of 27 communities (of varying ethnicity and varying levels of marginality). The team concluded, among other things, that boundary negotiations were deeply political, with the definition of conflicts, differences in capacity or power, and serious marginalization of groups (despite apparent comparative homogeneity among groups) as particularly important. They found agreements to be fluid and best considered as partial and temporary (the more intense the underlying struggle, the more fluid the interests, agreements, and coordination). They found power relationships to be central to understanding conflicts and concluded that what worked best was a phased process that allowed layers of conflict and awareness of changing political conditions to unfold and be identified. Many leaders were not accountable to their communities, so broad consultation was important. In general, special effort was needed to encourage participation and representation of disadvantaged groups, with explicit attention to capacity and power differentials, giving the disadvantaged groups earlier access to information and other forms of preferential treatment. Coordination had to be built on a strong foundation that reflected legitimate operating principles and outcomes. Although there were many pitfalls in the process of participatory mapping, very real gains were also made in empowering local communities to begin the process of asserting claims to their territories and of establishing debate about rights associated with those claims. A process was started that communities, government, and companies are now keen to complete (Chapter 1).

8. Use of Cross Visits

In an effort to strengthen community members' understanding of the implications of forest management for timber, community members from Guarayo, Bolivia, were invited to visit La Chonta, a well-known and nearby example of community timber management. BOLFOR[11]/ACM invited 10–12 people from each village on the trip, requiring that at least 5 women be included. In both field trips, twice that many women participated, and more were ready and willing to go had there been space to accommodate them. The women were especially excited to attend, since none of them had ever seen how timber extraction takes place. The women's participation provoked long discussions on training, project planning, and monitoring (on topics ranging from sawyer efficiency and vine cutting to safety and controlling food use at the communities' logging camps). These visits were so successful that BOLFOR incorporated the practice as a standard extension activity for each regional office in other parts of Bolivia (see other, related activities described in Chapters 13 and 14). Cross visits by women were also found to be useful in the Philippines (Arda-Minas 2002).

The ACM field teams approached gender and diversity issues from a variety of perspectives, using different methods, and succeeded in stimulating greater involvement of various marginalized groups in local forest management. Most importantly, the activities they undertook were fashioned in response to the contexts in which they worked, and were responsive to local conditions.

Authors and Chapters

Our concerns about polyvocality, mentioned above, mean that the characteristics of the authors contributing to this volume are also of interest. We come from various countries, experiences, and disciplines. The group is dominated by anthropologists (Akwah, Bolaños, Campbell, Chicchón, Colfer, Cronkleton, Diaw, Matose, Nemarundwe, Piland, Porro, Schmink, Sithole, and Stone). Other social scientists are also contributors (Anau, Cayres, Chitiga, Kusumanto, Mutimukuru, Nguié-bouri, and Wollenberg). All of the ACM teams are interdisciplinary in nature, and some of our other disciplinary representatives have opted to contribute. We have ecologists (Tiani and van Heist), an agronomist (Limberg), a wildlife specialist (Hakim), and foresters (Dangol, Iwan, Nuñes, Nyirenda, Pokorny, and Sudana) represented in this book. The fact that most authors have worked collaboratively with foresters and ecologists has strengthened their abilities to phrase their findings in terms that are familiar to those who are not social scientists, which we hope will increase the utility of their results.

The authors all have long-term involvement in fieldwork in their areas. They write compellingly, based on the kind of intimate knowledge of daily life on their sites that comes only with time and the personal sacrifices and delights of long-term fieldwork. As a group we have taken seriously the admonition that dealing equitably with women and other marginalized groups takes special effort.

Finally, the authors represent an unusually rich diversity of nationalities (Brazil, Cameroon, Colombia, Germany, Indonesia, Nepal, Netherlands, Peru, Senegal, United States, and Zimbabwe). Feminist scholarship has been frequently criticized for reflecting the views of white American and European women (e.g., Shiva 1989; Braidotti et al. 1994). In this collection, only 21% of the authors fall into that category; 31% are women from the developing world; 48% are men; and 69% are men or women from the developing world. Seven of the 14 lead authors are women from the developing world.

We begin this collection in Asia (Part I). Anau and others, in Chapter 1, describe the effort currently under way in Indonesia to shift from a top-down approach to forest management to a more pluralistic and collaborative approach. Their discussion of the history of Malinau is instructive, sharing many features with forested areas elsewhere in the world. The authors narrate a process of determining village boundaries in the Bulungan Research Forest in East Kalimantan. With the help of geopositioning (GPS) specialists, villagers were trained to map their village boundaries. The process involved serious and sometimes difficult negotiation among and within villages. This chapter examines local representation in the context of significant intravillage variability, during the process of negotiation. It draws conclusions about the conditions under which such negotiations can profitably and sustainably occur. It also argues for special attention to power differentials and for giving preferential treatment to marginalized groups in such negotiations (cf. Hildyard et al. 2002, *70,* who make a similar point).

Chapter 2, by Hakim, takes a slightly more macro-level view of similar processes in the Indonesian site in Pasir, East Kalimantan, focusing specifically on interactions between communities and district level stakeholders (government officials, timber company personnel). He provides a thorough discussion of PAR and collaborative management from a theoretical perspective, going on to describe the practical steps involved in bringing the various stakeholders together in an iterative process to improve forest management. He concludes with some comments on the strengths and weaknesses of PAR as an approach—emphasizing its utility for stakeholders in identifying problems, analyzing the experience gained, drawing conclusions, and planning new or improved actions, as well as the centrality of good facilitation in the process.

Dangol's chapter (Chapter 3) focuses on the importance of the participation of a wider range of stakeholders in local forest management, using Arnstein's "ladder of participation" as a conceptual framework. She presents the steps that were undertaken in her role as an ACM facilitator in Bamdibhirkhoria, Kaski District, in western Nepal. Her goal was to catalyze local initiative, to stimulate a management process that was more iterative and collaborative. She analyzes "before" and "after" scenarios from the perspective of various functions considered important for effective forest management and improved equity within the community. Her analysis provides useful hints on how to replicate what she has been able to accomplish.

Chapter 4 represents our most theoretical analysis, drawing on studies from Indonesia and Cameroon. Diaw and Kusumanto use their experiences in Ottotomo, Campo Ma'an, and Lomié in Cameroon and in Baru Pelepat in Indonesia, to illustrate how social science methods can go beyond simply extracting information from local actors. They use their own research experience to demonstrate how the data collection process can provide valid platforms for learning interactively and for negotiating meanings, power, and representation. The authors argue for the possibility of a "science in action" that neither takes sides nor claims neutrality, but rather accepts being an integral part of a social interface, being both influenced and transformed in the process.

Moving to Africa (Part II), the next chapter (Chapter 5), by Oyono, provides both historical and contemporary descriptions of the position and conditions of Pygmies in Cameroon. He shows the marginalized status of this sedentarized "forest and mobile people" (the first inhabitants of the Congo Basin), in interaction with the dominant Bantus. He focuses specifically on the inequities of Pygmy access to resources and their neglect in policy, even under recent, supposedly people-centered forestry legislation and a strong devolution policy at the national level. He argues that this marginalization, or "malintegration," carries with it the possibility of more extensive forest degradation, with Pygmies as meaningful actors. Rejected by the wider society, the Pygmies are being forced back into the forest—a process that contains a significant danger, since Pygmies now have more effective harvesting technologies and greater need to market their products. This they can do most easily in the service of external stakeholders whose interests do not include sustainability of forest resources. The inequity of the situation is dramatically clear from this account.

Tiani, Akwah, and Nguiébouri (Chapter 6) examine some of the impacts of the creation of a large national park (Campo Ma'an) in southern Cameroon on women and men in two communities. Although the park's creation has had negative impacts on both men and women, the strength of women's antagonism to the park was unexpected. One of the central bones of contention between the park and the populace revolves around hunting, a predominantly male activity. This study shows, however, that the enforcement of the prohibition on hunting by park guards significantly reduced women's access to cash that had previously been earned by selling wildlife products openly. Although the sale of wildlife products continues on a reduced scale, its current, clandestine nature has resulted in a shift from female to male sales. Tiani et al. conclude with some suggestions for improving community-park relations, based on suggestions from local women.

In Chapter 7, Nemarundwe talks about formal and informal decisionmaking platforms in Zimbabwe, where women's views are and are not represented. Her analyses are based primarily on qualitative methods, including Participatory Rural Appraisal (PRA) and long-term participant observation. She contrasts the decisionmaking processes relating to management of the woodlands, where men's views hold greater sway, to those in community gardens, where women have attained some formal rights to manage (and even own land, in contradiction to tradition). She also looks at the differences in gender representation within the communities, the governmental sector, and NGOs, demonstrating the significance of NGO involvement in securing land rights for women.

In Chapter 8, Sithole introduces the Mafungautsi Forest Reserve in western Zimbabwe. Within a general context of antagonism between the communities and the Forest Commission, women's roles are examined. She shows how women in formal positions of power tend to be there by means of their kinship relations with powerful men and how women in the public eye bear numerous undesirable consequences, like widespread suspicions of witchcraft and promiscuity, as well as disapproval for not fitting into the conventional female role. Sithole questions whether well-meaning attempts to involve women in the formal sector are in fact consistent with women's own desires. She argues that Zimbabwean women prefer to influence events informally, behind the scenes, rather than become directly involved in overt politics (cf. Cleaver's similar observation about a Tanzanian village (2002, *43*)).

Mutimukuru, Nyirenda, and Matose continue discussing Mafungautsi in Chapter 9, where they analyze how social learning affects collective action. They describe the use of Training for Transformation, workshops, and focus groups as empowering tools, in their attempts to catalyze an ACM process among three formal resource management committees. Specific groups focused on the management of thatch grass, broom grass, and honey. The authors show how the involvement of women and other marginalized groups was enhanced and how people's propensity for social learning was strengthened by more equitable interaction among diverse stakeholders.

Next, we move to South America (Part III). Campbell and others (Chapter 10) look at intracommunity variation in natural resource management and use in two similar areas: eastern Peru and western Brazil. Building on long term research in both sites, and using the contrasting methods of linear programming in Peru and indepth ethnographic methods in Brazil, they examine the different strategies families use to subsist in forested environments. The evolving developmental stages of domestic groups and gender, as well as broader contextual matters, emerge as important factors in the allocation of labor to different land uses. The authors' specific interests focus on the implications of these different strategies for conservation.

Pokorny, Cayres, and Nuñes (Chapter 11) are concerned with the relationship between researchers and local communities, fearing that local roles and values are too often overshadowed by the structural dominance of outsiders (a concern shared by Hailey 2002). Pokorny's team members began by clarifying their own perceptions about desirable changes, using criteria and indicators in a novel way. They then tested a series of participatory methods in three communities, including evaluating the methods in terms of their empowering potential. Finally, in formal efforts to address issues of intra-community heterogeneity, they were convinced that the local communities looked at their own heterogeneity in very different terms than did outsiders.

Porro and Stone, in Chapter 12, compare gender in the eastern and western Amazon. Porro describes the patterns of changes that have characterized the women babaçu palm nut collectors in Maranhão, while Stone focuses on the roles of women among the rubber tappers residing in an agroextractive reserve in Acre. They argue that the kinds of differences they describe mandate long-term attention and careful, site-specific treatment of gender in the diverse places where forest management is an issue. They warn against the common tendency to apply the kind of standardized gender frameworks put forth by the World Bank and UN agencies, and correctly identify the huge challenge in being responsive to local realities while at the same time enhancing equity.

Moving to Bolivia in Chapter 13, Cronkleton first describes the legislative changes that have legalized local Indian rights to their territories. The research site, Salvatierra, is part of the recently constituted Guarayo TCO (*Tierras Comunitarias de Origen,* or Indigenous Community Territory). He then describes the major shift under way from subsistence agroforestry to small-scale, commercial timber management under the guidance of a large, U.S.-funded forestry project (BOLFOR), in this Guarayo Indian community in the lowlands. He anticipates likely changes in gender roles and potential conflicts and describes the application of two techniques he has used to help the community prepare for the changes ahead. One is the sharing of visions for the future in a facilitated workshop; the other involves increasing transparency in the mechanisms for sharing benefits from the timber operation. In both cases, the views of women were heard to an unusual degree.

Chapter 14, by Bolaños and Schmink, describes the reactions of women and men in a minority community in lowland Bolivia to the same formal forest management project (BOLFOR). The people of Santa María are almost exclusively immigrants; and the community is within the Guarayo TCO. The authors first recount the division of labor between men and women; and then they examine the anticipated and differing implications of the forest management project for women's and men's roles. They conclude that more focused attention will have to be paid to women's needs, if adverse and inequitable impacts are to be avoided. Some specific suggestions for accomplishing this are suggested.

The final chapter concludes with pleas for increased attention to the involvement of all segments of local communities in forest management; and for training and mentoring of—and *trust* in—fieldworkers whose role is to serve local forest communities. Both the communities and the potential facilitators and fieldworkers need the authority and resources to plan and implement programs in ways that are responsive and appropriate to local conditions.

A Personal Perspective

In recent years, there have been repeated calls for authors to indicate their own philosophical and theoretical perspectives, as a means to help readers assess what is written and to acknowledge the plurality of orientations that exist legitimately in the world. Since I was ACM program leader from 1998 to 2002, my personal perspectives had some effect on the direction taken in the research reported here— despite our efforts to maintain nonhierarchical management.

My decision to compile this book on participatory action research and its impact on gender and diversity issues evolved within a particular vision of what researchers and practitioners of conservation and development should be doing.

We need to acknowledge and respect the plurality of positions and perspectives that exist. We need to acknowledge that there may in fact be several equally accurate ways of accounting for and describing the same phenomena.

We need to recognize that different people approach the world with different goals and different operating strategies. These need to be granted the same respect as one's own approach. When considering others' behavior relating to their futures, their perspectives are even more important (though one's own perspectives may contribute to and strengthen theirs, if offered with humility and desired by the others).

I believe that people everywhere sometimes exhibit what can be called altruism. This is a resource that should not be discounted. Whether such altruism is "pure," or is really self-interest that derives from the need to feel oneself a good person, or is related to anticipation of eventual return (generalized reciprocity), is immaterial in practical terms.

Rural peoples' capacities and potential contribution have been widely under-valued. Such people have diverse perspectives that can contribute to a richer un-derstanding of our world; they have knowledge about and perspectives on their own environments that others do not have. They also have creativity and energy that can make their own and others' lives better if unleashed.

Development, in a positive sense, should not be seen as a single path ("the main-stream"). Different individuals, groups, and cultures have their own trajectories, and the global diversity this implies is a valuable resource, as well as a type of insurance for humanity. People and groups evolve in different ways, along paths that follow from their own agency and structural settings.

I do not find convincing the common ideas that we are all involved in a capital-istic juggernaut, impotent to act in any way that will not conspire with this system, or that all our efforts to enhance human well-being are really ways of inadvertent-ly strengthening an oppressive system. I do not think that the future of humanity is predetermined, whether by class, prices, race, or any other overarching factor or entity. I think we all *can* act in our own (and our families' and communities') interest and that the task of development practitioners is to work toward building a context in which people's freedom to act is ensured.

I understand that many of the social structures that are in place at this time interfere with human well-being. Nonetheless, I believe that the "oppressed" and the "oppressor" can and sometimes do work together to improve such structures. I do not think it is constructive or even realistic to consider those in positions of power to be necessarily acting in ways that are antithetical to the interests of those with less formal power. I *do* believe it is constructive to retain an element of skep-ticism about their motives, initially.

Our search for means of empowerment is not a paternalistic approach—"we" cannot empower "them." Rather, we should be seeking ways that, together (with community members and others), we can alter contexts to make life more equita-ble, more conducive to human and environmental well-being—for the benefit of all. "Oppressors" are also oppressed by oppressive systems.

Notes

Special thanks go to the Adaptive Collaborative Management team, the administrators at the Center for International Forestry Research, and Cornell University's Department of Natural Re-sources and Mann Library, all of which granted me the space (both physical and emotional) and resources to put this collection together. I also thank the many people in the respective field sites who have patiently put up with our many intrusions into their lives, with equanimity, good hu-mor, and valuable insights.

1. Dr. Ravi Prabhu was the leader of the project entitled, Assessing Sustainable Forest Man-agement: A Test of Criteria and Indicators, and Dr. Eva Wollenberg was the leader of the project entitled, Local Livelihoods, Community Forests, and Devolution.

2. C&I is shorthand for a conceptual framework that includes principles, criteria, indicators, and verifiers in decreasing levels of abstraction (analogous more generally to wisdom, knowledge, information, and data). These elements, organized hierarchically, have been widely used to assess the sustainability of forest management (cf. Prabhu et al. 1996; Lammerts et al. 1997, Chapter 12).

3. Devolution is a purposeful process designed to transfer varying amounts of formal authority from "higher," "centralized" levels of government to "lower," more "peripheral," local level governmental and nongovernmental institutions. We are particularly interested in the bundle of policies that together serve this function.

4. CIFOR defines adaptive collaborative management as "a value-adding approach whereby people who have "interests" in a forest, agree to act together to plan, observe and learn from the implementation of their plans (recognizing that plans often fail to fulfill their stated objectives). ACM is characterized by conscious efforts among such groups to communicate, collaborate, negotiate and seek out opportunities to learn collectively about the impacts of their actions." This definition was initially crafted by Ravi Prabhu (Prabhu et al. 2001), and modified by the team in conjunction with a meeting of the program's International Steering Committee, in Manila, Philippines, October 2001.

5. This well established approach involves several features: a cyclical process of problem identification, joint planning, implementation, monitoring, and reflection on progress. This is then followed by alterations to the plan as needed, implementation of planned changes, and so on. The ACM program used PAR as a central tool in its efforts to work collaboratively with forest communities (see Cooke and Kothari 2002, for a valuable critique of this method; and Chapter 2, for a full description of our use of it).

6. CORMAS is a multi-agent system-based computer program designed to contribute to negotiation processes among stakeholders; CIMAT is a CIFOR program designed so that users can adapt existing sets of criteria and indictors for their own purposes; CO-View is a computerized tool for use by natural resource managers in articulating and exploring shared visions of the future; CO-Learn is a meta-tool which uses a computer-based map as a metaphor to guide users through a suite of learning support tools of use in natural resource management. Herry Purnomo and Ravi Prabhu have been leaders in developing these software packages, in cooperation with Mandy Haggith, Gil Mendosa, and scientists from the French CIRAD-Forêt. See other tools described in Colfer et al. 1999b; Wollenberg et al. 2000; Sithole 2002; and Hartanto et al. 2003.

7. Led by Ravi Prabhu and Frank Matose, and composed, in this work, of Richard Nyirenda and Tendayi Mutimukura as ACM facilitators.

8. Led by Cynthia McDougall, Laya Prasad Uprety, and Netra Tumbahangphe, there were four principal ACM facilitators: Sushma Dangol and Mani Ram Banjade, working with NORMS in the central part of Nepal; and Kalpana Sharma and Narayan Prasad Sitaula, working with New Era in the East.

9. Led by Stepi Hakim, the ACM facilitators were Suprihatin and Amin Jafar of the NGO, Padi.

10. Led by Eva Wollenberg, the team included Njau Anau, Ramses Iwan, Miriam van Heist, Godwin Limberg, and Made Sudana.

11. BOLFOR is a USAID-funded project on sustainable forest management, which has been operating in Bolivia for nearly a decade. ACM work there is also funded by USAID and is intimately integrated with BOLFOR's activities. Peter Cronkleton is in charge of ACM activities in Bolivia.

PART I

Asia

THIS SECTION INVESTIGATES the experience of four Asian locales with adaptive collaborative management of forest resources. Each locale is described briefly below. Three are in Indonesia—Bulungan and Pasir in East Kalimantan and Baru Pelepat in Jambi, Sumatra—and one is in central Nepal—Bamdibhirkhoria. The lack of attention to gender issues, compared with other kinds of human differentiation, in the Indonesian sites is a reflection of the relatively high and unproblematic status of Indonesian women.

Part I focuses on conflicts and their resolution in agreements negotiated within and between communities, including the influence of power differentials, the mechanisms for cross-stakeholder communication, and the strategies for strengthening the participation of marginalized groups in decisionmaking. The final chapter in this section (Chapter 4), which combines analyses from Indonesia and Cameroon, leads into Part II on Africa.

Indonesia

Malinau, East Kalimantan (Chapter 1)

Bulungan Research Forest is an area in Malinau, East Kalimantan (Indonesian Borneo), where CIFOR has been invited to conduct research. It is characterized by some of the most intact tropical rainforest remaining in Asia. Some areas are very remote and comparatively untouched, while others are rapidly changing and subject to strong outside influences. The indigenous people practice shifting cultivation (Dayak groups) and hunting and gathering (Punan groups), supplemented by the collection of nontimber forest products, hunting, fishing, and periodic wage labor. Gender roles are less rigid than in most other parts of the world. Important outside influences include timber companies, oil palm plantations, and mining companies. Located on the Malaysian border, it also has significant military and smuggling activities. Like other Indonesian areas, the local government and people are coping with a recent law devolving much authority to the *kabupaten* (or district) level. Local people still do not have legal rights to their traditional territories.

Pasir, East Kalimantan (Chapter 2)

This site includes two small, remote communities, Rantau Layung and Rantau Buta, in a single microcatchment in southern East Kalimantan. These communities are homogeneous, with only one ethnic group predominating in both communities. Although the people are Muslim, they share many aspects of their culture and livelihood strategy with the Christian Dayaks. The communities are reached by small canoe, with accessibility varying greatly depending on water conditions. Forest condition remains good. Timber companies have been active in the area for some time, and part of the village territory is technically part of a protected forest area. There are oil palm and transmigration projects not too far away.

Baru Pelepat, Jambi, Sumatra (Chapter 4)

Baru Pelepat is a small community located in the buffer zone of Kerinci Seblat National Park in the center of Sumatra. The indigenous community consists of the Minangkabau ethnic group, a Muslim, matrilineal group of swidden cultivators. Others have settled in the community recently from other areas of Jambi and from Java, either spontaneously or as part of a government resettlement scheme. The Orang Rimba are a hunter-gatherer group that also has close ties to this forested area. Relations have been problematic among these groups and with the neighboring park and timber and plantation companies. Like the Malinau and Pasir sites, Baru Pelepat is struggling to understand what decentralization will mean in their area.

Nepal

Bamdibhirkhoria (Chapter 3)

Located a few hours west of Pokhara in central Nepal, this comparatively accessible community has had a considerable amount of outside help from bilateral projects, nongovernmental organizations, and the government. Since the area was badly affected by a major landslide some years ago, it has been actively involved in conservation activities, including maintenance of a formal community forest, managed by a forest user group and recognized by the government. The community, which includes many caste and ethnic groups, has been governed by a traditional elite, which has, in cooperation with the state forestry officials, monopolized control of the community forest. Gender and caste roles are clearly differentiated and supported by a strongly patriarchal ideology.

CHAPTER ONE

Negotiating More Than Boundaries in Indonesia

Njau Anau, Ramses Iwan, Miriam van Heist, Godwin Limberg, Made Sudana, and Eva Wollenberg

*I*N MALINAU, EAST KALIMANTAN, Indonesia, the poor and the more powerful increasingly compete for the same land and forest resources. Swidden farmers, hunter-gatherers, timber companies, mining companies, and local government make diverse demands on the forest, yet coordination of forest management among these different land users has been weak.[1] During Indonesia's implementation of decentralization reforms between 1998 and 2001, when demands on the forest increased and local coordination was at its lowest, social conflict and forest degradation increased dramatically.

Malinau is not unique. Large forest landscapes everywhere are under increasing pressure from diverse and incompatible demands. In this chapter we argue that, unless appropriate mechanisms are in place for forest users to coordinate among themselves, large forest landscapes such as those in Malinau are at risk from escalated and entrenched social conflicts, increasing social injustice, open access competition for resources, and even willful destruction of forest resources. Because of recent reforms, stakeholders in Malinau face an additional challenge. They must make a transition between the top-down, more coercive coordination by forest departments in the 1970s to 1990s, when conflict was rarely openly acknowledged, to more deliberative[2] and pluralistic coordination, where self-organization, transparency in government, conflict management, and greater citizen participation guide decisionmaking (Anderson et al. 1999; DiZerega 2000; see also Chapter 2).

We present here the findings that led us to these conclusions.[3] We focused on village-to-village coordination, because it has received little attention, yet is fundamental to multi-stakeholder processes. We wanted to know whether the principles guiding more formal and complex multi-stakeholder processes were relevant to intervillage coordination, where fewer people were involved and they lived together with greater familiarity, more of a moral economy, and stronger kin obligations. We asked the research question, what conditions facilitate coordination of interests within and among villages? We were also curious to learn more about

local people's concepts of conflict and agreement and how these might be changing during the reform period. The work focused on village boundary demarcation
as a means of land use coordination and as a tangible source of conflict about
which agreements could be negotiated.

Below we present a review of current thinking about coordination processes,
our study methods, and a history of village-level coordination efforts in Malinau as
the context for our study. We then present the results of the study, with data about
sources of boundary conflict among communities and how they sought to overcome that conflict. We conclude with several recommendations.

Current Thinking about Coordination

What constitutes "good coordination" in forest management? During the last two
decades, proponents of community management have often advocated that the
state should decrease its involvement as the primary coordinator of local management (Poffenberger 1990; Sarin forthcoming). Where government coordination
has been weak, however, local entrepreneurs and strongmen often gain control over
the forest at the expense of communities (Barr et al. 2001; Dove 1993; Kaimowitz
et al. 1999). A new paradigm is emerging in which coordination occurs through
pluralistic processes that take into account the interests of different stakeholders. In
these multi-stakeholder processes, the central challenge is "how a society composed of formally equal citizens could be ordered, so that those having access to
more political resources, luck, or talent would not use their advantages to exploit
others weaker than themselves" (DiZerega 2000, 1).

To answer this challenge, current thinking indicates that coordination should
be grounded in negotiations that involve all relevant stakeholders, identify their
interests, facilitate effective communication and learning, create a neutral space for
interaction, and seek to achieve consensus (Allen et al. 1998; Borrini-Feyerabend
1996; Fisher 1995; Porter and Salvesen 1995; Röling and Maarleveld 1999; Röling
and Wagemakers 1998). Iterative cycles of conflict and adjustment are likely to
occur, and conflict should be managed (Lee 1993; Ramírez 2001).

Experience with decentralization and comanagement agreements between states
and communities indicates, however, that some of these recommendations are un-
realistic. Seeking to achieve politically neutral negotiations and collaboration can
work against politically weak communities and the vulnerable groups within them
(Anderson et al. 1999; Antona and Babin 2001; Baviskar 2001; Contreras et al.
2001; Edmunds and Wollenberg 2001; Sundar 2001; Wollenberg et al. 2001a).

We argue that a more realistic view of coordination requires modification of
the principles commonly used to facilitate negotiation. First, in contrast to certain
current beliefs about conflict mediation, evidence strongly suggests that it is difficult, if not impossible and undesirable, for people engaged in negotiation to define
stakeholders' interests clearly. Interests are many-layered, and we tend to construct
our interests in response to specific contexts and for strategic purposes (Baviskar
2001; Leach and Fairhead 2001). Especially where trust among groups is low, it

may be unwise for any group to reveal its true interests or to assume that other groups are communicating their interests honestly. Baviskar argues that we can best infer interests from how people act, not from what they say.

Second, as proponents of pluralism would argue (Anderson et al. 1999; Bickford 1999; Rescher 1993), consensus is impossible, and participants in a multi-stakeholder process should treat agreements as inherently unstable. Complete agreement is impossible because differences in experience prevent even two individuals from ever having the same desires (Rescher 1993). As only temporary states of coordination can occur, coordination is best thought of as a process of on-going accommodation and negotiation involving multiple actors. Agreements are not the end of the process, but rather are a set of guidelines providing legitimacy for new actions. People negotiating contractual agreements and management plans should therefore build in flexibility to accommodate adjustment and acknowledge these as temporary measures (Wollenberg et al. 2001b). Boundary agreements should acknowledge flexibility in rights allocated across borders. Facilitators of coordination should work with the plurality of institutions with which local actors interact, and not just through single user groups or local forest departments (Leach and Fairhead 2001).

Third, some parties consistently enjoy disproportionate control over coordination. Weaker groups' interests are routinely excluded, represented ineffectively, co-opted, or negotiated away (Anderson and Grove 1987; Hecht and Cockburn 1989; Parajuli 1998). The exercise of power is thought to depend on who assumed the convenor and facilitation roles (or controlled these roles), who was represented in the process, and who has the greatest capacities for communication and negotiation among participants (Ramírez 2001; Steins and Edwards 1999). Government agencies have often assumed this role in forestry by working in an "expertocratic" mode that relies on opinions of professionals rather than wider citizenry (Rossi 1997, 237). In these situations, the interests of disadvantaged groups are often masked under the guise of agreements (Edmunds and Wollenberg 2001; see also Chapter 11). Well-intentioned efforts to expand participation in forest management by including marginalized groups can actually work to their detriment, unless certain checks and balances and accountability measures are used. Multi-stakeholder negotiations are likely to be more just if they acknowledge existing power relationships—rather than assuming that negotiations can be neutral—and enable weaker groups to work politically in more effective ways.

From the points documented above, we suggest that more strategic principles[4] for multi-stakeholder processes are necessary. These require facilitators to manage in ways that are sensitive to power differences. Participants need to be aware of their options and willing to demand them. Any group or coalition that facilitates will seek to meet its own self-interest to some extent, so it is necessary for the participants to collectively agree on norms, rules, and sanctions that encourage socially responsible facilitation. Basic principles include the following:

• Improve the preconditions for disadvantaged groups to participate and negotiate effectively. This principle includes (a) seeking out possibilities for alliances among select stakeholders, rather than trying to achieve an apolitical agreement among all stakeholders; (b) enhancing the power, urgency, or legitimacy associated with certain stakeholders to increase the likelihood of their being noticed and involved in decisions (Ramírez 2001); and (c) enabling excluded stakeholders to work through parallel arenas to challenge decisions.

- Ensure accountability of coordination decisions to interest groups through effective representation (facilitating proximity of leaders to their constituencies, elected leaders, and delegates and fostering an ideology of civic dedication), transparency (third-party monitoring, public meetings and reporting, participatory processes), and checks on power (legal appeals to existing decisions, separation and balance of decisionmaking power across several authorities, enabling civic education and social movements) (Ribot 2001).

- Evaluate the legitimacy of negotiation processes, decisions, and agreements. This means analyzing the reasons for each group's participation or nonparticipation in negotiations, determining each group's form of representation, identifying the roles of convenors and facilitators, and understanding the historical context for such agreements. It also means treating legitimacy as partial and contingent rather than assuming that an unproblematic legitimacy is ensured through open negotiations.

- Acknowledge the fluid and complex nature of interests, agreements, and coordination processes and encourage people to communicate, debate, and negotiate.

To test the applicability of these principles and refine them, we examined the extent to which they were relevant to village-to-village coordination about land claims in Malinau. We report our findings below.

Methods

The Center for International Forestry Research (CIFOR) used action research to examine negotiation among stakeholders over forest land claims and coordination of land use in the 27 villages of the upper Malinau River in East Kalimantan (see Figure 1-1). Action research enabled us to conduct research that generated local impacts, thus mutually benefiting CIFOR and local stakeholders, and also enabled us to observe directly how these impacts occurred. The methods and focus of the work evolved in response to local needs in iterative stages.

In 1999, CIFOR conducted a systematic survey of stakeholders, land tenure, and forest-related conflict in the 27 villages of the watershed and organized a five-day community workshop involving representatives from all villages to identify a mutual agenda for collaboration between communities and CIFOR (Anau 1999; ACM-CIFOR 1999). The survey and workshop demonstrated that all the communities had a strong interest in demarcating village boundaries through mapping. They also revealed high levels of conflict among villages and between villages and companies about a range of forest-related issues, especially land claims.

In response to the high interest in documenting boundaries, CIFOR trained village-level committees in participatory mapping and, from January to July 2000, facilitated conflict mediation and mapping among the 27 villages. CIFOR created a core team of nine trainer-technicians that included six Malinau community members.

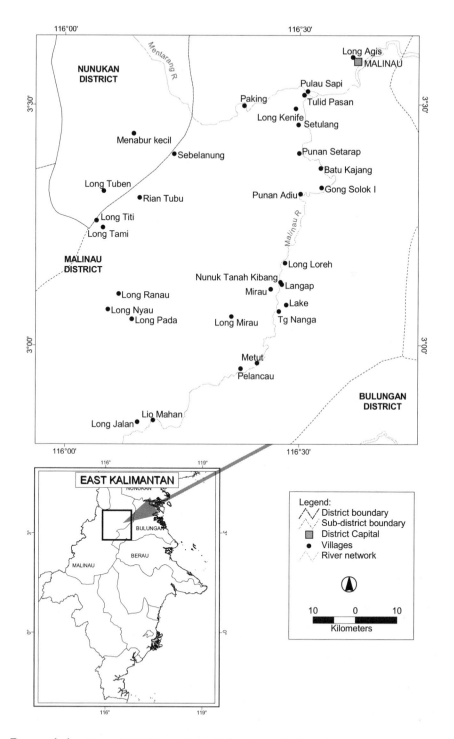

Figure 1-1. *Map of Malinau, East Kalimantan, Indonesia*

This team documented and analyzed the types and causes of conflict over boundaries (Tim Pendampingan Pemetaan Partisipatif 2000). Communities frequently asked the team to mediate their boundary conflicts, which allowed the team to directly observe the types of conflicts that occurred and the strategies communities used to achieve agreements. The team also examined the types of agreements produced to understand who was involved in producing the agreement and whether the agreement was legitimate and acceptable to community members.

Twenty-one villages produced draft maps. Because of the long-term nature of boundary adjudication and the role of local government in approving boundaries, CIFOR and the Malinau District government agreed that local government would be better placed to continue and complete the mapping. Since 1999, CIFOR has worked with communities to facilitate their negotiation with the district government. As of December 2003, however, the local government had yet to even begin demarcating village boundaries.

A History of Coordination Efforts among the Villages

Struggles over land and forest in Malinau have been longstanding, even if the reasons and means of managing them have changed. We can trace the shifting authorities that worked to overcome these struggles and served to coordinate control over land during the last several centuries.

Traditional Authorities

At least until the early 1900s,[5] intergroup warfare was common and a major influence on territorial control. Minor customary leaders (*kepala adat*) and their circle of close advisors controlled access to land, managed conflicts, and coordinated decisions within their ethnic group, while major customary leaders (*kepala adat besar*) helped to coordinate these matters at larger scales. Minor and major leaders formed alliances to oppose common enemies, especially unfriendly groups that ventured into their territory. Evidence suggests that in the upper Malinau, the *kepala adat besar* for the Merap ethnic group[6] was the reigning local power, together with the local Tidung sultanates, for most of the 20th century. Settled rice farming communities like the Merap sought control over defined territories, while others (e.g., the Punan) were more concerned with access to key resources, such as sago palm.

Where they existed, territories seem to have been conceptualized in terms of a central settlement point or a river and its watershed, with less emphasis on the exact location of the outer bounds. Individual property also existed. Major customary leaders controlled rights to valuable resources, such as birds' nest caves, located within the territory of a sultanate.[7] The sultanate in turn levied taxes on the traded products. Interestingly, the descendents of these customary leaders have used the letters of tax payment in recent years to establish ownership over the caves.

Customary leaders used hierarchical social controls within their villages and periodic consultation—especially with a close circle of influential villagers, or *tokoh masyarakat*—to manage conflict within the group. Unresolved conflict was handled through the fission of the group, with one faction moving to a new settlement or, in the case of external groups, through warfare. Around 1900, Kayan groups repeatedly attacked Merap groups on the Malinau to reclaim caves taken over by the Merap.

Rights of access to village territories were based on ethnic, marriage, and trade alliances, as well as respect for customary authority. Outsiders entering another group's territory were formally expected to request permission from the customary leader, although in practice people often casually entered watersheds to hunt or collect forest products without permission. Some leaders required not only permission, but also fees from outside groups to collect forest products in their area (Sellato 2001).

Colonial and Governmental Authorities

As warfare declined, the Dutch,[8] and later in the 1950s the government of Indonesia,[9] constituted additional layers in the institutional hierarchy in what was first the Bulungan Sultanate and later the *kabupaten* or district of Bulungan. The new Indonesian government established an infrastructure of centralized control. The role and direction of accountability of customary leaders became muddled as many were appointed by outside officials as village government heads and thus became upwardly accountable to a *camat* (subdistrict head), *bupati* (district head), governor, and the president. Local social institutions were severely weakened with the delegitimation of customary laws and leaders. Government maps of villages bore little relation to actual settlements and their claims. Local people became increasingly disenfranchised and politically distanced under this system. Most matters of village concern continued to be settled by customary leaders, especially those who also worked for the government as village head. These leaders made decisions among a group of ethnically homogenous people. Access to land and forest continued to be managed as before.

In addition to establishing a new overlay of administration, the state staked extensive claims to forested territories starting in the 1960s as Indonesia's timber boom began. Nearly 95% of what is now the district of Malinau was designated as state forest land, and in the late 1960s the central government allocated all the state forest land to timber concessions. Suddenly villagers found themselves sharing the forest with logging companies and were told that the land belonged to the government of Indonesia. The state's assertion of authority over land through the logging companies' presence openly challenged local sovereignty and claims to land in a way that sultans and the Dutch had never done.

Villagers accommodated the concessions, largely out of feelings of intimidation. Military officials usually accompanied logging company staff or otherwise harassed potentially troublesome villagers. The gradual degradation of forest, loss of wildlife for hunting, and lower water quality that villages experienced were partially offset by the construction of roads, provision of transportation, generation of employment (albeit limited), and occasional contribution to a village project. A few villagers also benefited from illegal logging opportunities. Local concessions did not strictly enforce hunting and burning prohibitions and allowed swidden agriculture in some forest areas in an effort to maintain good relations. During this time, the most common forest-related conflicts were intervillage quarrels about access to agricultural plots and for a few individuals, disputes over claims to birds' nest caves. The state simply did not allow conflicts with the government or concessions.

The authority of village customary leaders eroded further and land claims became more complex as ethnic groups began to share territories. Government resettlement programs of the 1960s through 1980s, and an ad hoc case of government-sponsored resettlement in 1999, reallocated most formerly Merap[10] lands along the upper Malinau River to other Dayak groups from more remote areas. The newcomers did not, however, always sever ties with their former territories. As a result of these programs, two to four ethnic groups now reside collectively in 9 of the 16 settlements (*lokasi*) of the upper Malinau River. Population pressure has also increased substantially. Local groups now claim multiple territories, with overlaps common. The role of customary and government authorities in settling these claims has not been clear. Because of the resettlement programs, the upper Malinau River area suffers more from these sorts of multiple claims than other parts of the Malinau District (see also Chapter 2).

Recent Developments

The final and most recent major development occurred with three overlapping phenomena: decentralization reforms; new access for villagers to monetary payments for timber and land; and the creation of the new Malinau District. With the initiation of decentralization and the associated uncertainty, local people from all sectors of society have sought to seize their share of Malinau's resources. Even before decentralization policies were formally implemented at the district level, de facto decentralization began taking place with villages making claims to customary (*adat*) lands and negotiating directly with local investors (Rhee forthcoming). Villagers made demands for compensation or benefits from timber and mining companies more freely and requested larger amounts than ever before.[11] Since former President Suharto stepped down in 1998, villagers say they can express their discontent without fear and have been much freer about speaking out against their leaders and the government. Military officials only rarely accompany timber companies or government entourages any more. New political associations have formed among different ethnic groups. Village leaders can be seen as often in the central town of Malinau, meeting with government or company officials, as they are in their own villages.

Fuelling the race for resources was the offer of payments by companies for harvesting timber. In 1996, the exploitation of coal in the Loreh-Langap area resulted in payments to some 10 to 20 households and to the customary Merap leader of Langap for rights to excavate their fallowed swidden fields. In 2000, the governor of East Kalimantan passed a provision (stimulated by the new Basic

Forestry Law 41, 1999) enabling communities to claim compensation from timber companies for logs harvested in their areas.

The most lucrative deals, however, were the IPPKs, or *Izin Pemungutan dan Pemanfaatan Kayu*. Beginning in April 2000, the district head began allocating these small-scale logging permits for 100 to 5,000 ha each to hastily formed small local companies. With decentralization, the district became responsible for generating its own income and could also keep a larger proportion of the income it generated than ever before. The incentives for intensive resource use were therefore high. The presence of the Kayan Mentarang National Park increased the pressure on the district to make more intensive use of the remaining areas, such as the upper Malinau River, which is more accessible and has better infrastructure. The result has been extraordinarily high levels of timber extraction among more diverse groups and increasing conflict among nearly all parties, including small-scale timber harvest permit holders, concession holders, villagers (themselves often forming factions), and mining interests (Barr et al. 2001). Forty-six IPPKs were issued, granting access to more than 60,000 ha in Malinau since April 2000. Underlying the logging deals have been negotiation of territorial claims and speculation about the value of these lands for future claims or compensation. A laissez-faire, frontier atmosphere has emerged in which the importance of making money far outstrips the value of being lawful or fair.[12]

The symptoms of trouble have been clear: community protests against the investors for not paying expected fees or wages to local harvesters; complaints among villagers about opaque deals struck between leaders and investors; and forest logged in areas where permission was not granted by villagers. In most villages now, few people know the content of the evolving law or are aware of their entitlements. Usually only a select elite close to the village head are involved in district matters and negotiations with investors. Many villagers are excited about trying new economic options, but lack information about how to consider trade-offs in livelihood security and long-term resource use. The communities' euphoria about receiving several thousands of dollars now will most likely be forgotten in a few years when both the forest and their money have run out.

Furthermore, the formation of the new Malinau District in October 1999 (one of three new districts formed from the district of Bulungan) resulted in a one-and-a-half-year period of temporary leadership, when the district government was not accountable to a local assembly. Not coincidentally, this was also the time when the bulk of the IPPKs were issued. Many government offices, including the forest service, were not filled until 2001. Another significant aspect of the new local government was its dominance by locals. For the first time, people from the district (or married to someone who was) filled most government posts. Previous officials were mostly from Java, Sulawesi, or other parts of Kalimantan. The "Dayakization" of local government has meant that authority is now rooted in the local politics of more than 18 different ethnic groups. Local relationships of power are more intertwined and complex than ever before.

These evolving relationships have affected how struggles over land play themselves out. Current alliances in Malinau reflect a set of fluid, interlocking networks of ethnic affiliations, economic interdependencies, strategic kin relationships, and even historical alliances from the headhunting period. Kenyah, Lundaye, and Ti-dung groups have been the most politically aggressive in recent years and domi-nate Malinau's new local government. These groups, together with the Merap, have also worked most aggressively to consolidate their claims to land. Punan groups, meanwhile, have had little representation in the district government, and are always the weaker partner in alliances with other ethnic groups (cf. the situa-tion of Pygmy groups in Cameroon and Orang Rimba in Sumatra discussed in Chapter 4).[13] Individuals from all groups have maintained an opportunistic atti-tude toward building alliances and have sought to strike new deals as they may, making it difficult to know at any one time precisely who has control where. Unfortunately, only a relatively small group of leaders and their circles has enjoyed the benefits of these deals and exerted any real influence over decisions.

Decisions made in the next several years will have huge consequences for who will control land and how that land will be used in the medium term. Current trends indicate very real threats of rapid deforestation, disenfranchisement of the Punan, and loss of long-term economic opportunities by most local groups. Be-cause Malinau is one of Asia's largest remaining expanses of continuous forest and home to the largest group of Punan in Borneo, it is vital that coordination be improved to encourage a longer-term, more equitable, and more integrative view of how the area's forest can be managed to benefit local communities. Local stake-holders feel these challenges intensely.

Setting Village Boundaries

In the context of current theories of coordination, available methods, and these recent developments, CIFOR's action research on intervillage boundary conflict proceeded. Below we report on the lessons learned about the sources of conflicts encountered, the factors influencing how communities reached agreements, and the results of the boundary demarcation process.

Sources of Conflict

Most conflicts between villages over boundaries involved overlapping ownership or use of agricultural lands (swidden fields, wet rice fields, and perennial gardens) and a history of mistrust and noncooperation. Other conflicts arose over rights to timber, valuable nontimber products like *gaharu* (*Aquilaria* sp., a fragrant resin) or birds' nests, and land containing coal deposits. Every village, however, experienced its own unique constellation of specific conflicts (Table 1-1). In the far reaches of the upper Malinau, where only Punan groups lived, conflict focused on access to forest products, in addition to the sources mentioned above. In the central portion of the watershed, where rich coal deposits occurred, conflicts emerged over com-pensation claims against the coal-mining company for the use of cultivated or fallow fields. In the lower stretches, problems focused on access to agricultural lands and problematic relationships.

Table 1-1. *Sources of Conflict Affecting Boundary Negotiations*

Village boundary	Coal deposits	Agricultural lands[a]	Nontimber forest products[b]	Timber	History of poor relations
		Lower Watershed			
Lidung Keminci–Sentaban	—	X	—	—	X
Sentaban–Setulang	—	X	—	—	X
Setulang–Setarap	—	X	—	—	—
Setarap–Batu Kajang	—	X	—	—	—
Batu Kajang–Gong Solok	—	X	—	—	X
Batu Kajang–Adiu	—	—	—	—	—
Gong Solok–Adiu	—	X	—	—	X
Adiu–Loreh	—	X	—	X	—
Adiu–Nunuk Tanah Kibang	—	—	—	—	—
		Middle Watershed			
Long Loreh–Gong Solok	—	—	—	—	—
Long Loreh–Nunuk Tanah Kibang	—	—	—	—	—
Long Loreh–Langap	X	X	X	—	X
Langap–Seturan/Punan Rian	X	X	X	—	X
Langap–Nunuk Tanah Kibang	X	X	—	—	X
Langap–Laban Nyarit	X	X	—	—	—
Langap–Tanjung Nanga	X	X	—	—	X
Laban Nyarit–Mirau	—	X	X	—	—
Laban Nyarit–Halanga'	—	X	—	—	X
Laban Nyarit–Tanjung Nanga	X	X	—	—	X
Laban Nyarit–Metut	—	—	—	—	—
Laban Nyarit–Pelancau	—	—	—	—	—
Laban Nyarit–Long Lake	—	—	—	—	—
Tanjung Nanga–Seturan	—	—	—	—	—
Tanjung Nanga–Metut	—	X	—	X	—
Metut–Pelancau	—	X	X	X	X
		Upper Watershed			
Pelancau–Long Lake	—	—	—	—	X
Long Lake–Long Jalan	—	—	X	—	X
Total conflicts	6	17	5	3	13

[a]Swidden fields, rice fields, perennial gardens.
[b]Birds' nests, gaharu, etc.

Disparities in economic or political status among villages exacerbated mistrust and lack of cooperation. These disparities affected how a conflict over boundaries manifested itself, as well as the possibilities for resolving the conflict. As we discuss further below, the larger the discrepancies between villages, the less likely it was that villagers were able to reach agreement about boundaries.

Although many of these conflicts were longstanding, villagers noted that the intensity of conflicts increased as outside parties sought to exploit timber and coal and offered lucrative compensation payments. The promise of significant extra income raised the stakes of the conflict and made people determined to protect or expand their claims to timber- or coal-bearing areas. The high stakes brought to the surface more latent, long-term conflicts related to intervillage differences or rights to agricultural land, which further fuelled the intensity of the immediate conflict.

Factors Influencing the Negotiations

Community Participation. Community participation in negotiations was lower than CIFOR expected, even though, as others have noted, generating adequate participation is a central challenge of populist approaches (Rossi 1997). Although villagers asked us to conduct the mapping during a period of low agricultural activity (April to July) when they would have more time and CIFOR's team actively sought to stimulate broad community participation through meetings and informal interaction, decisions tended to be controlled by only a few individuals. We have observed this decisionmaking pattern to be typical in villages of the upper Malinau River for most matters at the village or intervillage level.

Participation and representation were ineffective within villages, as well as in meetings between two or more villages. Within villages, participation in meetings was low. In the Loreh site (four villages) only 50 people out of 1,000 ever attended public village meetings. Only 20 people from the Loreh villages were later involved in the final boundary negotiation with the neighboring village, Langap. Of the 60 people interviewed in the Loreh villages after the mapping had been completed, only a small proportion knew that the negotiations and mapping had taken place.

Factions were common in even small villages. Representatives of these factions were frequently not present in meetings, either because they had not been invited or because they purposely did not attend. Boycotts of meetings were a common means of quiet protest against the group calling the meeting. Village leaders usually only consulted a small circle of influential colleagues among the *tokoh masyarakat* and never actively sought the views of different groups, let alone represented them. Women rarely participated in meetings, and if they did, rarely spoke. Predictably, village politics led to some groups giving more weight to their own preferences, while marginalizing others.

In intervillage meetings, one to six influential village members (*tokoh masyarakat*), including among others the village head, members of his staff (*aparat desa*), and customary leaders, represented a village. Decisions often could not be reached if a key person was absent. In Langap, for example, a decision could not be taken without the endorsement of the Merap customary leader. In Metut, the absence of the village secretary completely stalled negotiations with Pelancau.

In cases where the village leader only needed to reaffirm an existing agreement, we observed that the participation of a few individuals was sufficient for ensuring the acquiescence of other villagers and the stability of the decision. A small delegation became problematic, though, where changes needed to be negotiated and consultations with key influential people and representatives of groups were needed before settling on a particular option on behalf of the village. People attending meetings on behalf of a village rarely reported back to the village about the outcome of their negotiations.

One of the most important factors affecting participation was the location of a meeting. Time and transportation expenses limited the number of people willing or able to travel the often significant distances between neighboring settlements (one hour to one day). For example, in a meeting in Langap, 21 people from Langap attended, while 0–3 people from each of the eight neighboring villages attended. Similarly in a meeting held in Setulang, 30 people attended from Setu-lang, compared with 3 from the neighboring village of Setarap. If negotiations are held repeatedly in the same village (because of better facilities, ease of access for the facilitator, etc.), other villages may be compromised in their ability to partici-pate fairly.

Representation and participation among the Punan was especially poor (cf. Chapter 4 on Orang Rimba, Chapter 5 on Pygmies). The Punan faced special constraints to participating in meetings called in villages. First, Punan families frequently went to the forest for long periods of time, with men additionally going into the forest to look for *gaharu* for weeks or months at a time. As a result, they would often not know about meetings in advance and lacked time to consult with other community members before attending a meeting. More dominant groups also did not always invite them to meetings, and information from meet-ings was not always shared with them.

Second, the Punan often did not feel comfortable expressing themselves freely in the presence of more dominant groups. Among the nine locations where Punan villages coexisted with other ethnic groups, participation of Punan groups was extremely weak in three communities (Seturan-Punan Rian, Tanjung Nanga-Respen, and Gong Solok I-II).

Third, in at least Langap, the Punan living in neighboring Long Rat and Punan Rian had a historically subservient relationship with the Merap, having been giv-en land locally to facilitate their work as birds' nest collectors for the Merap.

A final reason for weak Punan participation was that in four settlements (Pelan-cau, Long Lake, Metut, and Long Jalan), members of the village were scattered in several locations, making it difficult to involve representatives from all groups and distribute information to everyone.

Internal Village Processes. Often villages did not undertake adequate internal consultations to reach agreement among themselves. In every village, the village head coordinated whether these consultations occurred or not, sometimes to-gether with the customary village head (*kepala adat*). Of seven pairs of villages that engaged in broad consultations within their respective villages before negotiating with their neighbors (Setarap-Setulang, Setarap-Batu Kajang, Batu Kajang-Gong Solok, Tanjung Nanga-Langap, Langap-Laban Nyarit, Langap-Loreh, and Metut-Pelancau), five resulted in agreements. Internal preparations ensured that the ne-gotiated decision would be acceptable to the broader community. They also helped community representatives explore different options and have more information at hand to be able to negotiate better. Nonetheless, only 11 out of 27 villages held formal community consultations. Others held small informal meetings. Aspects of internal consultations that seemed most important in producing a stable, broadly acceptable outcome included:

- Transparency, indicated by the holding of a community meeting attended by a majority of the families. Where transparency was lacking (e.g., Metut, Sentaban, and Laban Nyarit), people within the village later challenged the agreement determined by the village head.

- Community capacity to work together and trust each other (community cohesion), indicated by a history of low factionalism and high cooperation at the village level and support for the village leader. Such capacity was high, for example in Tanjung Nanga and Setulang. Where people did not work together, negotiations were less effective. Langap representatives, for example, negotiated demands from Tanjung Nanga that were not supported by other Langap villagers (where at least four factions exist). As a result, when the mapping team tried to identify boundaries, the agreement was rejected.

Negotiations between Villages. In observing the negotiation process, we sought to understand how negotiations were organized and what factors influenced their outcome. Although we initially encouraged parties to reach agreement[14] quickly and described such an outcome as a successful negotiation, we soon learned that most of the agreements were short-lived and partial in their support. An agreement reached quickly enabled communities to conduct the mapping of their territory, but we fear this occurred too often at the expense of a more socially inclusive process that would have probably resulted in more stable results. We learned that we should have evaluated the process underlying how a village reached an agreement as a basis for proceeding with the mapping, not just whether an agreement had been reached.

Villagers used two approaches in their negotiations: meetings between village heads or meetings between selected village representatives. Meetings between village heads usually occurred where there had been no previous village consultation. As noted above, this occurred where both parties already accepted a boundary and the boundary only required affirmation (Laban Nyarit-Pelancau, Laban Nyarit-Metut, Laban Nyarit-Long Lake). In these cases one meeting was sufficient to agree on boundaries. Where there was a disagreement about the boundary, however, community members consistently rejected agreements reached by only their village heads. For most villages, negotiations commonly involved one to five meetings. One set of villages held 19 meetings! As noted above, village heads and other representatives were only partly accountable, if at all, to their broader village constituency.

Five factors appeared to help communities reach agreements in these negotiations. First was consultation with the other village. Among the 27 villages, 8 held consultative meetings with neighboring villages as part of their preparation for the mapping. Six of these villages successfully negotiated agreements. Good relations did not predispose these villages to reach agreements, since half of the six were communities with long-standing difficulties with their neighbors.

Second, family relations among villages encouraged compromises that led to more rapid agreement. Six villages (Long Jalan, Long Lake, Pelancau, Metut, Laban Nyarit, and Langap) sought agreements based on compromises. Although these communities may have wished to expand their territory to take advantage of the changing value of resources, they ultimately decided to maintain existing boundaries, rather than invite conflict, because they were all members of the same extended family.

Third, financial incentives encouraged speedy resolution. Potential compensation payments by the coal company or sharing of benefits from small-scale

Table 1-2. *Difference in Capacities and Power Status between Two Negotiating Villages and Nature of Agreements Reached*

Difference in capacities and power status	Score	Agreement reached[a]		Stability of decision[b]	
		No	Yes	Stable	Not stable
No difference	0	0	6	5	1
	0.5	1	7	6	1
Moderate difference	1	2	2	1	1
	1.5	1	5	2	3
Large difference	2	1	1	0	1

[a]Number of villages.
[b]Stability was counted only for cases where agreement was reached.

timber harvesting (through IPPK holders) promised concrete benefits that encouraged villagers to reach agreement quickly, get on with mapping their lands, and secure additional income. The uncertainty of decentralization policies also pushed people to quick settlements. Many people adopted a first-come, first-served attitude, fearing that someone else would benefit from the resource if they did not make use of it first or that benefits would no longer be available if the policy changed.

Fourth, villages with similar institutional capacities and power were more likely to reach agreements than villages that differed.[15] In a number of cases, especially the case of Langap and its weaker neighbors Long Rat or Paya Seturan, more powerful villages presumed themselves entitled to exert their will about a boundary decision and disregarded the need to build agreement with a weaker village. Weaker villages often passively resisted these decisions by the more aggressive villages. This pattern is evident with the application of a simple scoring system,[16] the results of which are summarized in Table 1-2 and shown more fully in Annex 1-1. Even where villages reached temporary agreements, those villages having lower capacity-power differentials were more likely to reach stable agreements. The more similar the villages, the more likely they were not to challenge boundary agreements.

The fifth factor influencing outcomes was the opportunity to share benefits across villages. CIFOR assisted villages to reach agreement in several cases by encouraging villagers to treat the boundary not as a fence excluding nonvillagers, but as a set of rules about sharing access or benefits. In seven cases, villages negotiated agreements enabling neighbors to maintain their swidden fields, perennial gardens, or hunting rights (Langap-Loreh, Langap-Seturan/Punan Rian, Langap-Nunuk Tanah Kibang, Langap-Laban Nyarit, Laban Nyarit-Tanjung Nanga, Metut-Pelancau, Long Lake-Long Jalan). In five of these cases villagers reached agreement. Langap and Nunuk Tanah Kibang agreed to share future compensation payments from the coal company. Langap and Long Loreh reached agreement about an area under which lay valuable coal, by acknowledging that Loreh could continue to use existing cultivated plots in the Langap territory.

Where these five factors were not present, villages with conflicting boundary claims were not able to reach agreement. These villages were ultimately not able to

Table 1-3. *Results of Boundary Negotiations among Villages of the Upper Malinau River*

Village boundary	Agreement reached	Documented in writing	Agreement stable[a]
Lidung Keminci–Sentaban	—	—	—
Sentaban–Setulang	—	X	—
Setulang–Setarap	—	—	—
Setarap–Batu Kajang	X	X	X
Batu Kajang–Gong Solok	X	X	—
Batu Kajang–Adiu	X	—	—
Gong Solok–Adiu	X	X	X
Adiu–Loreh	X	X	X
Adiu–Nunuk Tanah Kibang	X	—	X
Long Loreh–Gong Solok	X	—	—
Long Loreh–Nunuk Tanah Kibang	X	—	X
Long Loreh–Langap	X	X	X
Langap–Seturan/Punan Rian	X	X	—
Langap–Nunuk Tanah Kibang	X	X	X
Langap–Laban Nyarit	X	X	X
Langap–Tanjung Nanga	X	X	—
Laban Nyarit–Mirau	X	—	—
Laban Nyarit–Halanga'	—	—	—
Laban Nyarit–Tanjung Nanga	—	—	—
Laban Nyarit–Metut	X	X	X
Laban Nyarit–Pelancau	X	X	X
Laban Nyarit–Long Lake	X	X	X
Tanjung Nanga–Seturan	X	—	X
Tanjung Nanga–Metut	X	—	—
Metut–Pelancau	—	X	—
Pelancau–Long Lake	X	X	X
Long Lake–Long Jalan	X	—	X

Note: X indicates yes; — indicates no.
[a]Stability means that the two villages involved had not challenged the agreed boundary as of July 2000.

sustain a supportive political base. We observed in particular that community members in these villages frequently did not support agreements produced by their leaders and in several villages refused to map the suggested boundary. Two villages also had the practical problem of not being sure whose territory adjoined their own because their borders were far. They had not prepared for negotiations with these neighbors, and preliminary agreements had to be renegotiated (Gong Solok–Long Loreh and Batu Kajang–Adiu).

Results of the Boundary Demarcation Process

Of the 27 boundaries among villages in the upper Malinau, villagers negotiated 21 agreements (Table 1-3). During the period of the mapping, most villages relied on written agreements between villages, which for many was a new development. Verbal agreements had previously been more common for boundaries. Written agreements were produced as public announcements signed by two parties and sometimes further signed by the local subdistrict leader (*camat*). In the past, one village had attached the signatures from attendance at a meeting to a statement of supposed agreement and produced a map showing their own version of the boundary. Trust in written agreements appears to be increasing despite such past abuses. Trust in verbal agreements has certainly declined, perhaps because they are seen as less legitimate and no longer binding.

Negotiations conducted transparently with written agreements were more stable than those that were not. Of the 21 boundary agreements, 14 were stable, while 7 changed within the seven-month period of mapping. The 14 stable agreements were based on more transparent negotiations in which negotiators shared information about the process or contents of their meetings; 10 of the 14 stable agreements were written.

Within six months, however, nearly all villages requested changes even to previously stable boundaries. We attribute these demands to the increasing activity of the small-scale timber permit (IPPK) holders during the latter half of 2000 and the introduction of a new provincial provision enabling villages to claim compensation from timber companies for timber previously harvested. Both changes led villages to increase their land claims even further. Also contributing to the fluidity was the lack of involvement from a higher institution with the authority to provide formal recognition of boundaries and control ad hoc revisions. With decentralization, just where this authority lies is not clear, although many have assumed it is now with the districts. The establishment of the new Malinau District has delayed the local government's involvement in the boundary demarcation to date. As the district asserts its authority and endorses boundary agreements, we can expect to see more stable results.

Conclusions

What does the Malinau experience in boundary demarcation indicate about the kinds of conditions necessary to facilitate better coordination with other stakeholders and improved negotiations by communities? What do they add to our understanding of emerging principles related to multi-stakeholder processes? We summarize our observations below and draw conclusions about the lessons that could be generalized to other settings.

Understanding the Complexity of Local Power Relations

Boundary negotiations in Malinau highlighted the deeply political nature of co-ordination efforts, even among seemingly homogeneous community groups. Portraying agreement-building as apolitical or neutral would ignore fundamental power relations that were influential in Malinau. Those power relations were expressed in the way conflicts were defined (e.g., mistrust between villages), the inequalities that made agreements hard to reach or less stable (e.g., differences in capacity and status among villages), and the lack of representation and attention to negotiations (e.g., among weaker or more marginalized groups).

The Malinau case showed the problems of basing negotiations solely on people's voiced interests, as leaders covertly discussed among themselves and changed their minds about previous agreements. Instead, we suggest that a fuller understanding of the diverse political relationships among groups can facilitate coordination and negotiation. Such an understanding is necessary for dealing fairly with differences in power among stakeholder groups, particularly in selecting representatives, choosing forums for decisionmaking, and identifying subjects of negotiation that deal with the different aspects of the conflict.

Special effort is needed to encourage effective participation and representation of weaker or disadvantaged groups. We suggest that at a minimum facilitators of coordination pay attention to these differences and give certain advantages to weaker groups. These include, for example, distributing information to them earlier, giving them priority access to resources, and facilitating their preparations for negotiations. More significant measures for longer-term empowerment include organizing community members, assisting them to mobilize resources, and helping them to develop strategic alliances. All of this depends, of course, on the desire of the group in question to receive such extra attention, and facilitators should take care not to create an identity of disadvantage that prevents the group in question from empowering themselves. Facilitators also need to take care not to alienate more powerful groups in the process by creating unfair advantages or overprotecting the group in question.

Dealing with Fluidity

In Malinau, the more intense the underlying struggle, especially in the absence of a third-party authority, the more fluid the villagers' interests, agreements, and coordination. Periodic opportunities to claim resources—as with the advent of decentralization, the changing monetary value of local resources, the creation of the new district, and the introduction of the mapping activity—directly increased these struggles. Villagers sometimes actively avoided reaching stable decisions in part because they lacked knowledge of an appropriate solution, given the rapidly changing conditions.

During periods of fluidity, a focus on managing conflict to maintain constructive debate is likely to be more productive than forcing an agreement. Instead of investing in formalizing and implementing agreements, facilitators can anticipate conflict and seek only tentative agreements that require a testing period. In this way, agreements reached quickly can be tested for their loopholes. Villagers can also use this time to develop a shared understanding of what the agreements imply. Investment in implementing the agreement could occur after evidence of reasonable stability.

Where constant modifications of agreements cause frustration and incur high costs, we observe the need for negotiators to build more supportive political constituencies from both the top and bottom. In Malinau, transparent decisionmaking and consultation with communities were key to achieving and then keeping an agreement with villagers. The presence of a third party with authority and legitimacy above the level of the village to set the criteria for resolving conflicts and to validate and enforce legitimate agreements could have ended much debate. Such top-down measures probably need to include districts making use of provincial and national policies and agencies to reinforce local decisions.

Improving Accountability in Decisionmaking

In Malinau, only a handful of people were involved in negotiations. These representatives, if the label is even apt, were weakly accountable or unaccountable to their communities. Networks, communication, and trust were frequently strong among selected leaders, or between leaders and companies, but much weaker between leaders and their constituencies. These conditions made it difficult for conflict to be managed in transparent ways, which kept disagreements from being acknowledged and agreements from being implemented.

Abuses of power are likely to persist unless certain checks are put in place. Broader consultation with factions in communities and better reporting back to communities can assist with building transparency. Complete accountability to villagers would be difficult to implement, however, with groups such as the Kenyah and Merap, because of their tradition of hierarchical control by the aristocratic class. In Malinau, the history of upward accountability of government representatives, the hierarchical nature of customary leaders, the strong local networks, and the pressures for striking quick deals have led to regular abuses of power that will not change easily.

Increasing Legitimacy and Coordination

Given the on-going struggles and highly unequal power relationships in places like Malinau, the potential is great for abuse of power in multi-stakeholder coordination processes. Such abuse can cause existing conflicts to escalate, particularly under the conditions of greater openness enjoyed now in Indonesia, and result in protests involving disruption of work, degradation of forest resources, and destruction of property, as has already been seen in Malinau.

In Malinau, the present institutional gap has left no clear customary or government authorities for settling conflicts. Self-interest and the close relationships between many government officials and local customary leaders have made many citizens question the legitimacy of these authorities. There is thus a need to build coordination upon stronger foundations of governance about what citizens and authorities consider legitimate operating principles and outcomes.

Conclusions

Our experience in facilitating boundary demarcation in Malinau marked only the beginning of a long and multistranded process for achieving better coordination among the very diverse stakeholders interested in Malinau's forests (see Annex 1-2 for some preliminary impacts). The research demonstrated the current vulnerabilities in coordination and agreement-making in Malinau. The political support for coordination is often fragile; few safeguards exist to ensure fair negotiations for weaker groups; and no clear authorities are in place to support and endorse these processes. Very real gains have been made, however, in empowering local communities to begin the processes of asserting claims to their territories and of establishing debate about rights associated with those claims. Communities, government, and companies are now keen to complete those processes, and by heeding the lessons of Malinau's experience, they may succeed in generating their own local brand of democracy.

Notes

The project was conducted by the following members of the Bulungan Adaptive Collaborative Management and Core Mapping Teams from 1998–2000: Salmon Alfarisi, Sargius Anye, Njau Anau, Ramses Iwan, Pajar Gumelar, Miriam van Heist, Godwin Limberg, Made Sudana, Nyoman Wigunaya, Asung Uluk, and Lini Wollenberg. We wish to express our thanks to and acknowledge the support of the following parties: the local government of Malinau District; Roem Topatimasing and INSIST; Carol J. Pierce Colfer, Kuswata Kartawinata, Steve Rhee, David Edmunds, Yurdi Yasmi, and Herwasono Soedjito from CIFOR; Jalong Lawai, and Paulus Irang of Long Loreh; Samuel ST Padan and WWF-Kalimantan Action Network; WWF-Kayan Mentarang; Ade Cahyat and Konsortium Sistem Hutan Kerakyatan-Kaltim; Niel Makanuddin and Plasma; Franky and Yayasan Tanah Merdeka; Amin Jafar and Yayasan Padi; H. Sayo and Pemberdayaan Pengelolaan Sumberdaya Alam Kerakyatan; Mairaji and Lembaga Pemberdayaan Masyarakat Adat; and Jon Corbett and the University of Victoria. The work was jointly funded by ITTO (primary donor) and the International Fund for Agricultural Development.

1. Coordination refers here to decisions that seek to achieve an aim on behalf of a group in light of the many self-interests of individual group members. Coordination can be self-organized (DiZerega 2000; Ostrom 1999) or imposed from outside. We assume here that coordination is likely to be more successful where it can balance self-determination by group members with institutions at the group or supragroup level that maintain authority and legitimacy to make and enforce decisions on behalf of the group.

2. We use Rossi's (1997) interpretation of deliberative here: dialogue and discussion that operate in an "engaged mode, somewhere between mere respect and confrontation" (*205*); deliberative democratic decisionmaking refers to a process in which individuals seek to go beyond their self-interest—although such interests might be part of the dialogue—and make decisions based on their perception of the common good.

3. Sections of this chapter are drawn from the CIFOR report *Pemetaan desa partisipatif alat penyelesaian konflik batas: Studi kasus desa-desa daerah aliran Sungai Malinau, Januari s/d Juli 2000, Tim Pengelolaan Hutan Bersama secara Adaptif*, 2001.

4. These recommendations are drawn from Edmunds and Wollenberg 2001 and Wollenberg et al. 2001c, the latter of which is a synthesis of other papers.

5. For the earliest periods we can only draw evidence from historical documents and oral histories and try to extrapolate from conditions observed in more traditional villages, although the latter is risky (see Sellato 2001 for a historical overview of Malinau during the past 150 years).

6. The Kayan, Merap, Kenyah, Punan are all local ethnic groups, as are Tidung, Berusu', and Lundaye (see below). The Punan are a hunter-gatherer group, traditionally, whereas the other groups survive by shifting cultivation. The Tidung are Muslim and were organized into a sultanate in the past.

7. Fox (forthcoming) characterizes overlapping sovereignties in the late 1800s in Thailand, writing that sovereignties were "neither single nor exclusive" (2), but rather (citing Winichakul 1994, *88*) "capable of being shared—one for its own ruler and one for its overlord—not in terms of a divided sovereignty but rather a sovereignty of hierarchical layers."

8. According to Sellato (2001), Dutch control in the Bulungan sultanate began in 1850 with a *Politiek* contract, was furthered in 1877 with an agreement for the Dutch to handle some of the sultanate's affairs, and was formalized in the late 1880s as part of the Dutch colony. In the early 1900s the Dutch forced the sultan to turn over control of the remoter regions of the Bahau River, Pujungan River, and Apo Kayan. They also worked with the sultanate, for example, to put down a Dayak rebellion in 1909 in the Tidung lands (which include the current Malinau River area).

9. According to Sellato (2001), in 1950 Bulungan became a *Wilayah Swapraja* (autonomous territory) of Indonesia after the Japanese occupation, and then in 1955, a *Wilayah Istimewa* (special territory). In 1959, after the last sultan, Jalaluddin, died in 1958, the sultanates were abolished, and Bulungan became an ordinary district (*Daerah Tingkat II* or *kabupaten*).

10. Prior to the Merap, it is believed that the Berusu' and Punan occupied the area (Kaskija 2000; Sellato 2001).

11. Even though demands for compensation had been made previously, villagers received few, if any, benefits in response.

12. This is not to imply that conditions before decentralization were always lawful or fair. There are numerous examples of small-scale illegal logging and other illicit and unfair activities from the prereform era.

13. Historically they have lacked the strong social cohesion of groups such as the Kenyah or Lundaye and have lacked effective institutions for representing their interests. Only in the mid-1990s did the Punan in Malinau organize the appointment of a Punan customary leader.

14. We defined agreement as concurrence between two villages about the location of their boundary.

15. We used *strength of leadership* (economic status of leader, e.g., food surpluses; quality of home construction; access to significant or regular cash income; possession of productive assets, such as rice mills, or luxury items like satellite dishes; alliances with powerful external groups; support of leader by community; and level of leader's education), *cohesiveness of community* (economic status of community, e.g., similar to leader but more broad-based; internal loyalties and mutual supportiveness; alliances with powerful external groups; skills and education levels; support of leader by community; and level of leader's education), and *access to information* (transparency of mapping process within village; knowledge of their territory) as indicators of a village's institutional capacities and power.

16. A score of 0, 0.5, or 1 was assigned for each of the three dimensions above.

Annex 1-1. *Scores of Village Capacity and Power Status*

Village	Strength of leadership	Strength of community	Access to information	Village score
Sentaban	Weak	Weak	Medium	0.5
Setulang	Medium	Strong	Strong	2.5
Setarap	Medium	Medium	Medium	1.5
Batu Kajang	Strong	Strong	Medium	2.5
Gong Solok	Weak	Weak	Medium	0.5
Adiu	Medium	Strong	Medium	2
Long Loreh	Weak	Medium	Strong	1.5
Langap	Weak	Weak	Strong	1
Seturan	Weak	Medium	Medium	1
Nunuk Tanah Kibang	Medium	Medium	Medium	1.5
Laban Nyarit	Strong	Medium	Weak	1.5
Mirau	Weak	Weak	Weak	0
Halanga'	Weak	Weak	Weak	0
Tanjung Nanga	Strong	Strong	Medium	2.5
Metut	Weak	Medium	Medium	1
Pelancau	Medium	Medium	Medium	1.5
Long Lake	Medium	Medium	Medium	1.5
Long Jalan	Weak	Strong	Medium	1.5

Note: The village score derives from strength of leadership, community strength, and access to information (value: Weak = 0; Medium = 0.5; Strong = 1). Strength of leadership is measured by economic status, external connections, community support, and education. Strength of community is measured by internal loyalties, economic status, human resources, and external connections. Access to information is measured by knowledge of their territory, knowledge of the mapping activities, and community discussion.

Annex 1-2. Impacts of Mapping

As of November 2001, more than a year after the completion of the mapping, we observed several important impacts of the negotiation process and mapping activity. First, a new awareness emerged among all stakeholder groups of the location and extent of different villages, as well as the value of mapping as a means for making claims to land. This awareness can be considered a necessary basis for coordinated landscape management. Although CIFOR did not distribute maps of a village to others and clearly marked maps as drafts, villagers themselves often shared them (especially with local investors), and CIFOR displayed the maps in several meetings with other stakeholders.

Second, new types of boundaries emerged as some villages (e.g., Tanjung Nanga and Langap) reconceptualized boundaries as straight lines or along roads rather than natural features. For most groups, the conceptualization of territory subtly shifted, from being defined by a center settlement point or a main river of a watershed to having an emphasis on outer boundaries. For some groups, especially the Punan, we suspect that the mapping has reinforced the historical trend of gradual territorialization of previously nomadic or shifting groups, a trend accompanied by an increasing tendency among inland groups to want to register their land as property, and even to seek private rather than communal property. It is too early to tell whether such changes are significant and benefit or harm Malinau's populations and forest, but they do signal changing attitudes and values related to land. Developments in the policy environment related to *adat* and IPPK claims will strongly influence how these changes play out.

Third, community capacities for mapping and negotiation improved. Small teams of people in each mapped community gained experience with the methods and equipment necessary to geographically reference and plot a series of points in their village on a map. Understanding of maps—including scales, legends, orientation, and their uses—became stronger in each village. Through the process of negotiation and with input from CIFOR, communities have greater understanding of representation and the need for building a wider political base of support for reaching an acceptable agreement. This understanding increased among a broad range of community members. Many community leaders were savvy enough to know, and decide when to use more (or as was usually the case, less) participatory approaches. Since the project began, community members have increasingly demanded that their leaders use more transparent, inclusive processes for consultations and decisionmaking.

CHAPTER TWO

Dealing with Overlapping Access Rights in Indonesia

Stepi Hakim

THE MANAGEMENT OF FOREST RESOURCES in tropical countries is a major political issue because many stakeholders claim such resources, and many conflicts arise over the benefits derived from forests. In some cases, stakeholders understand that cooperation is necessary to manage forest resources effectively and efficiently, and they agree to collaborate in the interest of everyone (Borrini-Feyerabend et al. 2000). Fisher and Jackson (1998) have pointed out that action research has a lot to offer such efforts to implement collaborative management of protected areas. However, the link between participatory action research (PAR) and collaboration in multi-stakeholder environments has not yet been studied much. This chapter describes some experience and lessons learned using PAR to enhance collaboration and effective forest management in a multi-stakeholder environment in Pasir, Indonesia.

What Is Participatory Action Research?

Many authors have presented definitions of PAR (see Fals-Borda and Rahman 1991; Rahman 1984; Maclure and Bassey 1991; Selener 1997; Chein et al. 1948; Wadsworth 1998). Fals-Borda and Rahman (1991) define PAR as "action research that is participatory, and participatory research that unites with action [for transforming reality]." It is not merely a research methodology on the relationship between subjects as they evolve in social, economic and political life, but also social activism committed to promoting people's practical application of knowledge, or praxis (ibid.). Wadsworth (1998, *16*) noted that "PAR is research which involves all relevant parties in actively examining together current action (which they experience as problematic) in order to change and improve it. They do this by critically reflecting on historical, political, cultural, economic, geographic, and other contexts which make sense of it." Many authors consider PAR as a type of action research (see Chein et al. 1948; Masters 1995; Selener 1997).

The concept of action research is recognized from the work of social psychologist Kurt Lewin in post–World War II America. Lewin described action research as "proceeding in a spiral of steps, each of which is composed of planning, action, and the evaluation of the result of the action" (Kemmis and McTaggart 1988). In practice, a group identifies problems of mutual concern and consequence. The group then decides to work together on a "thematic concern" to solve the problems. Kemmis and McTaggart (1988) defined thematic concern as "the substantive area in which the group decided to focus its improvement strategies." Then the group decides to carry out planning, action, and observation (individual or collective), and reflection together. From the reflection, the group reformulates (by analyzing the facts) new actions from the lessons learned. This sequence implies that the group may change plans for actions as it learns from experience.

Selener (1997, 35) pointed out that this process of a dialectical relationship involving theory and practice (praxis) would guarantee that the problem to be addressed is not just theoretical, but arises from practical experience. However, Moser (1975 as cited in Finger, undated, 261) argued that it is difficult to unify theory and practice and noted that the relationship between the two is the central problem of the dominant positivist social science.

There are two goals underlying this approach. One is to increase the correspondence between the actual problems encountered by the group and the theory used to explain and resolve them. The second is to assist the group in identifying and making explicit fundamental problems by raising the group's collective consciousness (Holter et al. 1993, cited in Masters 1995). When the group gains an increased understanding of the issues and analyzes them, this shared knowledge becomes the basis for implementing specific actions. Selener (1997, 41) noted the following four iterative steps: description of current perceptions of the issues; questioning of the representation of the issues or problems; reformulation of the problems, which includes building strategies of actions; and finally implementation of actions, which will shift the reality as initially analyzed. As new issues arise, these in turn need new analysis and new solutions (Selener 1997, 42). The research is an on–going process, and successful results can lead to innovations (Jackson 1993, 10).

Action research develops into participatory research when the action unites researcher and relevant stakeholders. Selener (1997, 17) noted that there are three principal activities in participatory research: research, education, and action. It is research in that the group is actively involved in identifying problems. It is education because the group together analyzes and learns about the causes of and possible solutions to the problem addressed. It is action because findings are implemented in the form of practical solutions (Selener 1997, 17).

In a participatory process, members of the group are not simply involved as data providers or recipients of research findings. They actively participate by implementing all activities during the research process. "They are the main actors in collectively identifying the research problem, the way said problem should be studied, the methods chosen to analyze data, the implementation of the research activity per se, and the transformation of results into action" (Selener 1997, 17). Although

PAR is quite close to a common-sense way of "learning by doing," it is very hard to achieve the ideal conditions for putting it fully into practice (Wadsworth 1998:16).

Several obstacles make PAR difficult to put into practice. One is the issue of the group's investment. Ideally, the group that is to take the action must be involved from the very beginning, and therefore the group needs to have the willingness to be involved in the process. This investment occurs if the group realizes that it needs the action program. Without such investment, research diagnoses and recommendations for change brought in from outside groups tend to stimulate insecurity rather than motivate changes (Chein et al. 1948, 5). Another obstacle to PAR, noted by Jackson (1993,6) is that it is more difficult to conduct than conventional research, because it requires researchers to be flexible in approach, to continually challenge assumptions and views, to call upon interdisciplinary skills, and to use a wide variety of methods to collect data. Chein et al. (1948) added that another obstacle is the limited capacity of lay people to participate in scientific research. The need for action may often conflict with certain research requirements, such as the need for simplification of the research process and the need for integration of the work of the specialists with that of lay collaborators (ibid.; see also Chapter 11). Some of these obstacles will be discussed further in the context of the case study described later in this chapter.

PAR, Social Learning, and Collaboration

Buck et al. (2001,12) noted that "participatory research can stimulate social learning by bringing different groups together through a conscious and deliberate cycle of inquiring, observing, reflection, planning and acting." In other words, the process generates learning within the groups (see Asanga 2001). This joint or social learning (sharing of knowledge and experience) fosters perceptions of interdependence and mutual appreciation. Thus, it facilitates "working together toward agreed-upon goals, generating confidence between and within the group in further efforts at collaboration" (Buck et al. 2001,2). The Romwe case in Zimbabwe shows that a social learning approach to management is one way to overcome differences in institutional priorities, sources of support, legitimacy, and culture and to encourage effective collaboration (Nemarundwe 2001,93; see also Chapters 7 and 9).

McNeely (1995) presented various forms of collaboration in resource management in many countries. Collaboration, by definition, is built on each collaborator's having something to give to the relationship and something to gain from it. Various definitions of collaborative management have been presented (see McCay and Acheson 1987; Berkes et al. 1991; Borrini-Feyerabend 1996; Brown 1999). Borrini-Feyerabend (1996,2) defined collaborative management as a partnership in which various stakeholders agree to share among themselves the management functions, rights, and responsibilities for a specific area or a set of resources. Collaboration emerges as a result of participation. A collaborative project could not be implemented properly in the field, if representative key stakeholders were not involved in all stages of the process. Participation of key stakeholders thus fosters collaboration by generating understanding and mutual trust.

Moreover, Brown (1999) noted that the principle of collaborative management requires the involvement of local communities, as well as the state, in forest management. He pointed out that the imbalance in power between industrial and nonindustrial users and the political will of key government agencies create barriers to meaningful community participation in forest management. The following case study outlines how PAR is able not only to create learning opportunities among stakeholders, but also to support collaboration among them in addressing forest management issues.

Case Study: Overlapping Access in Forest Management

The villages of Rantau Layung and Rantau Buta are located in the western part of Lumut Mountain, Pasir *kabupaten* (district), East Kalimantan, Indonesia. Administratively, both villages belong to *kecamatan* (subdistrict) Batu Sopang (about 200 km from the city of Balikpapan). Parts of these villages are located within the forest conservation area of Lumut Mountain, a 35,350-ha protected area declared by the government of Indonesia, through *SK Menteri Pertanian* (Decree of Minister of Agriculture) No. 24/Kpts/Um/I/1983. About 17% (5,892 ha) is located within the village of Rantau Buta, which covers 16,546 ha and includes 500 ha for shifting cultivation (Dinas Perhutanan dan Konservasi Tanah 1999). Data on the area of conservation forest within the village of Rantau Layung are unavailable, but the village covers 18,913 ha and includes communal forest, protection forest, wild gardens, shifting cultivation, and human settlement. Areas of production forest (outside of the conservation forest) are located in both villages and form the basis for the communities' livelihoods.

Demographics

According to the village *monograf*, a descriptive, official document available for all Indonesian villages (Anonymous 1999), the population of Rantau Layung is 50 households, or 213 people. The sex ratio is balanced: there are 113 males and 100 females. Rantau Buta, based on the village *monograf* (Anonymous 1999), has 20 households, or 82 people, of which 42 are male and 40 female. Both villages have similar educational levels. About 92% of Rantau Layung residents have graduated from elementary school, but only 2% have graduated from junior and senior high schools. About 6% of the people have had no formal education at all. In Rantau Buta, 90% of residents have graduated from elementary school, and 6% have no formal education. Because only one elementary school serves Rantau Layung (as well as the neighboring village of Rantau Buta), community members have to leave the village if they want education beyond the primary level. Faced with the inconvenience of river travel as the only means of transportation and the expense of living away from home to attend secondary school in the subdistrict capital, most community members prefer to stay in their village (based on interviews with key informants in the village and results of focus group discussion).

Economic Activities

When Rantau Buta and Rantau Layung residents estimated their calendar of activities[1] in one year, they determined that shifting cultivation took the greatest amount of their time. Second place was rattan collection, and third was lumbering. The fourth place for Rantau Buta community members was cultivating other crops (gardening), whereas for Rantau Layung residents, it was hunting.

The area of Lumut Mountain is rich in biodiversity and forest resources. Local harvestable plants (those with diameter exceeding 10 cm) include *buni* (*Aglaia* sp.), *wayan* (*Aglaia tomentosa*), *terap* (*Artocarpus elasticus*), *nato* (*Madhuca sericea*), and *red meranti* (*Shorea leprosula*). According to a government survey (Dinas Perhutanan dan Konservasi Tanah 1999), about 312 m^3/ha of harvestable plants are located around Rantau Buta. In addition, the area has a potential rattan production of around 1.2 tons/ha. Other forest products that are useful for the community are honey and *gaharu* (*Aquilaria* sp.). Given the area's rich biodiversity, there is no doubt that many stakeholders are keenly interested in the area.

Stakeholders and Their Interests

Our observations have identified 20 stakeholders in the research sites. These are listed in Table 2-1. Central issues in the area are the rights of access to forest, land, and products and the authority and responsibilities of stakeholders to manage the forest. Inconsistent policies on forest management have produced overlapping access rights and even some unrestricted or open-access situations, creating a "tragedy of the commons."

From the Pasir District government's perspective, the permissible activities of communities surrounding a protected forest are limited, in that no timber extraction is allowed. Nonetheless, the government issued a district policy in 2000 to provide licenses for communities to extract timber outside the protected forest

Table 2-1. *Lumut Mountain Stakeholders*

On-site	Off-site
Community of Rantau Layung	• Neighboring villages (Kasungai, Batu Kajang)
• Farmer groups	• Owner of the sawmill
• Youth (women and men)	• Subdistrict formal government (Batu Sopang)
• Forest workers	• Forest workers from neighboring villages
• Elderly (men)	• Logging companies (PT Telaga Mas, CV Teguh
• Village elite (formal government and informal customary institution)	Maronda Prima)
	• District Forestry Service
	• Regional Planning Agency
Community of Rantau Buta	• District Environmental Impact Agency
• Farmer groups	• NGO (Padi Foundation)
• Youth (women and men)	• Research institutions (CIFOR, Center for
• Forest workers	Social Forestry and Environmental Studies
• Elderly (men)	Center, Mulawarman University)
• Village elite (formal government and informal customary institution)	

areas. But, a license is granted only if a community owns the area as production forest. This condition stimulates logging companies to invest capital in the area by building partnerships with individual communities. The terms of the partnerships are questionable, however, since the communities lack financial capital, information, and legal clarity about their rights. Although communities in some sense own the area and receive a small fee for timber extracted, they have little power to monitor the use of forest resources (despite improvements with the passage of the recent Forestry Act 41/1999).

As a result, the logging company PT Teguh Maronda Prima, or TMP, holds the power in the management of forest activities. The communities have no power to monitor the amount of forest resources taken out by TMP. Moreover, communities understood that TMP would also take over responsibility for forest rehabilitation, an arrangement that was not in fact specified in the regulation issued by the local government. Degradation of forest resources has increased dramatically since TMP, having no sense of belonging to the area, took over management of the forests. Even though in April 2001 the local government withdrew its policy of granting extraction licenses in preparation for revising and reissuing policy guidelines it has made no formal announcement of a new policy terminating the activities in the field. As a result, TMP has been carrying on its timber exploitation.

Communities use the forests mainly for fulfilling their basic needs, that is, opening areas for shifting cultivation[2] and gardening (rattan and vegetables). They have managed forests sustainably for centuries, and their farming management systems are adaptive (see Colfer et al. 1997). In other words, they manage not to overuse their soils or convert large areas to *alang-alang (Imperata cylindrica)* or other weeds. Since local communities have used and maintained the areas for centuries, there is a strong argument to be made that ownership of the areas (for use purposes) belongs to the communities. However, the status of ownership is debatable. The government of Indonesia has claimed that it recognized the indigenous people living in and surrounding the forest, but in fact it has failed to provide secure access for poor people.[3]

Another overlapping access issue is the status of the forest concessionaire, PT Telaga Mas, or TS, in the area. The forest (outside the protected areas) has been occupied by TS as a concession since 1973. Inside the concession area, however, TMP is extracting timber. Forestry Act No. 5/1967 (revised as Forestry Act No. 41/1999) states that, to avoid overlapping areas, demarcation of the forest for production purposes has to be clear before the forest is utilized. In other words, utilization of the forest by more than one company in the same area is not allowed. In this case, however, policy conflicts between the district and national governments create overlapping access to the area's forests. It is understood that TMP has a right to extract timber in the area through a decree by the district head, whereas TS has a right to utilize the forest through a ministerial decree.

In solving the problem of overlapping access, a collaborative approach is needed to develop a shared understanding of what policies are in place and how they affect access rights. Thus, the main stakeholders (TMP, TS, communities, and local governments) need to collaborate to define rights and responsibilities for forest management in the area. The Center for International Forestry Research (CIFOR) and Padi Foundation have been carrying out collaborative research in the area since 2000. A central research question developed by CIFOR is:

Can collaboration among stakeholders in forest management in Rantau Layung and Rantau Buta—enhanced by processes of conscious and deliberate
social learning—lead both to improved human well-being and to the maintenance of forest cover and diversity? If so, under what conditions?

The next section outlines how we used PAR to address this issue.

The PAR Process

PAR encompasses the cyclical steps of reflection, planning, action, and evaluation.

Reflection 1

Through community meetings and discussions with researchers and field facilitators, it became clear that forest degradation is increasing. Many stakeholders expressed concern about forest exploitation, but discussions also revealed confusion
about rights and responsibilities in forest management. The community challenged
researchers to simplify CIFOR's research questions on adaptive and collaborative
management and to deliver them to relevant stakeholders. The community also
expressed its concerns about the local economic situation and its interest in addressing such issues.

Reflection 1 and Plan 1

Identify stakeholders and analyze their interests. Conduct assessments and collect
relevant information about local socioeconomic and forest policies.

Action 1

Participant observation and literature review were undertaken, and the results of
these were discussed among different stakeholders through meetings.

Evaluation 1 and Reflection 2

The research team observed that stakeholders initially could not collaborate on
forest management issues. During meetings, communities could not speak up to
other stakeholders, which was understandable since the communities had gotten
very little information on forest issues and had become accustomed to comparative powerlessness in interaction with external stakeholders.

Although it was theoretically possible to address the central forest management
issue directly by inviting TMP, TS, communities, and local governments to multi-
stakeholder meetings on that topic, we feared that bringing so many stakeholders
together would lead to power struggles and confrontations. Since most stakehold-

ers barely knew each other, there had been difficulties in building solid relationships and mutual trust. Thus, the researchers, facilitators, and communities decided to work on developing collaboration by first addressing local issues over which there was less disagreement, rather than directly tackling central problems that could give rise to more dangerous conflicts.

Plan 2

Identify and prioritize local issues that involve less conflict among interested stakeholders. To avoid direct confrontations or conflicts of interests among stakeholders in one forum, carry out discussions separately at the community level. Then organize district level discussions among all relevant stakeholders (only key community representatives would attend this discussion). Use facilitators to disseminate results from discussions at the village and district levels to all relevant stakeholders before the workshop at the district level.

Action 2

Discussions and meetings were conducted at the village and district levels to identify and prioritize local issues. The results of these were shared among all concerned stakeholders at the district level,[4] building scenarios of collaboration with interested stakeholders to address local issues. Facilitators helped to keep discussions on track, ensuring that local communities had a significant voice and avoiding misunderstanding of the issues among stakeholders.

Evaluation 2 and Reflection 3

Collaboration among different stakeholders is possible if the local issues addressed are not conflict laden. Collaboration in our sites occurred by building consensus among interested stakeholders to address local issues.[5]

Plan 3

Address the locally defined issue of using areas previously devoted to shifting cultivation for alternative livelihoods by seeking funds to rehabilitate the areas. The areas used in the past for shifting cultivation were considered degraded lands, so they were a concern not only for the community but also for the District Forestry Service (DFS), which wanted to rehabilitate them.

Action 3

The communities and researchers made a proposal to DFS jointly. Agriculture training in the communities was carried out jointly with DFS, the Department of

Estate Crops (DEC), and the Department of Agriculture and Horticulture (DAH). The communities of Rantau Buta and Rantau Layung made agreements among themselves to form farmer groups based on existing swidden clusters. In Rantau Layung there are three groups and in Rantau Buta, two groups. These groups also prioritized the most useful types of crops to be planted. The communities decided that each household would rehabilitate 1 ha of land in a former shifting cultivation area. The area would be used for on-farm agricultural research. Each farmer would observe the plants that survived in the first-stage planting process. Each group agreed to conduct regular meetings to exchange information among members.

In September 2001 the local government approved funding for this proposal. However, the communities recognized they might have to wait a while, because the money from the central government had not yet been delivered to the district government.

Evaluation 3 and Reflection 4

Because the money had not arrived yet in the district government, the rehabilitation activities were not going smoothly. However, the communities realized that local government was eager to collaborate with communities to rehabilitate village forests and lands. The provision of trainers by DFS, DEC, and DAH made it clear that these institutions were committed to supporting the people's efforts.

The private companies also began trying to rebuild a relationship with the communities. The company TS offered seedlings of white teak (*Gmelina* sp.), although these were refused because the communities did not see the benefit of planting such trees. The communities asked the company to provide fruit trees instead of white teak. And so the process continues.

Discussion: Leading to Collaboration?

The activities described have led to three general outcomes. First is the gradual increase in the involvement of stakeholders in addressing local issues. Second is the improvement in the ability of the local communities to make their voices heard by other stakeholders (TMS, TS, and district government). Third is the gradual development of a consciousness among relevant stakeholders to seek opportunities as a group for collaboration, communication, and learning from the actions they plan.

Previously, local communities had very limited involvement in planning and decisionmaking for addressing local problems. Problem-solving was a top-down process dominated by the subdistrict government. It was no surprise that genuine local problems could not be addressed effectively. The subdistrict had a strong influence on the governance of the village: we observed that communities rarely challenged information from the subdistrict and believed that the subdistrict had the authority to decide which village could receive funds (which is not true). Local communities were afraid to question the subdistrict officials. These fears resulted in less participation by communities in the process of addressing their problems. Forest management is an example of this.

Since PAR was introduced in Rantau Buta and Rantau Layung, however, local communities have become more aware of the problems affecting the forest resources in their village (see Reflection 1). They have recognized as well the need to participate in identifying problems and determining alternative actions to address the problems. They have found that by participating in village meetings they were able to realize, identify, and prioritize the problems. We observed that the communities grew more confident when they collectively identified their own problems. This confidence is a potential asset for communities in building collaboration with outsiders.

Previously, communities did not know the outsiders (TMP and TS) exploiting their resources. They recognized only the names of the companies, but did not know the people who were in charge of the activities, largely because collaboration among stakeholders did not exist in the area (see Reflection 2). However, by participating in meetings, discussions, and workshops, community members gradually recognized the outsiders. We found that these meetings are the best place to build relationships among stakeholders and also the best setting for other stakeholders to hear the voices of community members. The role of the facilitator is very important in this context. Before all the relevant stakeholders were able to sit in one forum to address an issue, the facilitator had to disseminate information to all of them. This helped to build relationships and avoid miscommunication. Facilitators had to be able to maintain information flows and to support discussions among relevant stakeholders.

The steps described above closely parallel the research cycle in action research. Attempts among relevant stakeholders to be involved and contribute to addressing the issues turned out to make the research more participatory, that is, to make it participatory action research. PAR is easier said than done, however. Our PAR activities face several obstacles:

- Avoiding false assumptions: Our early context studies and our ongoing involvement with the communities revealed a number of inaccurate assumptions that we held, and which could have caused serious problems.
- Poor coordination and commitment: At the outset, stakeholders were unaware or distrustful of each other and unaccustomed to working together.
- Political uncertainty: Some actions planned by stakeholders had to be postponed because of the national political situation.
- Bureaucratic delays: Government procedures for local communities to gain approval of their actions were long and cumbersome; see Reflection 4).

Coordination and commitment among relevant stakeholders is the biggest obstacle in PAR. To achieve successful coordination and action, representatives of all key stakeholders have to participate in setting goals and identifying constraints, as well as being involved in planning and negotiating for the future. Fisher and Jackson (1998) noted that mutual understanding and trust among relevant stakeholders could occur only if there is a genuine commitment to meaningful collaboration. Token involvement of less powerful stakeholders will not be enough.

There had been no previous collaboration in forest management among the relevant stakeholders in Rantau Layung and Rantau Buta. Thus, although it was not possible to address directly the main issues through collaborative group action, participants consciously saw there was an opportunity to build collaboration. Through learning gained from discussions and actions, the stakeholders were able to modify the approach (by first addressing less controversial local issues) to develop collaboration and mutual trust.

Conclusions

The objective of this chapter has been to show that PAR offers options for building collaborative management of forests in the confusing context of Indonesian forest management. Our experience in collaborative management in Rantau Buta and Rantau Layung remains incomplete. There is still no official community forest management—forests are still legally owned and managed by the government. But there is widespread interest in making Indonesian forest management more equitable. Our work is part of that effort. We are in the process of building collaboration among all the relevant stakeholders. We are in a negotiation phase in which local communities are working with officials in the district government and private companies to identify and agree upon the best approaches to sustainable community forest management. If greater collaboration occurs among the stakeholders, questions remain about the type of collaboration that will prevail: How equitable will it be? Will it be real collaboration or just a fake? Such questions will have to be addressed to achieve our hopes for collaboration.

Evidence so far suggests that PAR is feasible and provides a way to support conditions where collaboration among stakeholders could occur. As a type of research, our experience leads us to conclude that PAR helps stakeholders identify problems, analyze the experiences gained, draw conclusions, and plan new or improved actions. Through PAR, members of the Rantau Buta and Rantau Layung communities, private companies, and the district government learned many lessons in addressing local issues. PAR provides opportunities to participate and to learn from one's actions, which can lead to better future actions.

To achieve collaborative management among relevant stakeholders, we noted that there must be a genuine commitment among them to collaborate. We noted that meetings and discussions are the best venue for developing mutual understanding and trust among stakeholders. We believe that external facilitators can contribute significantly to the process of collaboration, although the question still remains whether meaningful collaboration can be maintained when external facilitators are phased out.

Notes

This chapter has been produced with financial support from the Asian Development Bank under RETA Grant 5812 and from CIFOR. The views expressed herein are the author's and do not necessarily reflect the official opinion of the Asian Development Bank or CIFOR.

1. The data were acquired from focus group discussion and open discussion through the application of a pebble distribution game. To allocate the time communities spent on various

activities over a year, we provided empty sheets for a seasonal calendar of activities each month. Members of each group filled the sheets with all their activities related to forests and lands for each month. Each group allocated its working time by placing pebbles on every activity conducted in one year. The number of pebbles distributed represented the amount of time allocated to each activity. Subsequently, the results of each group were discussed openly among groups to obtain responses and suggestions for improving the accuracy of estimates of total time allocated for each activity over the year.

2. The local community residents who live within the forest area mainly practice shifting cultivation, in which activities are conducted only once a year and rotate from one piece of land to another. There are two cycles of cultivation in Rantau Buta and Rantau Layung communities: the short cycle (5–6 years) and the long cycle (10–15 years). The length of rotation is based on soil fertility: when soil fertility has already declined, the long cycle is applied. The rationale for this practice is to provide opportunity for the land already cultivated to restore its original fertility. Generally, after planting various fruits and rattan on land with low levels of fertility, the community leaves the land fallow for some time (some fallow land is not planted with anything because it is located far from the village and thus would require much time and effort to harvest the products). The fallow land left with plants growing on it still requires regular maintenance and supervision. However, land that has been fallow for a long time may be left without regular maintenance, resulting in an unsatisfactory harvest for the community—though ecological functions are maintained.

3. In the late 1990s about 15,000 Indonesian households lived in and around the Lumut Mountain forests (Forestry Planning Agency 2001), but these people could not manage their areas to meet their livelihood needs because the forests are considered state land and regulated by state law.

4. The Indonesian governmental hierarchy is village, subdistrict, district, province, nation; the district level has gained in authority with recent political changes.

5. Local issues identified include: (a) need for infrastructure building, community involvement in forestry development, and marketing of forest and garden products (rattan, timber, fruits, honey, coffee, rice); (b) need for development of main commodity trees by planting teak, rubber, durian, and rambutan and annual crops in areas of past shifting cultivation; (c) need for teachers for elementary schools and financial support for further education; (d) need for development of human resources (due to low educational level); and (e) need for development of agriculture (on dry land) and agroforestry.

CHAPTER THREE

Participation and Decisionmaking in Nepal

Sushma Dangol

C OMMUNITY FORESTRY HAS BEEN defined as active management of forests through direct participation of local people (Arnold 1992). In Nepal, a community forest is part of the national forest, which the District Forest Office hands over to community user groups for development, protection, use, and management, in accordance with a work plan. The user groups are free to fix the prices of forest products and to sell and distribute them for the groups' collective benefit and welfare (Shreshtha et al. 1995, 2).

In Nepal, the community forestry program was originally based on handing over barren and degraded forest land to user groups. But now emphasis is on devolving all protection, management, and other responsibilities to the users, with proper technical support from the Department of Forestry.

Traditionally, the people of Nepal were dependent on forests for fuelwood, fodder, and timber. As long as population pressure was low, the local supply of forest products was sufficient. Increased human and livestock population and government policies on land registration contributed to the gradual depletion of forest resources. The participation of local people in community forestry became essential after the failure of conventional approaches, where attempts to manage forest resources were carried out without the people's participation.

The current approach is bottom–up. It involves all the users who directly depend on the forest for their livelihood, including disadvantaged groups whose survival depends on use of forest resources. The approach posits that, if people from disadvantaged groups are involved in decisionmaking about rules and practices, their needs and interests are more likely to be taken into account. This study deals specifically with differences in access and voice between the elite and the less advantaged, including lower castes, ethnic groups, women, and poor people.

In my view, community forests can be sustainably managed only if all concerned stakeholders participate in making decisions about how the forests are used.

Therefore, it is important to understand some participatory approaches that enhance active learning processes and thereby empower the users themselves.

The Ladder of Citizen Participation

Sherry R. Arnstein (1969) has identified a number of ways in which the term "participation" is used. The main difference in these uses is the level of decisionmaking involved. Arnstein has presented the framework in the form of a ladder with eight rungs, in which the level of participation increases as one goes from the bottom to the top. Here, I am using her framework to analyze the change in the level of participation by different stakeholders in participatory action research on the adaptive collaborative management (ACM) of community forests, which aims to ensure the sustainability of the forests and the well-being of their users. The elements of this framework are shown in Figure 3-1.

The bottom two rungs of the ladder, Manipulation and Therapy, represent non-participation that is contrived to substitute for genuine participation. The real objective at this level is not to enable people to participate in planning or conducting programs, but to enable powerholders to "educate" or "cure" the participants. Rungs 3 and 4 (Informing and Consultation, respectively) progress to levels of tokenism that allow the "have-nots" to hear and to have some voice. When powerholders proffer these token levels as the total extent of participation, citizens may indeed hear and be heard, but they lack the power to ensure that their views will be heeded by the powerful. When participation is restricted to these levels, there is no follow-through, no "muscle," and hence no assurance of changing the status quo. The fifth rung, Placation, is simply a higher level of tokenism in which the ground rules allow have-nots to advise, but the powerholders retain the right to decide. The sixth rung, Partnership, enables have-nots to negotiate and engage in trade-offs with traditional powerholders. At the uppermost rungs, Delegated power (7) and Citizen control (8), have-not citizens obtain the majority of decisionmaking power or full managerial power. Obviously, the eight-rung ladder is a simplification, but it helps to illustrate the point that there are significant gradations of citizen participation.

8	Citizen control	Degree of citizen power
7	Delegated power	
6	Partnership	
5	Placation	Degree of tokenism
4	Consultation	
3	Informing	
2	Therapy	Nonparticipation
I	Manipulation	

Figure 3-1. *"Ladder of Citizen Participation"*

Source: Arnstein 1969.

Disadvantaged groups in the context of Nepal's community forest management, with which this chapter is concerned, are those stakeholders who depend heavily on forest resources for their livelihood but have insufficient voice in decisionmaking about the management and use of those resources. Lower-caste people, poor people, and most women are often marginalized in the community forest management system and are therefore referred to here as disadvantaged groups.

Methods

Participatory action research (PAR) as part of CIFOR's ACM program has three broad processes: interaction among stakeholders; communication and learning among stakeholders; and collective action. These processes result in changes or adjustments to management for the greater sustainability of natural resources and the welfare of the communities (Prabhu 2000). This chapter is based on field experience at the ACM research site of Bamdibhir community forest (Bamdibhir and Bamdi are short forms of Bamdibhirkhoria) in Nepal. During our research, we used various methods such as key informant interviews, self-monitoring assessment records, and meeting minutes. We also used participatory rural appraisal (PRA) tools including: wealth ranking, Venn diagrams, social resource mapping, interaction within and between diverse stakeholders, cross-site sharing, self-monitoring tools, and heterogeneity analysis. Information was collected from disadvantaged groups including the marginalized, "untouchables," the poor, and women's groups.

The Bamdibhirkhoria Community Forestry User Group

The Bamdibhir site (one of two ACM sites in Kaski District) is situated in Phewa Tal Watershed, Chapakot Village Development Committee, in the Western Development Region of Nepal. It is in the forest governance area of Pumdi Bhumdi Range Post. The range post office is about a four-hour walk from the Bamdibhir Forest User Group (FUG) site. It is a 30-minute flight or five-hour drive from Kathmandu to Pokhara and about one hour by bus from Pokhara to the Bamdi settlement, where the FUG office is situated. The forest is located between Wards 3, 5, and 6 of the Chapakot Village Development Committee to the northwest of Lake Phewa.

The Bamdibhirkhoria forest had earlier been severely affected by a landslide toward the southwest side of the forest, resulting in dramatic erosion. The landslide area is covered by eroded stone deposition. The total area of the community forest is 48.5 ha, including both natural forest and plantation. The forest is composed primarily of tree species: *Katus (Castonopsis indica), Chilaune (Schima wallichii), Mahuwa (Engelhardia spicata), Uttis (Alnus nepalensis), Mallato (Macaranga postulata), Tindu (Dyospyrus montana), Angeri (Lyonia ovalifolia), Kaphal (Myrica esculanta), Semal (Bombax ceiba), Painyu (Prunus cerasoides), Kyamun (Syzygium operculata), Tooni (Toona serrata), Ankhitare (Trichilia connaroides)*, and *Siris (Albizia julibrissin)*. The following shrub species are also important: *Ghatu (Clerodendron viscosum), Archale (Antidesma bunius)*,

Dhayero (Woodfordia fruticosa), and *Ainselu (Rubus ellipticus)*. Significant herb species include: *Dattiwan (Achyrium bidentata), Ban Tarul (ioscorea alata), Siru (Imperata cylindrica), Pani Amala (Nephrolepsis cordifolia), Amliso (Thsanolaena maxima), Rudhila (Pogostemon benghalensis), Githa (Dioscorea bulbifera), Guhe Amala (Polystichum lentum), Seto Kanda (Rubus rugusus), Kukurdaino (Smilax aspera), Bamboo,* and *Gurju*. Among wild animals there are deer, rabbit, *kalij,* jackal, *kokal,* leopard, monkey, squirrel, porcupine, jungle cat, different types of birds (parrot, crow, *titra,* vultures), reptiles, and snakes, as well as butterflies.

The first forest protection committee was formed in 1979 at the beginning of the Phewa Watershed Project. The FUG, with the legal right to manage the forest, was formed in 1993 during the final period of the Phewa Watershed Project. The FUG comprises 134 households of various caste and ethnic groups, including Brahmin, Chhetri, Magar, and the deprived caste groups labeled as untouchables (e.g., Kami, Damai, Sunar, Sarki, and Chandara). The total population of the Bamdibhir FUG toward the end of the PAR phase was 722, with 384 males and 338 females. The largest population is Brahmins, followed by the so-called untouchables (Biswokarma, Damai, and Sarkis). The Magar ethnic group is third, and Chhetris, who are dominant within the national population (at 16%) according to the 2001 census, are fourth. Rais are minority in-migrants, along with the Bhujels. The average household size is five members.

Gender dynamics differ by ethnicity and caste group. In households of ethnic minority groups and low castes, the labor and decisionmaking are often shared more equally among men and women than in high-caste households. In all households within hill districts such as Kaski women spend more than 12 hours a day on farm and domestic work activities, compared with a much lower amount for men (see Bajracharya 1994, Gurung 1995). Women's role in household decisionmaking is less obvious and is affected by many variables other than gender, but is found to be higher in households of low castes and ethnic minority groups (Adhikary 1987).

In the Magar and Brahmin communities in our ACM site, both men and women are involved in domestic work and collect forest products as well. Fodder collection is the job of the women, as is the collection of most nontimber forest products. In lower castes, such as Damai, Kami, and Sarki communities, only the females are involved in domestic chores and, contrary to the general pattern described above, they have relatively less influence in household decisionmaking. In all caste and ethnic groups, timber extraction is the job of men.

Despite women's active engagement in day-to-day forest use and management, their participation in formal management has been low, in part because of their lower social status and the widely held perception that such activities are outside of women's domain (cf. Chapters 8 and 14). Gender norms and women's own reluctance to speak publicly and assume leadership roles have hampered their ability to participate equally with men in committees governed by formal rules and procedures (Denholm 1990).

Gender relationships within rural Nepal are changing in response to policies and programs of government agencies and nongovernmental organizations (NGOs) that target women for agriculture and natural resource management activities, leadership training, and empowerment. Another key source of change is an increased level of long-term outmigration of males, leaving women as de facto heads of households and therefore responsible for more decisionmaking and labor.

The major sources of livelihood and income for households in the Bamdibhirkhoria FUG are agriculture, livestock-raising, business, and service jobs in government offices, NGOs, and schools. Some people depend mainly on remittances from Qatar, Saudi Arabia, and Malaysia.

The socially marginalized castes such as Damai, Kami, and Sarki follow their traditional occupations of tailoring, blacksmithing, and shoemaking, respectively. In the past they depended completely on these occupations. As payment their clients provided food grains at each harvest. Today, because of ready-made and industrial products, their occupations are at risk. A number of them have small plots of private land for agriculture. The poor and marginalized caste members with little or no private farmland work for wages. They plow, weed, plant, sow, and reap others' grain fields, and they repair and construct houses. The local daily wage rate is Rs.150 for women and Rs. 200 for men. (The exchange rate was Rs. 78.6 to the US dollar in May 2003.)

Most of the people in the Bamdibhirkhoria FUG spend nine months of the year engaged in agricultural activities. They are involved in the complete cycle of farming, from sowing and planting to weeding and harvesting. The main cereal crops are rice, wheat, and maize. Farmers spend the remaining three months repairing and constructing houses, collecting firewood, and visiting relatives. They plant potato, onion, garlic, ginger, and other vegetables in their free time. A number of young people have been recruited to the Indian Amy. Others have gone to Qatar and Saudi Arabia and other countries to work. Some of the young men go to Pokhara and Kathmandu for higher education. Thus, unlike the situation for the entire village (which consists of nine wards), in the three wards where we focused our efforts, there are more elderly people and women.

During the context-study phase of our research, we conducted well-being ranking and committee discussions in different hamlets. The Bamdibhirkhoria FUG members were divided into high, medium, and low wealth groups. Information on land-tilling status and food sufficiency of the caste and ethnic groups was collected in each ward and later verified by the committee members. A large number of Brahmin households were found to cultivate their own land exclusively. Most of the Brahmins and Magars also cultivated some land that they rented from others. Besides renting from others, some Magars cultivated part of their own farmland and rented the rest to others to cultivate.

Later, during the PAR phase, the FUG committee members and each hamlet carried out well-being ranking on the basis of their own local criteria (see Table 3-1). Only three households among the Bamdibhir FUG members were classified as rich. While most of the households fell in the medium category, a significant share (one-fifth of the households) was considered to be poor. Brahmins and the Magars were generally considered better off: they had a smaller percentage of poor members, and among the three rich households, two were Brahmins and one was Magar. These two groups were seen to compete somewhat for power in Bamdibhir. The Chhetris were found to be in the minority and weaker in Bamdibhir than they are in Nepal as a whole. The marginalized caste groups such as Biswokar-

Table 3-1. *Distribution of Households in Bamdibhir Forest User Group, by Wealth and Caste or Ethnic Group*

Caste or ethnic group	Rich households No. (%)	Medium households No. (%)	Poor households No. (%)	Total households No. (%)
Brahmin	2 (3.70)	47 (87.04)	5 (9.26)	54 (40.0)
Magar	1 (3.22)	27 (87.10)	3 (9.70)	31 (23.0)
Chhetri	—	3 (75.00)	1 (25.00)	4 (3.0)
Biswokarma	—	19 (87.86)	9 (32.14)	28 (21.0)
Damai	—	6 (50.0)	6 (50.00)	12 (9.0)
Rai	—	2 (100.00)	—	2 (1.5)
Sarki	—	—	1 (100.00)	1 (0.7)
Bhujel	—	—	2 (100.00)	2 (1.5)
Total	3 (2.24)	104 (77.61)	27 (20.15)	134 (100)

Source: Local Heterogeneity Analysis, updated in July 2002.

mas were mostly found either cultivating others' farmland or doing wage labor; these groups, including the Damai, Sarki, and Bhujels are typically marginalized and poor all over the country.

Using ACM to Encourage Disadvantaged Groups

There has been increasing recognition of the important roles that researchers and facilitators play in processes such as ACM. For that reason, I describe my own role and the steps involved in the ACM process as practiced in Bamdibhir:

I am of Newar ethnicity, born in Kathmandu, Nepal. I earned my Bachelor's degree in Forestry at the Pakistan Forest Institute, Peshawar. My work experience started in October 1995, when I was employed as an extension organizer in the DANIDA/Regional Training Center Dhangadhi in the far west of Nepal. In this capacity, I organized, designed, and facilitated several awareness programs and training for forest dwellers and government staff. Since May 2000, I have been working for the Center for International Forestry Research (CIFOR) as a field researcher, conducting participatory action research on the adaptive collaborative management of community forests in Nepal. The research aims at investigating, promoting, and developing approaches, strategies, tools, and methods in a collaborative environment that ensures sustainable forest management and human well-being. As a researcher, I am responsible for communicating and negotiating the research agenda with local people and diverse stakeholders and for assessing, analyzing, and reporting on the ongoing research.

On site, I played the role of agent of change, acting as observer, facilitator, supporter, initiator, analyst, advocate, and catalyst in partnership with the forest users. When ACM first began in the research area, I was principally an observer. I tried to create a friendly, comfortable environment in which all users, especially women, the poor, and "untouchables," would feel free to express their feelings, experiences, interests, and ideas, either individually or in groups. During the context study, I spent a significant amount of time in each FUG, gathering information informally while building trust with the less vocal sectors of the community. I mobilized the disadvantaged groups using various PRA tools, such as wealth-ranking, Venn diagrams, social-resource mapping, histo-ecological matrices, historical time lines, key informants, informal interviews, group and informal discussion, and simply through encouragement.

Later, I attended the Forest User Group Committee (FUGC) meetings and interacted with the different groups organized by women, men, caste and wealth groups, and NGOs. I also observed the FUG's activities for planning, monitoring, making decisions, sharing benefits, enacting rules and norms, and mobilizing resources. I supported and advised the FUGC in formulating mechanisms for implementing the planned actions and reviewing and planning further actions. If conflict arose, I played a conciliatory role.

In monthly meetings and informal discussions, I routinely noted who spoke and who did not, who seemed interested and who did not. Afterward, I was able to make a plan for mobilizing the users. I visited homes and discussed the issues in a friendly way with the people. I started to know about the users' problems and followed their suggestions. As a result, users felt increasing ownership of the ACM program.

PAR in Bamdibhir FUG started with a participatory self-monitoring workshop held March 6–9, 2001. At the workshop, community members participated in activities, such as mirror and network games, to analyze the existing situation, develop a vision and indicators to measure changes, and prepare action plans for six prioritized issues. After the PAR interventions, I initiated discussion in the hamlets (*tole*) to lay the groundwork for regular meetings and facilitated the FUGC and *tole* meetings, indicators assessment, *tole* committee meetings, FUGC and *tole* representatives meetings, implementation process, constitution and operational plan preparation, and general assembly. With the PAR intervention, the FUG plans were identified and implemented, the reflection process occurred in a more participatory way, and the level of action increased. The FUGC and disadvantaged groups now monitor their activities by heterogeneity analysis and periodic reassessment of self-monitoring indicators. Community members prefer discussing upcoming issues in their hamlet before coming to the monthly meeting. Users from both disadvantaged and elite groups feel that they can meet me whenever they want to discuss and share their thoughts.

Comparative Analysis of the Effectiveness of the ACM Process

In this section, I compare the level of participation of disadvantaged groups in six ACM interventions, using before and after scenarios. The important examples of interventions, described below, include participation in forest management decisionmaking, participation in meetings and training, policymaking, information-sharing, participation in activities or implementation, and benefit-sharing.

Decisionmaking

Before PAR. The Bamdibhir FUG made decisions and managed the community forest at several levels. A general assembly was supposed to be held every six months.

In fact, it was rarely held once a year. When I arrived the assembly had not been held for three years. The Forest User Group Committee was formed or changed every two years, and it too did not meet regularly. There were no regular hamlet-based meetings on forestry issues. The FUGC chair was dominant in making decisions about sharing costs and benefits related to forest resources. Most of the time, other people kept silent and accepted the chair's decisions. There was top-down decisionmaking in the FUG and limited communication from the FUGC to FUG members. There was very limited input from FUG members, particularly the poor and marginalized, in decisionmaking (see Table 3-2).

There were two main decisionmaking forums: the FUGC meeting and the general assembly. When the committee called a general assembly, FUGC members would inform the households in their hamlet about it. The committee meetings were held once a month, and committee members attended them and discussed the agenda items presented by the chair. The chair was from a higher caste, rich and powerful. He had been an active member of the district development committee (DDC) and liked to deal with all the forest-related issues. In meetings, the chair's plans strongly influenced the decisionmaking process. On a number of occasions, the chair decided and wrote the decisions in the minutes, and other members of the committee simply signed. The committee meetings and general assembly were often postponed due to the lack of a quorum.

Users gave limited input to the committee in the decisionmaking process. Few committee members or users were involved in forest-related plans and activities. General users tended to be unaware of their roles and responsibilities and failed to carry out responsibilities assigned to them by the assembly to enhance the community forests. Magar and Brahmin men and women attended meetings to give individual input on decisions more frequently than did those from the poor and marginalized groups. There was no planned mechanism to include input from hamlet-level decisions to the FUGC.

Biswokarmas (a lower caste) were completely absent from decisionmaking. Attendance of women and marginalized castes in meetings or assemblies was extremely low. There were only 3 women (2 Brahmin, 1 Sarki) in the FUG committee

Table 3-2. *Distribution of Bamdibhir Committee Members, by Gender, Wealth, and Caste or Ethnic Group, before Participatory Action Research Intervention*

Caste or ethnic group	Number of males				Number of females				Total	
	Rich	Medium	Poor	Total	Rich	Medium	Poor	Total	No.	%
Brahmin	1	2	0	3	0	2	0	2	5	45
Magar	0	2	0	2	0	0	0	0	2	18
Biswokarma	0	3	0	3	0	0	0	0	3	27
Sarki	0	0	0	0	0	1	0	1	1	9
Damai	0	0	0	0	0	0	0	0	0	0
Chhetri	0	0	0	0	0	0	0	0	0	0
Bhujel	0	0	0	0	0	0	0	0	0	0
Rai	0	0	0	0	0	0	0	0	0	0
Total	1	7	0	8	0	3	0	3	11	100
Percentage	9	64	0	73	0	27	0	27	100	

Note: Percentages may not total 100 because of rounding.

out of a total of 11 members. Moreover, the women's voices in these forums were often ignored. They did not like speaking in public, and if they spoke they were often not heeded. Even if they spoke about practical forest management activities (like thinning and pruning), the men took their remarks lightly. The men's behavior disheartened the women.

After PAR. Decisionmaking for all activities is now more process-oriented. Monthly meetings and the annual assembly have become the main forums for dealing with forest-related issues such as benefit-sharing, conflict resolution, income and expenditures, work–plan approval, illegal cutting, leadership development, information-sharing, and training. The hamlet (*tole*) meetings now also provide an important forum for decisionmaking, planning, and conflict resolution, in addition to the FUG committee meetings and general assembly. The annual work plans, the constitution, and the operational plan are now prepared through these forums by involving a variety of stakeholders and socioeconomic groups. More hamlets have also started to participate in the committee meetings. There are four *toles* in the FUG (Raikar Magar; Maidanpokhari; Okhaldunga; and Bamdi), and each *tole* has elected a *tole* subcommittee. The *tole* committees have started to conduct monthly meetings.

Records indicate that FUGC meetings are held on a regular basis. At the monthly FUGC meetings, men and women members participate more comfortably and actively in discussions and decisionmaking. Women have gained representation on the FUGC; of the 11 members of the executive committee, now 6 are men and 5 are women (see Table 3-3). Poor people are also actively participating in meetings and decisionmaking forums. Now people from different hamlets discuss forest-related issues in their own hamlets and make a preplan, which is then discussed in the monthly meetings.

People from higher castes and the rich are pleased with the introduction of *tole* meetings and have expressed satisfaction with the decisions that are made, including distribution of costs and benefits. Poor people are also happy with the

Table 3-3. *Distribution of Bamdibhir Committee Members, by Gender, Wealth, and Caste or Ethnic Group, after Participatory Action Research Intervention*

Caste or ethnic group	Number of males				Number of females				Total	
	Rich	Medium	Poor	Total	Rich	Medium	Poor	Total	No.	%
Brahmin	0	3	0	3	0	2	0	2	5	45
Magar	0	1	0	1	0	1	0	1	2	18
Biswokarma	0	2	0	2	0	1	0	1	3	27
Sarki	0	0	0	0	0	0	0	0	0	0
Damai	0	0	0	0	0	0	1	1	1	9
Chhetri	0	0	0	0	0	0	0	0	0	0
Bhujel	0	0	0	0	0	0	0	0	0	0
Rai	0	0	0	0	0	0	0	0	0	0
Total	0	6	0	6	0	4	1	5	11	100
Percentage	0	55	0	55	0	36	9	45	100	

Note: Percentages may not total 100 because of rounding.

practice of meeting separately in their own hamlets, but they express less satisfaction with the decisions. Poor people think that they are spending more time in meetings and getting nothing immediately, but they also feel that they have increased their knowledge about different forest activities and possibilities for the future.

Women from all segments of the community now participate more actively in decisionmaking. Conflicts with their existing family responsibilities remain a difficult issue, but now they are more aware of community forest potential. They show interest in all activities and have created relationships with different stakeholders, shared information with them, and increased the level of mutual understanding. The conduct of ACM's context studies and initiation of PAR involved the use of various PRA tools: self-monitoring, heterogeneity analysis, and system analysis, which encouraged women, especially those from lower castes and the poor, to participate in discussions.

While women were initially unable to express their interests and opinions in public decisionmaking forums, for fear of showing disrespect to their male relatives or being ignored by them, now women are observably more involved. Respected people are aware of the involvement of women in meetings and in community forestry activities, and women are also interested enough to participate in every monthly meeting. They have the opportunity to share their feelings and opinions, communicate with each other about new occurrences, and to stimulate more effective decisionmaking with their input.

Participation in Meetings and Training

Before PAR. Previously, mainly FUGC members participated in training opportunities, and the chair of the FUGC was dominant in choosing who would get training. The government line agencies and the development projects used to inform the chair or the FUGC first about training, workshops, and study tours. The person who got the message would take advantage of the opportunity to participate in the training without any decisions in the FUGC. Often the one who received the message would offer such opportunities to his close relatives as well. Little consideration was given to equitable representation by gender, caste, or ethnicity or to implications for general well-being in the selection of individuals to be offered such opportunities. For example, the District Soil and Water Conservation Office provided nursery-foreman training, which might have benefited the FUGC nursery foreman, and the District Forest Office provided record-keeping training, which could have helped the FUGC treasurer. However, the chair did not inform these individuals. He quietly participated in all types of training and workshops himself.

Although the users objected to his behavior, they were too afraid of him to challenge his decisions. Furthermore, disadvantaged groups did not have much free time for participating in meetings, study tours, and training. Thus, they missed all the capacity-building and skill-acquiring chances. They also missed important information about such matters as community forestry policy, rules, and regulations. The few people who did participate in training opportunities hesitated to share the information out of fear of the elites.

After PAR. After the self-monitoring workshop, the participants assessed their FUGC's strengths and weaknesses. They noted the FUGC chair's dominant role in decisionmaking and proposed incorporating the opinions of users from the various *toles.* They made an action plan that involved *tole*-based user groups contributing to decisionmaking in the FUGC, including the selection of participants in training, workshop, and study opportunities. The participants developed selection criteria for committee membership, to be followed at the next assembly, including balance by gender, caste, ethnicity, class, and *tole.* The *tole* committee has been formed and *tole*-based issues have been brought up in the FUGC meetings by the *tole* committee representatives.

Now, people from disadvantaged groups are very interested in receiving training and participating in study tours. They appreciate these opportunities. Committee members have started to think that women should be involved in different meetings, training, and tours. For example, in the hamlet meetings and self-monitoring reassessment, the ratio of men to women participating is 60:40. A few women and poor people are able to express their opinions on important decisions in the community. Women are encouraged to attend the forest management activities and training. There were 13 women participants out of 16 at the three-month long bamboo craft training; none of the trainees was in a higher than medium wealth group. Indeed, four were classified as very poor, and three more as poor.

Policymaking

Before PAR. Nepal permits the District Forest Office (DFO) to hand over any part of the national forest area to a locally organized Forest User Group to manage and utilize the products of the forest. Nepal now has more than 11,000 FUGs, which are established and legally recognized by the government. In Kaski District, the Phewa Watershed Project started its activities in the Bamdibhir area in 1977. This project played an important role in carrying out activities to control landslides in the area.

Phewa Watershed Project staff and village leaders consulted the DFO and the ranger (DFO staff) to form the FUG, so that local communities would be able to use and manage the local forest. At that time only a few people from the user group gave input to prepare the required operational plan (OP). Users from Bamdibhir community forest did not know about the OP or the constitution that was prepared by the Phewa Watershed Project staff. Even committee members had no idea about the OP's contents because of their lack of involvement in its preparation.

After PAR. Presently, the FUG starts the process of revising the operational plan, and members of every hamlet are involved. Most of the committee members have realized that the assembly, which finalizes the OP's revision, should be preceded by a self-monitoring exercise. Discussions to achieve more gender balance in the committee are under way. The FUG is planning to have an assembly this year to approve the revised operational plan. DFO staff also attended the self- monitoring review meeting of FUG members.

After the self-monitoring workshop, the FUG recognized in their action plans the need to revise the constitution and operational plan. After the next regular FUGC meeting, the FUGC members and *tole* representatives agreed to discuss the contents of the constitution and the operational plan at the *tole* level. In the *tole* meetings, they discussed how to link the self-monitoring indicators under the different main headings of the constitution and operational plans. One *tole* also organized a mothers' group meeting, and that group kept written minutes for the first time. The FUGC representatives from this *tole* played an essential facilitating and organizing role. After joint discussion (with ACM facilitation) between some FUGC members and *tole* representatives about how to conduct the *tole*-level meetings, it was agreed that the *tole* members would examine and suggest revisions to the constitution and operational plan as needed.

In the Okhaldhunga *tole* meeting, participants gathered at the Temple of Shiv. The secretary of the FUGC facilitated the session. He highlighted the importance of the *tole* meetings for developing a viable self-assessment process and realizing the group's own program of community forestry. The vice-secretary of the FUGC also focused on the importance of self-monitoring indicators and the need to involve people to incorporate self-monitoring into the process of revising the constitution and operational plan. Another FUGC member further stressed the importance of involving the disadvantaged (poor, women, and "untouchables") in the *tole* meeting, as well as in community forest management more generally.

Tole meetings focused on some visioning and on the annual work plan, with periodic attention to the operational plan. The participants, as part of the self-monitoring reassessment workshop, stressed the importance and necessity of further improvement of visioning and the action plans. In this way, the FUGC representatives organized the other three *tole* meetings and one mothers' group.

The discussions held by the *tole*-level subgroups on the contents of the constitution and operational plan are significant outcomes that have helped to raise awareness among all FUG members about forest-related activities. There is a real change from a top-down decisionmaking process, mainly dominated by the FUGC or its chair, to a more bottom-up process that incorporates the views of a much wider and more representative group of community members (Table 3-4).

The Bamdibhir CFUG held a general assembly on March 9, 2002 to amend their constitution and operational plan, which resulted in the preparation and approval of a new 10-year operational plan and constitution. The community's vision, as well as indicators to monitor whether they are proceeding in a direction consistent with their vision, was included in the constitution and operational plan. The constitution provides that the indicators used for monitoring will be reviewed every six months. Subsequent planning will be based on what is learned from the monitoring. The annual work plan planning process followed by the Bamdibhir CFUG is summarized in Figure 3-2 (from McDougall 2001). The arrows at the top of the figure are intended to show the iterative and collaborative nature of this process.

Table 3-4. *Participation in Tole Meetings about the Constitution and Operational Plan, by Gender, Wealth, and Tole*

Tole	Rich participants		Medium participants		Poor participants		Total participants		Total
	Male	Female	Male	Female	Male	Female	Male	Female	
Bamdi	2	0	9	7	3	2	14	9	23
Raikar	0	0	14	17	4	1	18	18	36
Maidan Pokhari	0	0	5	3	0	2	5	5	10
Okhaldunga	1	0	10	7	1	6	12	13	25
Total	3	0	38	34	8	11	49	45	94

Information-Sharing

Before PAR. The formal forums to share information between members were monthly committee meetings and the annual general assembly. Information was also exchanged informally among members in teashops, small gatherings, and work places. If any members got training, participated in a workshop, or went on a study tour, they communicated their ideas informally to those who were interested. No system was developed to share information systematically.

Although the committees and assemblies existed, the decisions and discussions were largely led and influenced by the FUGC chair. The chair consulted with the related government and development institutions and prepared the FUG's annual plans. He called the assembly to rubber stamp his decisions. Information flow regarding training, tours, or workshops was not transparent, even within the committee. The chair used to go himself or send his cronies to benefit from such opportunities. After the training the participants did not share their new ideas or knowledge with other FUG members. There was no mechanism in the FUG for general user members to give input to the committee regarding the community forest's management. Thus, the decisionmaking and information-sharing processes in the FUG were individualistic, centering on the actions of the FUGC chair.

After PAR. The formal forums for information flow between members are monthly committee meetings, *tole* committee meetings, and the annual assembly. The committee calls the general assembly once a year. *Tole* meetings and FUGC meetings are held once a month. Committee decisionmaking has been delegated to two committee members from each *tole,* who conduct *tole* meetings, inform *tole* users of all decisions, and present all *tole* decisions at the FUGC meetings. Every *tole* representative and FUGC member simply describes his/her group's agenda items. These items are then prioritized and discussed in turn, and decisions are taken. The *tole* representatives subsequently inform users in their *tole* what has been decided in the FUGC meetings.

Participation in Activities or Implementation

Participation here refers to the involvement of disadvantaged groups in making decisions and in forest management activities, before and after ACM. Initially, I used a pebble distribution method (Colfer et al. 1999a) to analyze the level of participation; once the PAR process was under way, I used heterogeneity analysis, a moon phase self-monitoring tool from Criteria & Indicators (Pandey 2002), and Arnstein's "Ladder of Citizen Participation" (Arnstein 1969).

Before PAR. Participatory decisionmaking processes were not in place before ACM. Some elite persons and technical staff controlled the decisionmaking. The chair of the FUGC basically did what he liked. In the general assembly, and sometimes in meetings, the elites, who dominated the general assembly, discussed what the committee chair presented and approved it. If marginalized castes, people of low economic status, or women attended the meetings or assembly, they could not

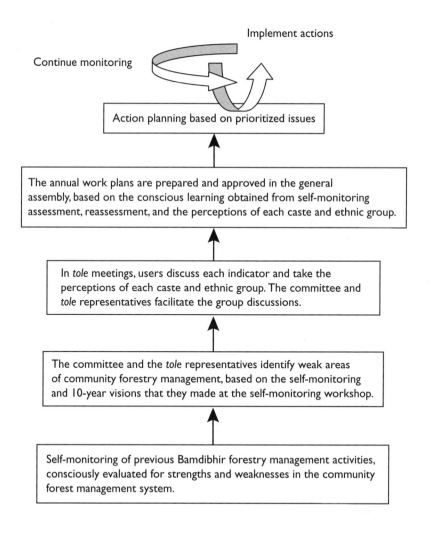

Figure 3-2. *Annual Work Planning Process Followed by the Bamdibhir Forest User Group Committee after Participatory Action Research (PAR) Intervention*

express their views. They were ignored or patronized by the local elite. The only purpose of the assembly was to form a new FUGC under the chair's control.

Because "untouchables" and the poor members have to work all day long to earn wages for the evening meal, it was thought they would be unable or unwilling to participate in public meetings. Similarly, women were thought to be busy with household work. If some women attended a meeting, they could not express their interests and opinions for fear of showing disrespect to the men in their families.

After PAR. Since the PAR intervention the level of participation among disadvantaged groups (women, "untouchable" castes, and other users) in the Bamdibhir community forest management system has been increasing. At the self-monitoring workshop, community members identified problems with representation in their management system. They initiated regular *tole* meetings, and planning and implementation of the annual work plans by *tole*s. Now *tole*s plan and make decisions about goals and work plans for the FUG. The processes in place for annual planning and revising the constitution and operational plan give equal opportunities to all castes and ethnic *tole* residents to make decisions on community forest management.

The meeting process also changed after the ACM intervention. Now, in committee and *tole* meetings, participants collectively work out the agenda for discussion on each issue, giving all opportunities to express themselves. Decisions are recorded in the minutes with due consideration of everyone's opinions. The committee chairs, secretary, and other *tole* representatives facilitate rather than lead the discussion, as I have observed at meetings and from reports by FUG members. The result of introducing ACM processes in the Bamdibhir community forest user group has been positive, increasing the power and access of women and marginalized groups in making forestry management and community development decisions.

The pebble distribution results reported in Table 3-5 show that after the ACM intervention, the reported participation of FUG stakeholders increased. The participation of committees, the occupational castes, *tole* representatives, and the poor in the forest management systems increased sharply compared with that of the traditional elites.

Table 3-6 shows the participation of each caste and ethnic *tole* in decisionmaking before and after ACM, using Arnstein's "Ladder of Citizen Participation." As noted above, the level of participation of disadvantaged groups of the Bamdibhir community forest has been increasing since the PAR intervention. The ACM process minimized caste discrimination in the FUG and increased the power and access of marginalized groups in making forest-related decisions. The level of participation of women and marginalized caste groups in the FUG increased from Arnstein's therapy level, indicating very little input, to the partnership level, indicating much greater equality in participation.

Benefit-Sharing

Before PAR. Benefit-sharing was based on equality in the FUG. The community forest opened once a year for timber to be collected by those who were in need

Table 3-5. *Changes in Perceptions of Participation Based on the Pebble Distribution Method, by Caste and Ethnic Group, Institutional Membership, and Wealth*

Group	Before PAR intervention (2000)	After PAR intervention (July 2002)
Brahmin castes	45	55
"Untouchable" caste	35	65
Bhujel	40	60
Rai	44	56
Magar	39	61
FUG committees	30	70
Tole FUGC members	35	65
Rich	48	52
Medium	45	55
Poor	30	70

Table 3-6. *"Ladder of Participation" in Bamdibhir Forest User Group, before and after Participatory Action Research Intervention*

Level of participation	Steps of ladder	Before PAR intervention	After PAR intervention
Degree of citizen control	Citizen control	User committee and Brahmin *tole*	—
	Delegated power		User committee and Brahmin *tole*
	Partnership		Magar, Bhujel *toles*, "untouchable" castes, women's groups, and poor
Degree of tokenism	Placation	Magar *tole*	—
	Consultation	—	—
	Informing	"Untouchable" castes, women's groups, and poor	—
Nonparticipation	Therapy	—	—
	Manipulation	—	—

Note: "Ladder of Participation" is based on Arnstein 1969.

and could pay a certain price to the FUG committee. The FUG members were satisfied with the benefit-sharing process for ground grass and for fallen branches for firewood, which they could collect freely any time. For timber, however, the royalty rate was the same for the poor and the rich. The wealthier people were satisfied with the rate, but the poor and deprived-caste FUG members were not able to pay the charges to extract timber. The rate had been fixed on the minimum level that middle-class people could afford, rather than on what the poorest could pay. Thus, the poor and disadvantaged people were not benefiting from the timber of the community forest at all. When they needed timber to construct or repair houses, they had to wait several years until they could accumulate the money to pay the fixed price.

Wealthy people with large livestock herds were benefiting more from the system of free collection of fodder from the community forest than those with no or few livestock. Poorer members of the FUG noted this inequity and said that timber charges should be reduced for the poor; they argued that, since the poor people did not own much land, they actually needed more forest products and deserved a more equitable distribution system.

After PAR. The distribution system is much the same as before, although each hamlet's people and committee members are negotiating to try to create a more equitable distribution system. It takes time though. The FUG has decided that fuelwood collection should be once a month. FUG members and I have begun to analyze their situation in terms of benefit sharing, collaboration, forest sustainability, and well-being. They have also started to plan using situation analysis, with implementation followed by monitoring and evaluation, and to assess their FUG's access to decisionmaking about sharing benefits in terms of class, caste, ethnicity, and gender (Tumbahangphe 2002). Benefit-sharing is among the trickiest issues in forest management (cf. Chapter 13).

Conclusions

Decisionmaking is an important function in community forestry user groups. To be successful, significant decisionmaking should take place at the grassroots level. Most of the Bamdibhir FUG members realize that community forests can be managed sustainably only with the participation of all users. Though comprehensive participation is routinely emphasized in theory, in practice disadvantaged groups are frequently marginalized, whether because of their own hesitation, ignorance, and illiteracy or because of the dominance of elite groups.

Women and poor people have real constraints on their participation in decisionmaking. Women are engaged in daily household work, and poor people must work for daily wages for their livelihood. These obligations interfere with their ability to attend meetings, and as a result they are less aware of the decisions made by the FUGC. Furthermore, women's voices are often ignored in such decisionmaking forums. There are few women on the FUGC, and when they speak male members tend to treat their input lightly, which further inhibits women's willingness to speak out.

The proliferation of forestry-related training, workshops, and study tours available through government, NGO, and donors' programs—although intended to benefit forest users, especially the disadvantaged—also tends to work against equitable participation in management activities. Women, marginalized castes, and poor people have little time for such activities. They are more likely to be found working: planting, thinning, pruning, weeding, and collecting forest products. In addition, development projects initiated by other sectors, such as agriculture, health, and education, also solicit women and others to participate in meetings, workshops, training, and study tours. Often villagers experience serious tension in balancing time allocation between household and development activities.

In the Bamdibhir site a more inclusive decisionmaking mechanism is occurring now in the hamlets, where disadvantaged groups are encouraged to participate and interact with other local stakeholders. The ACM experience has provided a good opportunity for these groups to collaborate more effectively. In monthly FUGC and hamlet meetings, disadvantaged groups including women, the poor, and lower castes, are now routinely involved in planning, decisionmaking, and monitoring, enabling them to participate in all forest-related activities. The traditional elite (position holders, high castes, the rich, and social workers) are increasingly recognizing the importance of involving more marginalized groups in meetings and forestry activities, as a road to improving their well-being and strengthening sustainable forest management.

Although it cannot be said that the ACM process has solved all problems between the marginalized and the elite in Bamdibhirkhoria, both groups have made significant improvement in their abilities to work together to solve shared problems. My conclusion is that this kind of process is effective in adding value to the community forest processes already under way in Nepal.

Notes

This chapter is the outcome of mental, physical, and financial investments from officials and individuals that deserve my sincere acknowledgment. I would like to express my cordial thanks and gratitude to my mentors, Robert J. Fisher and Jeanette Gurung, and my editor, Yvonne Byron, who perpetually encouraged me and gave valuable time, guidance, suggestions, and feedback until the chapter was completed. I also express my sincere gratitude to Carol J. Pierce Colfer, Ravi Prabhu, and Cynthia McDougall, for editing, giving feedback, encouraging and participating in CIFOR's ACM writing workshop in November 2001. I am deeply indebted to my parents for giving me permission to participate in the writing workshop. It would have been difficult to complete the chapter without their valuable support and advice.

CHAPTER FOUR

Scientists in Social Encounters

The Case for an Engaged
Practice of Science

Mariteuw Chimère Diaw and Trikurnianti Kusumanto

T HE END OF THE TWENTIETH CENTURY was marked by an exponential growth of knowledge and power networks, along with an unprecedented scale of poverty. As the cognitive gap between global agencies and local
societies steadily grew, the way knowledge is routinely extracted from local people
became increasingly questioned. At stake is the model of development by external
injection—a model unique in history—that tries to address inequality and poverty
on a world scale. The nature, object, purpose, and integrity of science as a human
agency is also in question.

Alternative modes of investigation (e.g., Kuhn 1996; Giddens 1987) contributed to this nascent epistemological crisis by exposing the myth that science is inherently neutral and objective. In the 1980s came the idea that more inclusive and
participatory approaches were needed to reduce the gap between givers and takers
of information. Yet, old issues about just how participatory the new methodologies
were (Guijt and Cornwall 1995) or how really new they were (Chauveau 1992)
never went away. Questions arose about the knowledge or powers that were really
distributed in participatory projects. Criteria of "good science" were attributed
shifting meanings and assumptions; notions of validity, reliability, or objectivity
were challenged to include the judgment of the people involved in the investigation (Lincoln 1990; Chambers 1997). How development science—an extractive
activity by nature—could possibly remain scientific while organizing a system of
knowledge generation with clear populist traits was an implicit concern.

Some critiques saw that participation was being abused by the continued neglect of the social processes taking place during and following the use of participatory methods (Cornwall et al. 1994). Others called for a deeper understanding of
the determinants of social change, while questioning the sometimes obscure in-
tents and assumptions of investigators (Richards 1995; White 1996 in Groot and
Maarleveld 2000). In separate but closely connected fields, difficult questions were

coming to the fore on the relevance of development studies (Edwards 1989) and the reasons for the "impasse" of development sociology. This questioning led to new debates and attempts to reframe the complex relations between theory and action, between structure and agency in development studies (Booth 1994; Long 1992; Buttel and McMichael 1994).

Research on adaptive collaborative management of forests, on which this book is based, is at the crossroad of many of these efforts to move beyond the theoretical impasses of development science and the hesitations of action research and participation. This chapter focuses on the encounter of scientists with other people in the course of participatory research. Its aim is not to make a wide sociohistorical observation of participation or the role of science in society, but to make the link between fine microprocesses and three key theoretical questions about science and participation: How participatory is participation? Can methods be empowering? and How scientific is "participatory science"? It then proceeds to address, through theory and case studies from Cameroon and Indonesia, the conditions under which these questions can be answered positively. The chapter's first key argument is that local empowerment (that is, the acquisition and redistribution of powers over local resources and benefits) is possible mainly through a practical epistemology[1] based on in situ generation of interactive knowledge for action. Its second argument is that there is nothing inherently contradictory between science—an activity that requires, at some point, externalizing real-life processes and treating them as "things"—and participation. What is wanting is a renewed understanding of how the floating frontiers of theory and history manifest themselves in action-based science; how does this latter serve and be served by participatory processes?

The chapter relies on a social-learning perspective based on the analysis of "interface situations" (Long 1989) and the "interplay of individuals and situational factors in generating human behavior" (e.g., Maarleveld and Dangbégnon 1999). It recognizes the paradox between the empowerment paradigm of participatory approaches and the extractive nature of scientific knowledge and explores options beyond that paradox. These options are defined in the context of forest management, but aim at the wider range of issues of power, knowledge, and benefits in social encounters. We begin with a review of the participatory and social learning approaches of the last decades and then link these to the debates on science, knowledge, and action in the sociology of development. This is followed by an account of interactive experiences in the field and by a discussion of the conditions under which an "engaged practice of science" is possible.

Knowledge and Power: Epistemologies of Social Change

More than two decades ago Michel Foucault's visionary work (1980) drew our attention to the mobile, pervasive, and "capillary" nature of power, and to its dependence on producing discourses of truth that subjugate other forms of knowledge. The continuous concentration of knowledge and imbalance of power in society are at the core of Agrawal's (1995) critique of *neo-indigenismo*—the well-intentioned project to conserve indigenous knowledge by archiving it in special places. It also underlies the critique of the transfer of technology model and its "transportational paradigm,"[2] which views knowledge dissemination as a one-sided transfer from one individual or social unit to another. The linearity of this model is, to this day, the principal basis upon which development science is organized.

But knowledge is not a commodity that is easily packaged and transferred through socially neutral channels. Knowledge is a battlefield (Long 1992) where struggles over meanings constantly take place. The need to account properly for these cultural transactions influenced the shift in methods of generating social sciences data. A renewed emphasis on local realities and inequalities, along with the development of investigative techniques departing from time-bound, resource-demanding surveys, marks this shift.

In perhaps too many cases the change amounted to little more than quick-and-dirty project validation activities; the centers of power that decide upon what constitutes knowledge and what modes of knowledge ought to be preferred are still largely located in a small number of scientific communities and resource-endowed institutions from the North (Davies 1994). On the whole, however, a new generation of methodologies reframing the complex question of the social construction of knowledge and inequality have been emerging and consolidating. This search for emancipatory forms of knowledge is justified by the realization that within prevailing power structures there are "opportunities for the powerless to exercise influence and for the powerful to have incomplete control" (ibid.).

Participatory Methods and Action Research

Starting in the early 1970s with the emergence of participatory action research (PAR), the new methodologies began to penetrate agriculture and natural resource management fields, in conjunction with the rapid growth and diversification of participatory rural appraisal (PRA). Probably the best known of these schools is the so-called farmer-first perspective (Chambers 1983), which brought together a family of ideas and methods seeking to increase the timeliness and social relevance of field research.[3] Rapid rural appraisal (RRA) and other rapid assessment techniques were tested and developed in the late 1970s.[4] By the late 1980s, RRA, which stressed observation, semi-structured interviews, and focus groups, had mutated into PRA, with a noted reliance on mapping, diagramming, and other visualization techniques borrowed from agroecosystem analysis (Cornwall et al. 1994; Chambers 1998). The methodological focus had also shifted from collecting data rapidly to facilitating farmers' own representation and analysis of their data.

The parallel development of PAR outlines the eclectic character of this methodological renewal.[5] Originally framed by the work of Kurt Lewin (1946) in the United States, action research is one of the oldest methodological articulations of science and action. It inspired several attempts to structure the relation between researching, learning, and acting (Fisher and Jackson 1998), including the project in the late 1940s to develop action anthropology as a legitimate field of applied science (Tax 1960). PAR, as such, grew mostly as an intellectual social movement in Latin America in the

1970s[6] (Molano 1998); it then moved on to influence PRA approaches before being co-opted into the mainstream of participatory-methodological transformations.

Inspired by the whole spectrum of critical and "heretical" epistemologies of the last two centuries, PAR has been "a constant search for scientific theories on participation" (Fals-Borda 1998); it shares common intellectual and political roots with the action research, action learning, and process management schools, which came into focus in Australia in the 1990s (ibid.), and with other populist methods of participation.[7] In June 1997 in Cartagena, some 32 schools of participatory action research, management, and systems sciences were accounted for at the World Congress on Participatory Convergence in Knowledge, Space, and Time (Fals-Borda 1998). The congress celebrated, in that sense, a global stream of "converging heterodoxy" for methodological innovation, ethics, and responsibility in science and development.

Social Learning

Since Paulo Freire (1972) showed some 30 years ago the fundamental link between knowing and learning, both social participation and natural resource management have progressively evolved to integrate social learning as "the interactive way of getting things done [by] actors who are interdependent with respect to some contested natural resource or ecological service" (Röling 2002 in Leeuwis and Pyburn 2002). Social learning entered the science of natural resource management in conjunction with the critique of reductionism and positivism in the 1990s, but can be traced back at least to the earliest moments of action research and PAR. It is concerned with methodologies that integrate complex, value-laden, resource management processes and systems of inquiry[8] that "seek the multiple perspectives of ... stakeholders, encourage involvement and action, and resolve conflicts for the common and future good" (Pretty 1994). Hence, aspects of learning are explored in terms of knowing, perceiving, sense-making, and acting. As a result, a set of concepts have arisen that all have collective cognitive processes as a common feature: comanagement, interactive policy development, participatory learning, knowledge system thinking, collective action, communicative rationality, negotiated conflict resolution, as well as social learning (ibid.).

Social learning phenomena are diverse yet they all share the "interplay of individual and contextual factors in determining human behavior" (Maarleveld and Dangbégnon 1999). Many have sought answers to the question of how social learning can catalyze adaptation in natural resource management. Key elements for facilitating social learning seem to be communication and experimentation, as well as the creation of spaces, metaphorically called "platforms,"[9] for negotiations about resource use. The role of such platforms in enabling joint learning and adaptation among actors and institutions is at the core of the emerging literature on adaptive comanagement and pluralism; it was also a lead theme in the critical self-evaluation of PRA in the mid-1990s.

Buck and others (2001) suggest that in facilitating social learning it is critical to understand the collective aspect of learning, as well as the political process among actors, notably, the ways they share knowledge, communicate, and build their relationships. Some authors caution against facilitation that aims at shaping "neutral" negotiation conditions, for fear that this may disadvantage the less powerful groups (Edmunds and Wollenberg 2001). Scholars have recognized in social learning a promising approach for facing society's natural resource dilemmas; central planning—with its belief in technology and markets—increasingly fails to govern development sustainably (Röling 2002; Röling and Woodhill 2001; and Woodhill 2002). Recently, social learning has entered into discourses of forestry development in relation to cognitive processes within communities, institutions, and project or institutional teams (Wollenberg et al. 2000; Buck et al. 2001; De Boo and Wiersum 2002).

Relevance, Theory, and Agency

There is a mutual challenge between participatory approaches and theoretical fields in the social sciences. The case for an engaged practice of social research was made during the so-called impasse and post-impasse debates (Booth 1994) in development sociology, a disciplinary field strongly attached to the central place of theory in science. Scott and Shore's (1979) landmark analysis, *Why Sociology Does Not Apply*, argued that sociologists were divorced from the policymaking process because their inward-looking perspective was preventing them from adapting sociological knowledge and methods to problems and variables outside the traditional ken of the discipline. This critique contrasted "knowledge for understanding" with "knowledge for action" and was echoed 10 years later by Edward's (1989) radical attack on "the irrelevance of development studies." The article made the case against "academics that had taken information and ideas from those whom they were researching without contributing to the lives of these people." It also juxtaposed "the continued existence ... of poverty and exploitation in the 'Third World' with the ever-increasing amount of development research, advice, and funding, supposedly undertaken to counteract these trends" (Edwards 1994).

Cernea (1995) also argued that incorporating social-science knowledge into development policies "is the most effective way of employing this body of knowledge." Otherwise, "studies and books produced by social scientists [keep] accumulating [with] little effect" on policies or grassroots processes. "Technocratic planning will continue to rule under the clout of cosmetic rhetoric," and participation will remain a "hot ideology lacking a social technology." But for Cernea, going "beyond explanation to action makes [of] applied sociology and anthropology a distinct enterprise." He thought it necessary, for this reason, to consolidate the cognitive identity of these disciplines to develop a model of "social engineering" attractive to policymakers, and with its own know-how and change tools (Cernea 1995, 24–36). This approach has a distinct flavor of action research, but it also opens itself to the perils of standardization and elitism. If science and policy are both socially constructed with power to "close controversies" (Keeley and Scoones 1999), the effects of their "mutual construction" on the "black-boxing of uncertainties" and diversity in social processes cannot be underestimated (ibid.).

In response to the "relevance" critique, development sociologists basically replied that the impulse to produce useful theories places a straightjacket on the intellectual growth of the discipline (Buttel and McMichael 1994); for them, relevant research is not applied research, which "often not only ignores the wider picture, but also the rather more local picture" (Bebbington 1994). The need to maintain the separation between development science and development practice (Buttel and McMichael 1994; Booth 1994) is the dominant view in this field, shared even by Long's school of actor-oriented analysis. Inspired by the works of Latour and Giddens, this school made a distinct contribution by reintroducing the concept of agency—the staging of "knowing active subjects"—in the understanding of the variable outcomes of social change (Long 1992; Arce et al. 1994).

Long's action-orientation is essentially theoretical. Concerned with the analysis of "discontinuities" (that is, "discrepancies in values, interests, knowledge, and power" in social interfaces; Long 1989), it "is not action research, but rather a theoretical and methodological approach to the understanding of social processes." The actor-oriented approach "must stand or fall by its analytical results"; different social actors can use it to assess their own life circumstances and strategies for action, but it "does not offer a recipe for 'getting development right'" (Long and van der Ploeg 1994). Long recognized that, as such, those in positions of influence or authority can use it against the poor and weak.

This, in our view, is a central problem for theoretical disciplines that shield themselves from direct social engagement. Their claims to relevance are certainly valid; the question is, Relevant to whom? Once generated, knowledge does not circulate randomly; how it subsequently redistributes cognitive powers among unequal actors remains a scientific and sociopolitical question wanting for transparent answers.

The Challenge of Adaptive Collaborative Management

With an early start in agriculture and community development in the 1960s, participatory approaches have more recently penetrated natural resource management fields, mainly in the form of participatory or collaborative management schemes. This collaborative emphasis (comanagement, joint management, etc.) is influenced by the common or disputed nature of ownership of many natural resources and by the "access power" that local communities and marginal groups hold over them (Diaw 1998). In the case of forest resources, both ethical and pragmatic considerations (Colfer et al. 1999a), related to local people's independent capacity to affect production or conservation objectives, have prompted basic professional and disciplinary consensus on participatory methodologies.

This consensus, linked with earlier concerns about development and equity in the history of agriculture, is at the origin of an emerging family of participatory forest management approaches based on the recognition of multiple realities and pluralism (Anderson 1998). To varying degrees, these approaches emphasize institutional issues of governance (Ostrom 1990), negotiation and learning-by-doing (Borrini-Feyerabend et al. 2000; Ramírez 2001), adaptive comanagement (Colfer et al. 1999b), and patrimonial mediation (Ollagnon 1991; Weber 1998). In all of them, the question of the social conditions needed to achieve sustainability and human well-being holds a central place. A shared assumption is that environmentally aware, self-confident, and empowered communities are necessary to the emergence of self-improving systems of forest management.[10] But achieving such conditions is both a methodological and social challenge. The following case studies are meant to provide insights in that regard.

Lessons from the Field: Social Interfaces and ACM

The social science methods (SSM) tested between 1996 and 1998 by CIFOR (Colfer et al. 1999a) are part of the general renewal of social-sciences epistemology. Initially used to fulfill criteria-and-indicators–based assessment and monitoring objectives (Prabhu et al. 1998), these tests[11] led to the realization that something had to be done to catalyze social capital in local forest management. Without direct involvement of research in the facilitation of adaptive collaborative forest management, it was thought, the capacity to develop social methodologies adapted to the complexity of forest situations would remain elusive. From this, came the idea of a participatory action research program, with occasional backing from conventional research methodologies. The aim was to codevelop with local and central actors self-sustaining sets of institutional arrangements, methods, and strategies based on conscious learning.

Field sites were established in a network of 11 countries, including Indonesia and Cameroon that are the focus of this chapter and where the SSM tests had taken place.[12] In each country and site, the entry of the research project was discussed with all relevant stakeholders and institutions to adjust the research framework to key policy issues. Entry was not done on a purely scientific or random base, although it retained basic cross-site consistency for comparative analysis. Rather, entry was negotiated and consciously fashioned to be policy-relevant and partner-dependent. This configuration of research and policy relevance set a local stage for ACM research that is very different from that of the SSM tests. Methods are also being used in a different action context that improves our understanding of how they interplay with interface processes.

Our first case study documents our involvement in the research and facilitation of community representation and decisionmaking in Sumatra, Indonesia. The second analyzes the multiple ramifications of using a simple pebble game for facilitating the negotiation of power and meanings in forest management in southern Cameroon. The cases are a sample of the many ways in which the interfacing of science and participation facilitates social learning and mutual discovery. They focus on two related, though different, processes—the first, broad and extended over time, and the second, punctual but comparable across sites—and provide an empirical context for the issues of power, knowledge, and methods that we have been discussing.

Case 1. Jambi, Indonesia: Learning and Cultivating Diversity

In May 2000, being part of a team of research-facilitators, we (one of the authors) made our first acquaintance with the community with whom we would become involved for the following two and a half years. Below, we share some of the lessons learned in participatory action research and in the interactive use of science methods across a variety of stakeholder groups and institutional hierarchies.

People's "Own Ways" for Development and Social Learning

Over the past three decades, the research site has persistently suffered from declines in local forest resources, state-fostered evictions of local communities, and severe social conflicts over access to the remaining resources. The government's resettlement schemes, beginning in the 1970s and continuing to the present, resulted in a mass relocation of populations from other regions to and within the island of Sumatra, under conditions that ignored preexisting local rights and seriously complicated the area's human ecology. The resulting conflicts and failures, among other reasons, pushed the government to embark on decentralization as a means of reform. The design of this reform is still being debated (cf. Chapters 1 and 2), as are questions about whether the goal is genuinely to empower local communities to choose their own ways to development.

The research in Jambi was intended to investigate this complex situation and facilitate local stakeholders' search for their own ways to development, that is, paths over which they exert influence and perhaps full ownership. Thus the research emphasized social learning as a means of empowering the actors. This emancipatory focus is in line with the current views, noted earlier, that regard social learning as an effective way of resolving conflicts over competing social and ecological demands on natural resources. A common hypothesis is that, because social learning takes place in everyday interactions, it can be fostered by opening spaces for discursive communication among local actors with different values and interests.

As a part of ACM's global research, our field team took up the challenge of exploring these spaces of social learning in a multiactor forest context. For us, the research strategy had an additional benefit besides the emancipatory purpose of social learning: the actors' social interface was a primary source of knowledge that we could integrate into our own research process and questions.

Divergent Realities and the Need to Learn Diversity

Our research took place in Baru Pelepat village[13] in central Sumatra, in the buffer zone of Kerinci National Park, one of the four largest conservation areas of Southeast Asia. The basic social fabric of Baru Pelepat is made up of a diverse ethnic population[14] comprising the original Minang people, who descend from the matrilineal Minangkabau of West Sumatra, as well as settlers from different ethnic groups and a nomadic group naming itself Orang Rimba—"people of the forest."[15] An active social plurality of women, youth, customary leaders, and village elites accentuates this diversity. Institutional diversity also contributes to the pluralism, whereby a customary institution coexists with a formal village government, women's groups, and a religious institution, each governing a specific facet of community life.

Before undertaking the action research, our team determined that social learning had been seriously lacking among actors and institutions; coordination of natural resources use and management was also found to be weak. Poor social relationships, low social capital, and distrust among actors or institutions had severely hampered the social exchange of knowledge and experiences, obstructing discursive communication. People were overwhelmed, unable to cope with social pressures, and in some cases apathetic about their environment and even their future. Diverging realities and value orientations had become problematic.

Issues of land and mutual perception well illustrate the problems. The perception among the original Minang that the land holdings allotted to the settlers by the resettlement project had been seized from their customary lands had made the Minang antagonistic toward settlers. The latter were consequently excluded from community decisionmaking and prohibited from planting perennials on customary lands, which denied the settlers a traditional means of establishing land rights. In most customary forest tenure systems (Diaw 1998), tree planting, along with tree cutting, is the primary way through which individual land rights are established on communal land.[16] In addition, both groups developed a range of prejudices against the other. Both groups realized to a certain extent that they had dissimilar realities, competencies, and norms, but were incapable of dealing with these differences. The stakeholder meetings that our team facilitated at the beginning of the research revealed, however, that all groups and actors wanted urgently to tackle these problems, which they perceived as complex and difficult to grasp.

Making Diversity Salient

Our facilitation of social learning and action in the community of Baru Pelepat was designed according to the needs and problems that the stakeholders considered important. Two overriding issues were identified: the improvement of community governance[17] and the recognition of formal rights to manage the village area.[18] Each of these issues then generated more questions. The collaborative identification of issues did not happen overnight; the community and the researchers took almost six months to set priorities and assess how best to organize the learning activities around the issues. Given that the actors had expressed a need to learn about their own diversity (cf. the similar process in Nepal, described in Chapter 3), these issues provided a space within which they could discuss, debate, and learn. In that sense, dissimilar experiences, histories, and cognitive repertoires had the possibility to meet and foster alternative valuing and acting.

Although we and the other stakeholders were unaware of it at that time, the design on which we agreed to facilitate learning and action fit relatively well with Groot et al.'s (2002) concept of "multiple learning trajectories." In this framework, each learning trajectory corresponds to a given actor's[19] domain[20] with specific learning characteristics, the smaller ones nested in larger ones and in turn embedded in a larger whole across a variety of social, institutional, and administrative levels. For this discussion, we will only pay attention to the issue of stakeholder representation that is part of the question of community governance, which we depict in Figure 4-1 on the basis of Groot's concept.

We agreed with the community stakeholders that choosing representatives to sit in a village decisionmaking body was a good opportunity for learning about diversity.[21] The following section summarizes the learning content and process before the election of community representatives, including the methods and facilitation techniques used and their impact.

Content and Process of Learning

In the months prior to the election of representatives, community stakeholders in Baru Pelepat started a learning experiment that focused on notions and issues of representation in community decisionmaking. The content and orientation of the learning mainly focused on building the group's awareness of the different perspectives of stakeholders and of the different contexts that gave them meaning. In joint meetings, no matter the discussion topics, facilitation was directed at enabling participants to understand such relationships; this was pivotal to the creation of mutual trust and openness to further dialogue and negotiation.

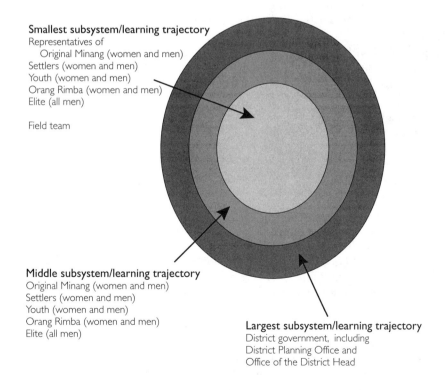

Smellest subsystem/learning trajectory
Representatives of
    Original Minang (women and men)
Settlers (women and men)
Youth (women and men)
Orang Rimba (women and men)
Elite (all men)

Field team

Middle subsystem/learning trajectory
Original Minang (women and men)
Settlers (women and men)
Youth (women and men)
Orang Rimba (women and men)
Elite (all men)

Largest subsystem/learning trajectory
District government, including
District Planning Office and
Office of the District Head

Figure 4-1. *Nested Subsystems and Corresponding Learning Trajectories in Jambi*

An example is the learning Minang women and women settlers went through to locate each other's perspectives in their wider context. Settler women had complained that the original Minang women discriminated against them in the women's savings and loan groups. Discussions revolved further around the historical and cultural context in which each group had developed its understanding of the savings and loan activities. The groups did not start by discussing present realities for fear that the actors might become locked into fixed interpretations and diverted from further dialogue (cf. the concern in Pasir, East Kalimantan, to avoid serious conflicts early on, Chapter 3). In the end, there were visible changes in women's attitude and communication, as well as a better appreciation for each others' views.

To both the field team and villagers it soon became obvious that the learning of diversity did not occur in a vacuum but rather in a micropolitical context in which significant power imbalances dictated relationships and influenced learning. An influential customary leader, for instance, immediately began campaigning for villagers to give him their vote. These maneuvers occurred both inside and outside the learning sessions facilitated by the team. They provoked tense and nonneutral conditions that hampered communication and learning among groups and actors. The field team countered this development by providing information about the coming election and about broad notions such as governance and leadership. Discursive processes occurred among the villagers directly involved in the learning, as well as those who were not. Information leaflets, pamphlets with drawings, or meetings after the Friday mosque prayer served to stimulate discussions. We cannot be sure whether this strategy worked, but the customary elite was not elected.

The learning of diversity followed an iterative pattern that helped to untangle seemingly fixed assumptions and values. We started with simple concerns that were easy to talk about, before proceeding into more difficult and deep-seated topics. Before starting a new topic, actors were encouraged to reflect back to the preceding topics that they had succeeded in untangling. By doing so, people built familiarity and certainty with the conditions for further dialogue. For example, during one meeting stakeholders drew up an agenda of activities for making preparations for the election. This agenda started with relatively neutral and "value-free" activities (e.g., registering the population entitled to vote or looking for voting locations) and ended up with the most value-laden ones (e.g., discussing mechanisms for the screening of eligible candidates). Although emotions ran high during some of the later meetings (one elite person walked out of a meeting), it appears that the familiarity and knowledge previously built helped actors, including the less articulate ones, to participate in the discussions. There were separate knowledge-building sessions for certain less powerful groups who needed extra attention to foster their self-confidence (cf. Chapter 2). For example, separate sessions were held just for women to discuss the coming election to enhance their understanding of issues and stimulate greater participation in joint activities with other groups. These targeted sessions complemented the joint stakeholder meetings, which emphasized sharing knowledge (e.g., the exchange of perceptions and information on community governance between the original and settler groups) and building relationships. Stakeholders' enhanced appreciation for each other's perceptions and interests appeared to be the result of exchanging ideas and knowledge, as well as taking part in joint activities.

Method and Tools for Learning

The method for "learning diversity" comprised a range of means to guide actors through their learning process. Special tools were used to facilitate particular learning processes during stakeholder and group meetings. Everyday, there were unstructured, yet to some extent focused, interactions between facilitators and stakeholders. These interactions were mostly helpful for those groups not familiar with structured settings, such as the women and the Orang Rimba. Future scenarios, or "stories that might be" (Wollenberg et al. 2000), were especially useful to help the actors analyze and understand the underlying value orientations and contexts of their different perspectives (see also Chapter 13). Narratives, pictures, and other visual aids supported the interactive building of scenarios. During a stakeholder workshop on village devolution, separate groups elaborated their scenarios of the future, bringing in their own perspectives and values.[22] The scenarios were subsequently shared and discussed with other groups. All groups were represented in at least one scenario to ensure that their perspectives were explicated and communicated in the discussions. Previously obscure assumptions and presuppositions were made explicit, and as a result seemingly intractable issues became discussable. This process of exchange, interaction, and confrontation formed a solid basis that made possible the subsequent creation of a future scenario shared by the whole community.

Impact

At least to some extent, the outcome of the election reflected the way the learning and its facilitation had affected the people involved. Although social processes other than learning could have influenced the outcome, evidence shows attitudinal and behavioral changes clearly related to better interactions between groups and actors.[23] Of the seven elected members, one was a woman, several were settlers from various ethnic groups, and the majority were members of the original Minang community. None of the customary elite was elected, nor was any Orang Rimba nomad. We drew three conclusions from these results. First, the election of women and settlers seemed to indicate that fixed prejudices or stereotypes toward women and settlers were receding, as villagers saw them as eligible to be their representatives. Second, it appeared that as groups learned to think, value, and act alternatively, they went through a process of empowerment that may have led to their refusal to elect powerful elites as representatives. Just a year before, it would have been unthinkable for communities not to elect their powerful leaders and elites to be their representatives. During early informal interactions between the field team and villagers, many initially conveyed that, while disapproving of the selfish and corrupt behavior of their leaders, they did not know how to oppose them. Third, the absence of any representative chosen from among the Orang Rimba demonstrated that the nomadic group hardly benefited from the learning experiment (cf. the Pygmy situation described in Chapter 5). In the period when the learning took place the group had left the community. Extra efforts will be needed to take account of this disadvantaged group in future learning processes.

Facilitation

At this point, it is worth touching on the issue of representation, a key element in the facilitation of diversity. Groot et al. (2002) identified three challenges often encountered in facilitating multiple actor settings: misrepresentation, powerless representation, and disassociated learning:

Misrepresentation. Misrepresentation is a situation in which the people elected do not truly represent the characteristics or aspirations of their formal constituency. Misrepresentation occurred in our case because of the various meanings actors ascribed to the notion of "representative." It was common practice before the learning intervention to appoint village representatives from among formally educated people or the traditional village elite. A way to tackle the issue of misrepresentation is for the facilitators and the local actors to devote sufficient attention to the process of electing representatives, a key learning aspect in itself. The field team chose to pay a lot of attention to actors from the elite group (hence, some form of "imbalanced representation" in learning). This group was assessed as being most in need of learning because its behavior was often a root cause of conflicts. The learning interactions between researchers (acting as facilitators) and the elite mostly focused on perceptions of good governance and effective leadership. In other words, knowledge development was needed among the powerful as well as the powerless, yet it concerned different learning content.

Powerless Representation. Powerless representatives are those involved in the change process yet lacking power. The issue of powerless representatives held true in our case: some elected community representatives obviously lack power, which will certainly have consequences for future facilitation.

Disassociated Learning. Disassociated learning happens when representatives grow apart from their constituencies. Disassociated learning occurred in almost all groups in our case. This was most probably due to the fact that representatives were not used to reporting back to their groups. Finding ways to involve as many group members as possible in learning processes required new workable mechanisms, such as having several representatives per group, or developing forms of "rolling" representation.

On Researching and Facilitating Diversity

Doing participatory action research in multi-actor settings poses a double challenge: understanding complex interactions in a social-ecological system and being part of the very system under investigation. This dual role of scientist and facilitator has implications for our discussion on "learning about diversity." A scientist is expected to develop understanding of the world by deploying certain methods and to have the potential for critical, self-reflexive learning (Bateson 1972 in Alrøe and Kristensen 2002.) A facilitator, in contrast, is supposed to play a social role and to foster change through communication. It is suggested here that in participatory action research, the scientific outcomes are directly communicated through the facilitation and constitute a basis for action and empowerment. Researchers have their own constructions of the world; maintaining self-critical and self-reflexive observations is thus a basis for maintaining sensitivity to the plurality of interpretations and social representations. This sensitivity, in turn, is the basis upon which to establish the trust needed for effective facilitation.

Case 2. Campo-Ma'an, Cameroon: Negotiating Powers and Legitimacy

At the start of its work in Cameroon, ACM initiated stakeholder forums in three sites, after prior consultations with local stakeholders, to develop criteria and indicators (C&I) for an ACM intervention.[24] The sites are large management areas comprising from eight to more than 150 villages. In all of them there were serious conflicts over management and tenure, as well as recurrent problems of power and exclusion. SSM tests had taken place before in the Campo-Ma'an area, the largest of the sites (see Figure 4-2).

The forums were intended to facilitate, through participatory system analysis and interactive games, a joint assessment of local collaboration and the rights and means to manage local forests. Various analyses of these exercises and of the processes that followed have been or are being documented.[25] This case focuses on the interactive games and the workshops' facilitation, which are directly relevant to this chapter's focus on meanings, power, and science.

At the Core of Conflicts: Tenure, Histories, and Power

In all the sites where ACM has conducted research, tenure rights and historical power relations between stakeholder groups have been found to be at the root of present-day forest management problems. This finding was confirmed by the three stakeholder forums we conducted, which all started with heated discussions about past and present wrongs done to one group by another with respect to tenure rights and access to resources.[26] This tension was stronger in the first two forums, for reasons directly related to the different evolution and configuration of these rights-and-power issues in the sites.

The first forum was held in the Ottotomo state forest reserve in the Yaoundé area. Surrounded by eight villages, the reserve crystallized many of the problems of state forest reserves. Off-limits to human activities since its creation in 1930, the reserve faced growing encroachments from the local Ewondo population, which claimed use rights in the forest. In the small group meetings that were set at different moments

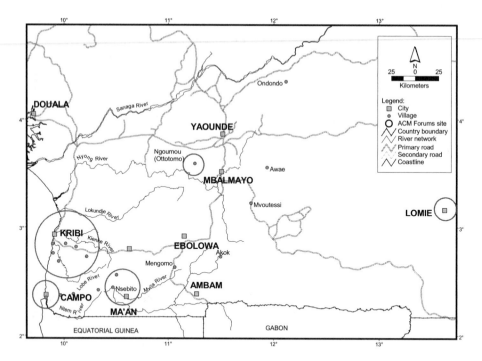

Figure 4-2. *Map of Cameroon ACM sites*

to prepare stakeholder-based positions and analyses, community representatives articulated these claims in detail, with an overriding demand for the right to carry out activities such as hunting and collection of forest products inside the reserve.

ONADEF, the reserve's managing agency and the key player in the area,[27] developed arguments that paralleled those of community representatives and that emphasized the population's "misunderstanding" of the situation, the "invasion of the reserve by slash-and-burn agriculture," and the need for control and protection of the resource. The third group of stakeholders was a mix of NGOs and intervening actors, including the Ministry of Agriculture. This group highlighted the "unsuitable production systems," the numerous tenure challenges, and the lack of material and financial means for forest management. As in all the forums, the researchers strictly maintained their roles as facilitators.

There were four stakeholder groups in the second forum,[28] which took place in the Campo-Ma'an National Park (see also Chapter 6). Through negotiation, the government, the World Bank, and the Global Environmental Facility set aside this protected area in 1999, as an environmental compensation for the Chad-Cameroon pipeline. Lodged at the heart of an 800,000-ha Technical Operational Unit, its overall area covers more than 150 villages in the historic points of settlement of local Mvae, Ntumu, Mabi, Iyassa, and Bakola-Pygmy populations, who claim rights of first occupancy in the region. The park is also surrounded by logging concessions and agroplantations and interacts with a marine military unit at the border with Equatorial Guinea. The park was run until mid-2002 by the Campo-Ma'an project (CMP), a tripartite arrangement between MINEF and two bilateral cooperation agencies. Local issues, which came out early during the workshop, include conflicts between biodiversity conservation and logging interests, concerns about poaching and transborder encroachments, disputes over traditional hunting and forest access rights, and compensation claims for land lost to the park and to the industries. Local stakes were high because the CMP promised potential logging revenues and alternative benefits from ecotourism and community forestry. The first two days of the workshop were marked by strong discursive battles around those issues.

The third workshop was held in the Lomié community forests area of East Cameroon. It gathered representatives from five community forests (Kongo, Koungoulou, Eschiambor, Bosquet, and Ngola), among the first to be legally recognized in Cameroon under the 1994 Forestry Law. Other actors included representatives from SDDL (*Soutien au Développement Durable de Lomié*), a project attached to the Dutch cooperation agency, SNV, which supported the creation of the community forests, as well as representatives of the logging industry. At the outset of the joint stakeholder analysis, and in contrast to the other forums, the Bantu and Baka Pygmy representatives stood separately for their respective constituencies.[29] Although tension was lower in Lomié than in the other two forums, probably because of the existing devolution in favor of communities, the Lomié forum started with a list of grievances and concerns related to land use and social well-being, including the size and extent of community forests, the divergent interests among local communities, loggers, and the project over logging in the community forests, and the influence of elites on decisionmaking. The map of tenure issues in the area seemed to have moved from the traditional state–communities confrontation to a more nuanced set of contradictions among a wider range of actors.

Actors at Odds: The Misgivings of Collaboration

The rhetorical stances that dominated the opening of the stakeholder forums indicated a weakness in local cooperative capital. The stakeholders were profoundly dissatisfied with each other and were all locking themselves into an impasse. Further evidence was provided by a simple game in which each group scored the intensity and quality[30] of its collaboration with the other groups. Collaboration could be nonexistent (0), bad (1), acceptable (2), or good (3); indicators such as the frequency of coordinated activities or information-sharing gave the assessment a concrete basis. The groups' scores were jointly interpreted in plenary, using paired off graphics to visualize the actors' mutual perceptions of collaboration. We organized the six figures of mutual perceptions that we found into four basic cases (see Figure 4-3, in which the base of the arrow represents the speaker, or giver of the score, and its tip the other actor concerned).

- *Open conflict* (2 cases; 13%): A says that its collaboration with B is bad; B says the same; the stakeholders agree that their collaboration is bad or nonexistent. These are mainly cases of "constructed adversity." We found such oppositions between promoters of community forests and industrial loggers in Lomié and between the project and indigenous communities in the Campo-Ma'an National Park.

1. Open Conflict

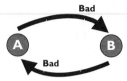

2. Self-Deceptive Conflict Situations

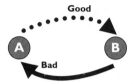

3. Slightly Biased but Acceptable Collaboration

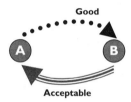

4. Fair to Good Collaboration Based on Congruent Perceptions

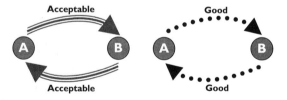

Figure 4-3. *Visualization of Interactions among Stakeholders*

- *Self-deceptive conflict situations* (7 cases; 44%): A says that the collaboration is good or acceptable; B says it is bad. We found such cases in all the sites (see Table 4-1). The most severe cases were the first two in Lomié, but the basic situation was the same for all of them: one actor (always the less powerful) was profoundly dissatisfied, while the other thought that there was little or no problem. We noted that the position of state and project actors found in Lomié was reversed in Campo.
- *Slightly biased but acceptable collaboration* (3 cases; 19%): A says the collaboration is good; B says it is just acceptable. This is typical of situations in which the actors recognize their need for each other, but the actor who is more in need overemphasizes the collaboration. Three instances of this pattern were found: Ottoto-

mo reserve managers with regard to third-party support for comanagement; Baka with regard to the project; and loggers with regard to community forest managers in Lomié.

- *Fair to good collaboration based on congruent perceptions* (4 cases; 25%); this situation, unique to Lomié, involved mainly the forest administration, in relation with the industry, the Bantu, and the Baka (the only case of two-way good collaboration in the whole sample). The project and the Bantu communities also perceived that they have acceptable collaboration. This perception reflects, to a certain extent, the goodwill liberated by the process of devolution in the area.

Overall, this analysis of collaboration in the three sites gives a sharply contrasting map of perceptual differences between powerful and less powerful actors and between protected areas (Campo, Ottotomo) and community forests (all but one of the acceptable or good cases). The average rating of collaboration by sites and actors clearly reflect this contrast, with a rating of 1.3 (bad) for Ottotomo and Campo Ma'an and 2 (fair) for Lomié.

Rights and Means: Discourses on Power and Legitimacy

During the forums, we used a pebble game to trace the correlation between mutual perceptions of legitimacy and power in tenure and management. The game brought new understanding to these issues and was a powerful tool for negotiating the conflicts of meanings that sustained them; it illustrates the way science and participation can mutually serve and reinforce each other.

Playing with Pebbles: Games, Polysemy, and Mathematics. The pebble game is a simple, user-friendly technique that we first applied during the SSM tests. It is played by distributing 100 pebbles among stakeholder groups, each pebble corresponding to one unit of rights or means (R&M) held by a group of actors. Despite its procedural simplicity, the pebble game has complex semiotic content requiring conceptual exploration. Rights and means are highly polysemic concepts. The Beti concept of *ngul* (rights), for instance, also means force while being the root word for authorization (*o va ngul);* at the same time, one of the terms for authority, *edzoe,*

Table 4-1. *Actors and Sites with Self-Deceptive Conflict Situations*

Actor A	Actor B	Sites
Administration	Project	Lomié
Industry	Baka	Lomié
Bantu	Baka	Lomié
Project	Administration	Campo
Administration	Communities	Campo
Administration	Communities	Ottotomo
Intervening actors	Communities	Ottotomo

also means power or responsibility. These words are therefore locked into a semantic kinship and continuity that we also find in French and English.

During the SSM tests,[31] there was much discussion in the field about whether rights could exist without means, whether rights referred to legal rights or some other form of legitimate rights, and whether rights also included responsibility or authority. The ACM forums, during which participants offered additional concepts of rights and means in Mvae, Ntumu and Mabi, were an opportunity to break this semiotic uncertainty.

The forums' stance was that the notion of rights essentially relates to the perception of legitimacy in society. When they talk about rights to manage, people are really talking about who, in their view, should have a say in management. Because the sources of legitimacy in society are plural, actors attribute legitimacy to actions based on their own, subjective ordering of those sources. It is a value judgment, in which are encoded the composite references and cultural hierarchies according to which people routinely appreciate what is "right" and what is not. In short, to do their scoring participants had to recognize that: (a) there are many sources of rights and legitimacy in society—moral, legal, traditional, formal or informal, (b) all sources of rights may be valid, and (c) each participant is free to integrate those sources of rights according to his or her own criteria, preferences, references, or hierarchies of legitimacy (cf. Introduction to this volume).

The notion of means refers to the actual, concrete power to do things and to the instruments of that power. These instruments include money, logistics, physical force, knowledge, and know-how, as well as the law (not as a moral category, but as a tool to have things done in a certain way). In attributing means, people make judgments about who, as a result of these various sources of authority, has the actual power to do certain things in forest management. This judgment of reality is composite and integrative, but yet relatively independent of normative hierarchies.

During the ACM forums, we came to the conclusion that adding rights to means, or simply averaging them, was not sufficient to determine the management interface, that is, the effective ability to limit, sanction, and resolve conflicts between groups. Despite all their entanglement, R&M are, in a way, like apples and oranges. We can say that one group of actors has more or fewer rights than means—more or fewer apples than oranges—but that doesn't give us a third product, a unique condensation of the two, which we can name, interpret, manipulate, or trade. We thus used the arithmetic base of the game to calculate that third entity. By subtracting the score of rights from the score of means, in particular, we obtained a measure of the R&M handicap of a group of actors, or stakeholder handicap.[32] This opened a whole new field of understanding and negotiation that we will detail later. The basic equation that we used is the following: $H = M - R$, where H represents the handicap, M the means, and R the rights. This equation leads to three types of situations:

No handicap: $H = M - R = 0$, where R&M are in equilibrium and stakeholder handicap is nil.

Power deficit: $H = M - R < 0$, where handicap, ranging from −1 to −100, is expressed as a power deficit.

Excess power: $Hp = M - R > 0$, where handicap is expressed in terms of excess power and a deficit of legitimacy.

Playing Hide-and-Seek: Collaborative Impasses in Management. Table 4-2 presents the sum of the handicaps distributed among stakeholders across three basic management roles: the R&M to define limits, to sanction, and to resolve conflicts. To obtain the total handicap of a management role, a site, or a stakeholder group, we added up the absolute values[33] of the related stakeholders' handicap and then weighted them against the tolerable level[34] of handicap defined by the stakeholders to get a management handicap index (MHI) by site and management role: In Table 4-2, an MHI of 1.0 means that the aggregate handicap concerned is tolerable. An MHI of 1.9 means that this aggregate handicap is almost twice the tolerable level. The equations defining the tolerable level[35] and the MHI are:

Tolerable level = 10 × number of score givers × number of score receivers

MHI = sum of handicaps/sum of tolerable levels

From Table 4-2, we see that, across the sites, the stakeholders thought that the aggregate handicap of management was not tolerable (1.7). Lomié stands again as the site with the lowest level of dissatisfaction, particularly on issues of sanction and conflict. As a group, the Campo stakeholders also thought the way the R&M to manage conflicts was distributed among them was acceptable. For the rest, we can only note the high level of concern of the stakeholders:

- With twice the tolerable levels of handicap across sites, the R&M to define limits was the most contested area, with the greatest accumulation of grievances.
- Among all sites, Ottotomo saw the most heated confrontations, with generally content state managers (score below the tolerance point of 90) and extremely unhappy communities (about five times the tolerance level).[36] At the time of the forum, the exclusionary policies of the reserve had prompted conflict between the reserve managers and the expanding neighboring population.
- In Campo, the R&M to define forest limits and to sanction were the main points of contention, with tolerance scores almost twice the acceptable level. The predominance of these issues is not surprising, given that Campo is the site of a newly created national park, where the demarcation of boundaries and control of poaching are the key conservation policies of park authorities. We also note that the project actors gave more weight to perceived problems of sanction than to issues of tenure. This was mainly a criticism directed at the state for its alleged failure to apply sanctions.
- Strikingly, project actors in Campo and Lomié stand out as the second unhappiest groups, with readings of handicaps that are about twice the tolerance level.

Givers, Receivers, and the Politics of Discourse

During the pebble game, groups of actors assigned R&M scores to themselves and to their protagonists in specific management contexts. In doing so, they were

Table 4-2. *Aggregate Handicap to Define Limits, to Sanction, and to Manage Conflicts, by Site and Giver*

Giver	Campo				Lomié				Ottotomo				Grand total
	Limits	Sanction	Conflict	Total	Limits	Sanction	Conflict	Total	Limits	Sanction	Conflict	Total	
Administration	90	50	30	170	110	40	50	200	30	10	40	80	450
Communities	80	50	40	170	70	52	20	142	130	200	130	460	772
Intervention agencies	60	100	50	210	120	140	180	440	70	70	130	270	920
Baka	—	—	—	—	150	80	40	270	—	—	—	—	270
Industry	—	—	—	—	0	0	0	0	—	—	—	—	0
Grand total	230	200	120	550	450	312	290	1,052	230	280	300	810	2,412
Tolerable levels	120	120	120	360	250	250	250	750	90	90	90	270	1,380
MH index	1.9	1.7	1.0	1.5	1.8	1.2	1.2	1.4	2.6	3.1	3.3	3.0	1.7

expressing their perceptions about order and disorder in their environment. In real life, perceptions cannot just be aggregated and turned into measures of central tendency; they must be acted upon—exposed, confronted, discussed, negotiated, and recognized for what they are—before the situation they support can be mediated. From the brouhaha of arguments an observer can assess a situation as bad, but cannot attempt to mediate the situation without getting all of the actors to communicate their perceptions and listen to others.

Tables 4-3 to 4-5 present the scores given and received by the stakeholders and an initial interpretation of their meanings. For each score received in a management area, we have several "givers" and several, often dissonant, viewpoints. These views form a discursive chain in which each score is only a fragment of a larger discourse. This discourse closely follows the positions, contradictions, and strategies of the players.

A few key figures of speech emerge from these discursive battles.

The Irrelevant Other. This is a tricky figure identified during joint analysis. One characteristic of the no-handicap equation ($H = M - R = 0$) is that an actor with no means and no rights will still be in "cohesive capacity," with no handicap. In that case, as we discussed in plenary, what the giver is telling the receiver is: "with zero rights and zero means, you are at your right place with no role to play; you have no handicap because you are irrelevant to that management issue." This was the communities' discourse toward the Campo-Ma'an project, despite the project's obvious responsibilities on borders, conflicts, and antipoaching issues. This was also the systematic attitude of the administration and industry groups toward Bantu and Baka communities in Lomié, that of the Baka toward the Bantu in that site, and that of most groups toward the industry in Lomié and Campo. In that last case, an interesting figure also emerged of the "invisible self." Of all the groups, only the industry[37] gave itself no rights and no means at all, while giving all the R&M to the state and the project and qualifying the whole situation as excellent. This denotes strategic behavior intended to overemphasize the subordinate position of the industry in relation to the other actors in power, in particular to the state from which the industry derives all its legal entitlements to local forests.

The Deprived Community. Except in Lomié, where Bantu and Baka claims of legitimacy were noticeably subdued, these communities saw themselves as seriously handicapped and despoiled by the state in matters of tenure and sanction (as well as conflict resolution in the case of Ottotomo). They gave more rights to themselves than to any other actor and claimed additional means to deal with those issues. Except for the Lomié project, which gave the communities even more rights and handicap than they themselves claimed in that site, non-community actors did not fully recognize this position. These actors accepted that in matters of border definition, communities had secondary rights (about 30%) and, to some extent, were deprived of those rights; and some of them, such as the administration in Campo extended this mitigated perception to all three management areas. Other state and project actors thought that communities had little business in matters of sanction and conflict resolution and tended to exceed their rights. Adding to the general perception that communities did not have a fair share of means to assert

Table 4-3. *Capacity and Handicap to Define Limits*

| | Stakeholders (Receivers) | | | | | | | | | | Discourse Statements | |
| | Admin. | | Community[a] | | Intervention | | Baka | | Industry | | | |
Capacity to define limits / Speakers	R	H	R	H	R	H	R	H	R	H	Rights holder(s)	Handicap
Campo												
Administration	65	−25	30	−20	5	25			0	20	The state is major rights holder, with a share to communities.	The project and the industry have powers exceeding their rights.
Communities	50	20	50	−40	0	0			0	20	We are the only co-holders of rights with the state.	The state and the industry usurp almost all our rights.
Intervention	75	−30	25	5	0	15			0	10	The state is the dominant rights holders and villagers have some.	The other actors have to make up for the state's weaknesses.
Lomié												
Administration	75	−55	0	0	25	25	0	0	0	30	Only the state and the project have rights and the state should be key.	The project and the industry take away the bulk of our (state) rights.
Bantu communities	50	−5	25	−15	0	15	25	−15	0	20	The state shares half of the rights with us and the Baka.	We are in deficit and the project and the loggers fill in the gaps.
Intervention	50	−30	25	−10	0	60	25	−20	0	0	The state shares half of the rights with Baka and Bantu communities.	We (the project) have the real powers.
Baka	100	−75	0	0	0	40	0	10	0	25	The state has all the rights.	The state is weak and the project and loggers have most of the power.
Industry	75	0	0	0	25	0	0	0	0	0	The state has the bulk of the rights; the project has some.	Everything is fine and balanced, as it is supposed to be.
Ottotomo												
Administration	50	10	30	−15	20	5					All have rights, but the state is the majority rights holder.	Only villagers lack capacity; the NGOs and we make up for it.
Communities	35	65	65	−65	0	0					We are the main rights holder, and the state has the remaining share.	Our rights are totally usurped by the state.
Intervention	60	−25	40	−10	0	35					The state has dominant rights, with the villagers having a sizable share.	We (NGOs) have good means to make up for their weaknesses.
Total	685	−150	290	−170	75	220	50	−25	0	125		

[a] "Communities" refers only to Bantu communities in Lomié, and to both Bantu and Bagyeli groups in Campo. These latter, however, were underrepresented.

Table 4-4. *Rights and Handicaps to Sanction*

| | Stakeholders (Receivers) | | | | | | | | | | Discourse statements | |
| | Admin. | | Community | | Intervention | | Baka | | Industry | | | |
Speakers	R	H	R	H	R	H	R	H	R	H	Rights holder(s)	Handicap
Campo												
Administration	30	10	30	5	10	10			30	−25	The rights are equally shared.	Only the industry is not delivering.
Communities	50	25	50	−25	0	0			0	0	The state and we have all the rights and share them 50/50.	But the state is exceeding its rights to our detriment.
Intervention	100	−50	0	30	0	20			0	0	The state is sole rights holder.	But the state doesn't deliver half of the time.
Lomié												
Administration	50	20	0	0	50	−20	0	0	0	0	Only we and the project have rights that we share equally.	But the project is not delivering accordingly.
Communities	74	26	13	−13	0	0	13	−13	0	0	The state has most of the rights; we and Baka have a little.	In addition, the state takes away the little we have.
Intervention	50	−20	25	−25	0	20	25	−25	0	50	The state has most rights and the Bantu and Baka share the rest.	Because all the rights holders are weak, the industry takes over and we fill part of the gap.
Baka	100	−40	0	0	0	40	0	0	0	0	The state is the sole rights holder.	But the state is not strong enough; the project fills in a big chunk of the gap.
Industry	50	0	0	0	50	0	0	0	0	0	Only the state and the project should have a say.	Everything is working perfectly.
Ottotomo												
Administration	90	−5	0	0	10	5					I am basically the rights holder.	Things are fairly as they should be.
Communities	0	100	100	−100	0	0					I am the sole rights holder.	The state is a despot; it takes away all our rights.
Intervention	90	−35	10	20	0	15					The state is the only significant rights holder.	The state can't do it all by itself; the community and we fill in some gaps.
Total	684	31	228	−108	120	90	38	−38	30	25		

Table 4-5. Rights and Handicaps to Solve Conflicts

| | Stakeholders (Receivers) | | | | | | | | | | Discourse statements | |
| | Admin. | | Community | | Intervention | | Baka | | Industry | | | |
Speakers	R	H	R	H	R	H	R	H	R	H	Rights holder(s)	Handicap
Campo												
Administration	45	5	30	-15	25	0			0	10	We share the rights with all stake-holders but the industry.	All are doing okay but the communities.
Communities	50	-20	50	20	0	0			0	0	The state and we share all the rights 50/50.	The state is not solving the problems; we actually do better.
Intervention	75	-25	25	0	0	25			0	0	The state is the key rights holder; communities have some rights.	State is not delivering; we make up for it.
Lomié												
Administration	50	25	0	0	50	-25	0	0	0	0	Only the project and we share the rights 50/50.	The project is not delivering.
Communities	30	0	30	-5	10	10	30	-5	0	0	State, Baka, and we are the co-holders of rights.	Fairly okay despite slight deficit in communities.
Intervention	20	-10	40	-40	0	15	40	-40	0	75	Communities are the major rights holders, with a minor role to the state.	Their capacity is nil, with the industry taking over and a little help from us.
Baka	50	-20	0	0	35	20	15	0	0	0	Only state, project, and we have a role to play; others are irrelevant.	We are just at the right place; the project makes up for weaknesses of the state.
Industry	50	0	0	0	50	0	0	0	0	0	Only the state and the project have a role to play.	Everything is fine and in balance.
Ottotomo												
Administration	60	20	10	-5	30	-15					The state is the main rights holder, with some role to intervention/NGOs.	We end up supplying weakness of the others.
Communities	35	65	50	-50	15	-15					We are main rights holders with roles to the state and intervention/NGOs.	State ends up imposing and taking over everything.
Intervention	100	-65	0	35	0	30					The state concentrates all the rights.	The state is weak and cannot do it without us and the villagers.
Total	565	-25	235	-60	215	45	85	-45	0	85		

their tenure rights, this discrepancy between communities and other actors partly explained the negative picture revealed by our previous game on collaboration (Figure 4-3).

State Powers: Token, Despotic, and Benevolent. The figure of the state in the discursive map of the pebble game is an elusive one. State actors in Campo and Lomié thought that environmental and community projects fundamentally deprived them of their right to define tenure. But they asserted themselves as the dominant power that manages sanctions and conflicts (and tenure in the case of Ottotomo). They also criticized the projects or the industry for shirking their share of the responsibility to put more law and order in the sector. The other actors threw back this criticism to the state and severely assessed its weakness and tokenism (the Baka and the projects in Lomié and Campo) or its despotism (communities, particularly in Ottotomo and Campo). Only the industry projected an image of power, benevolence, and tranquility onto the state, for reasons that can be decoded as essentially strategic.

Proxies and Civil Servants. The position of the projects and other intervening groups in the three sites is very peculiar. Strikingly, this group never made the "mistake" of giving itself any rights. This stance appeared to be dictated by a tactical awareness of their subsidiary position, as the projects did actually have legal and administrative rights given to them by the state. The projects perceived their rights as embedded in those of the state and better left unspoken. Nonetheless, this group accepted that it had moderate powers because of the weakness of legitimate rights holders. The projects were just filling in the gaps, as demanded by their civil-service mandate in forest management.

The Baka and loggers in Lomié agreed with an even more positive assessment of the projects. State actors were a bit more critical: first, they gave significant rights to this group of intervening actors; second, they said that the projects or NGOs either exceeded their rights or did not live up to them in a number of management responsibilities. Bantu communities disagreed all the way. In two thirds of the cases, they portrayed the projects as irrelevant entities with no rights, means, or role to play in forest management.

Overall, a picture of structural disagreements emerges from the R&M game, as well as the earlier assessment of collaboration in the three sites. The actors agreed that structural deficits hampered their effectiveness, but disagreed over the reasons. For a transformational paradigm such as ACM, the next logical question is, How is it possible to move beyond this adversarial rhetoric to problem-solving? In our view, the whole process of framing, measuring, and decoding these meanings provides a context for collaborative solutions. The next section recapitulates briefly this joint facilitation process and outlines its defining moments for further development.

Empowerment and Disempowerment: Social Facilitation and Negotiation

The basic setting for the discussions was established at the beginning of the forums through a series of facilitation techniques. The participants were given yellow cards that empowered them to control the facilitators and the most vocal actors in the forums; a steering committee, formed with representatives from each group, reported every morning to the plenary. By creating a more level, democratic playing field, these tools and the whole facilitation style helped relax the participants and build trust in the process.

The basic rules of the pebble game were set in an initial discussion of the concepts of rights and means, as described earlier. This first conceptual negotiation validated the actors' reference frames and claims to legitimacy or power.

The scoring in small groups was also a negotiated process. The stakeholder groups were of mixed ethnicity and social make up. The rules of the game compelled each group to develop a consensual position, that is, to trade meanings and negotiate positions internally. From our observation, the pebbles themselves were empowering to quieter individuals who did not have to voice their opinion, but could just walk up and rearrange the pebbles when they saw fit. Occasionally interspersed with justifying statements, these rearrangements would go on until the group arrived at a general sense of satisfaction. Although discussions were heated at times, the transparency of the process helped resolve these conflicts in a quick and fairly consistent manner.[38]

The next step was that of joint presentation and analysis in plenary. There were three moments in that process: the groups' presentation of their results; the definition of analytical rules; and the visualization, interpretation, and discussion of the results. The first was straightforward and without noticeable surprise; the other two need elaboration.

Definition of Analytical Rules. The R&M game was not invented on the spot—the scientists brought it in—but the method of analysis it required was not fully developed prior to the forums. We did not know how to turn the relationship of rights and means into that "unique condensation of the two" mentioned earlier. The scientists were forced to look for a solution because the actors had different competencies and needed simple, explicit ways to visualize the data. The puzzling R&M relationship had to be simplified, measured by a single indicator, and graphically expressed. This was done, first, through a clarification of the moral-action basis of the relationship:

> Q: What is the desirable balance between rights and means? Is it desirable to have rights without means, or means without rights, and why?

> R: The unanimous response was that the desirable situation was for someone to have means appropriate to his or her rights and rights appropriate to his or her means. For reasons of fairness and effectiveness, the relationship was considered balanced when rights and means annulled each other. Hence, the no-handicap equation $(M - R = 0)$. The other figures of handicap were derived from this equation: (a) the handicap of undercapacity and powerlessness and (b) the handicap of excess power and lack of legitimacy and recognition, where moral and external obstacles hamper effectiveness.

These notions were conceptualized, reviewed, and refined over the course of the forums.

Visualization, Interpretation, and Discussion of Results. The participants did not know the rules of joint analysis when they were doing their scoring. When the groups' scores started to be turned into statements, some participants were surprised, and did not recognize the viewpoint they had expressed. In one case, a community representative spoke out against "the numbers that confuse the minds of the peasants." Exchanges followed as to whether the mathematics were a mask or an eye-opener.

These legitimate interrogations were solved by going back to the data; people who were concerned about being misinterpreted could go back to the drawing board at any time and rearrange their data as they saw fit (which did not happen, because the data as much as their transformation rules turned out to be what they were supposed to be). In a few cases, people were actually saying more than they were ready to handle and were astonished by the radicalism of their own statements. At the same time, they were not comfortable taking back scores that reflected what they saw or wanted to say as a group about the distribution of rights and means. The significance of these discussions did not hinge on whether the statements were right or wrong. Rather, the significant outcomes were the revelation of ingrained attitudes of mutual rejection and criticism among social actors and the emergence of second thoughts and hesitations that expressed new opportunities for change.

The next step was, naturally, to push these openings for compromise a step further by using the graphics of stakeholders' handicaps to start negotiating powers and legitimacy. Figure 4-4 is a visual example of these drawings. Ideally, all actors would like to be at zero with the "right" amount of rights and means. We can see in the figure that the administration is demanding more powers for itself and for the communities to define forest boundaries. Following the rules of the game, these additional powers can come only from the actors in positions of excess power, that is, the industries (absent during the scoring) and the Campo-Ma'an project intervention group. In spite of its critical undertones, the scoring of the project intervention group comforts the administration; both agree that the administration must be empowered, which implies some voluntary disempowerment from the project itself. The figure also shows that, basically, the project and the communities are playing hide-and-seek; the latter are saying that the project is irrelevant, while the former says that communities should be content with what they have. Neither could really maintain its position when confronted with the reality of the site. In addition, by siding with the communities' demand for empowerment, state officials (involuntarily) killed four birds with one stone: they placed themselves in a position of referee, made it difficult for community representatives to maintain their adversarial stance (the state in huge excess of power), deflected part of this resentment toward the project, and consolidated their demand for more power-sharing with the project.

The communicative power of these drawings cannot be overstated. Even actors who have severe communication and collaboration problems, when suddenly confronted with graphic images that they themselves had helped to produce, could not deny that the images were demanding them to compromise. By showing that having lots of power is not the same as being an effective manager, and by making paths to empowerment and voluntary disempowerment clearly visible, the handi-

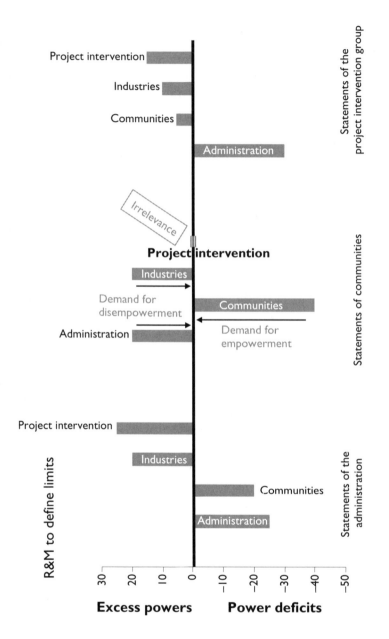

Figure 4-4. *Stakeholders' Perceived Handicap to Define Tenure in Campo*

cap figures enabled the actors to recognize that it is possible to trade powers legit-imately through negotiation. The images showed graphically that collaboration was needed and feasible, but implied that each actor would have to make a behav-ioral change to achieve balance.

This was not the end of the story, for no mathematical representation can give an exact measure of the power that should or could be traded. Genuine collabora-tion can result only from a negotiated, often protracted, process. The role of the actors in power is particularly important, as Chambers (1997) shows. In his explo-ration of "power as a source of error," Chambers gives telling evidence from the fields of development and psychoanalysis that "for learning, power is a disability." Further evidence became apparent in the collaborative efforts that followed the forums. Attitudes that compromised mutual adaptation and collaboration came from only one site, but from actors in key project positions who had opted not to be present at the forums. We are now making a fuller assessment of these interactive processes, but it appears on the whole that they contribute significantly to improv-ing forest management.

Encounters in a Mine Field: Conditions for an Engaged Practice of Science

Following the semiotic tradition, we can think of society as a vast trade of signs, in which cultural experiences and social meanings are constantly transferred and ne-gotiated. Science is a social construct. As such, it bears, encoded, both a social experience and a cultural perception of the world. By entering into a social field with their own social project, scientists and other development or conservation professionals open a new field in which new meanings are exchanged and expec-tations raised. Externally induced change is always about the addition or insertion of new meanings into a social and perceptual field. It is, at best, naïve to think that knowledge can be extracted from this intercultural field without affecting it, if only residually. Extraction and interaction are inherent to social encounters.

The defining character of extractive science is that it resists being altered by these encounters and, therefore, does not recognize what it does to them (see Figure 4-5). Thus, the myth of neutrality in science is entertained at the price of shirking social responsibility. During a field-based workshop on participatory systems analysis on the Wild Coast of South Africa in 2000, Whapi Siyaleko luminously expressed the meaning of extractive science for the "other side" of the encounter. This traditional healer, *qwele,* of the Xhosa community of Hubeni had previously contributed to studies of natural plants and processes, with no benefit to him in return; he told us this when we came asking—again—for the benefit of his knowledge:

I am poor; you study my poverty, you get fame—me, I remain poor;

I know; you study my knowledge and that of my neighbors, you now know more than all of us—me, I just know what I knew.

This statement and what it tells us of the social construction of knowledge inequalities and poverty best make the case for an engaged practice of science.

"Uniting research and practice, understanding and action, researchers and researched into a single unitary process" is one way of making research accountable to the subjects of its work (Edwards 1994). This process, however, is never unitary except, perhaps, in the classic dialectics of the unity of the opposites. The knowledge gained in social interactions is always integrated individually into previous knowledge and personal cultural experience.

Extraction is an integral part of colearning and social interfaces. The key problem is not extraction as such, which is inherent not only to scientific practice but to social encounters in general. What is problematic in purely extractive processes is the separation of agendas between givers and takers of information and the lack of "restitution." Extractive processes disempower the givers and reinforce existing social discrepancies in knowledge and transformative powers. Empowerment can happen only with three basic learning conditions: (a) mutual learning, (b) restitution, or "what we learned" accountability feedbacks, and (c) transparency of motives and language. Language is key, as it defines the cultural sphere within which the cognitive exchange or restitution takes place. These conditions play out differently, or do not play, in different types of research; not all "relevant research" is empowering to the poor.

Bringing Participation to Science: Lessons on Interactive Processes

In the case studies that we have narrated, we, as scientists, came to bring science to participation, only to discover, as it often happens, that participation was actually bringing a lot to our science and methods. The Jambi social learning experience stages community actors who willingly engage in a process of deliberate learning; this modifies the local political interface while bringing new insights

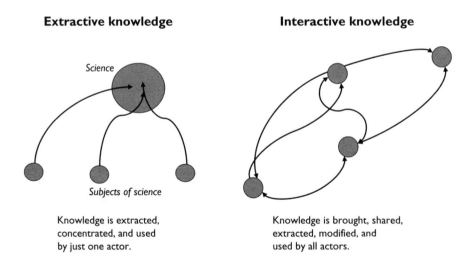

Extractive knowledge

Science

Subjects of science

Knowledge is extracted, concentrated, and used by just one actor.

Interactive knowledge

Knowledge is brought, shared, extracted, modified, and used by all actors.

Figure 4-5. *Models of Extractive and Interactive Knowledge*

on the nature of learning, the negotiation of perspectives, and the sensitivity of local political processes. In our experience with the pebble game, it is the insertion of the scientific method into PRA facilitation and visualization techniques that enabled the expression and negotiation of meanings among local actors. The participatory context set its own requirements on the actors' interplay and on our activity as scientists. It is the intense interactive context, as much as the requirements of joint analysis, that pushed the research-facilitators into finding ways to visualize the conflicts of perceptions and the "situated facts"[39] that supported them. To facilitate the emergence of a common language, the scientists had to talk among themselves and compare notes, while the main actors were occupied elsewhere.

Within the larger cognitive interface, solutions were not given but tested, and therefore modified by the interaction; problems and options were both defined within the facilitation situation. The R&M mathematical relationship that lifted the veil on unforeseen management disabilities and opened the door to the conceptual trading of powers and legitimacy emerged from the dynamics of the encounter. The process was not neutral, nor did it arise from below; rather, it came as a "middle way," where the confrontation of different types of knowledge and social demands opened new questions and offered new solutions and paths to explore.

The process was also not unitary; the actors had their own, different definitions of the problems and mobilized different capabilities and cognitive repertoires to struggle over them. Ultimately, each actor went away with some shared knowledge but also with his or her own interpretation of the encounters and degree of willingness to conform to the collaborative decisions. After the workshops, there was a backlash from the most powerful absentees in one site. They did not openly oppose the new collaborative commitments made at the workshop but managed in many subtle ways to undermine their credibility and interfere with follow up. In return, less powerful actors started using discursive arguments from the encounter to increase their bargaining position, notably, by opposing the attitudes of the scientists with those of the project and, at times, by taking rhetorical stances such as the demand for "criteria and indicators" of the project's social commitments. Such positions were found among community as well as state actors.

In another site, the reserve managers, who were key actors of the workshop, immediately interpreted a new case of encroachment as a consequence of the villagers having been empowered to do so by the workshop. This potential backlash was avoided through joint field investigation in the village concerned. It was found (with some luck) that the villagers involved could not have been influenced by the encounter because their representatives had yet to give them any feedback from the workshop, which the villagers were vocally unhappy about. These illustrations of "social processes taking place during and following the use of participatory methodologies" show that these latter have empowering qualities when appropriately used; they also express the contradictions and uncertainties of these processes and their dependence on larger institutional commitments that, at least for a time, are needed to support local collaborative capital.

Framework for an Analysis of Method

There has always been in the social sciences a desire to resemble the physical sciences, to produce general laws of society—higher-order explanations that, once "revealed" and formulated, could be reinjected into the social movement as predictive principles. Then, we discovered that, even in the physical sciences, this potential to predict, control, and engineer can be contingent and socially—historically—situated. We then turned to explanation as an approximation of proof, relinquishing the claim of innate superiority of scientific knowledge. Science, particularly social science, is one form of knowledge among others and is not unconnected to common sense and cultural endeavors. But if explanation, conceptualization, and error are constitutive of other modes of explanation, what then may be the defining characteristics of science?

Our contention is that, as a cognitive activity, science is distinguishable mainly by the way it approaches discovery and refutation. Its search for answers is deliberate, systematic, and structured to allow for surprise and refutation. Scientists do not work, at Popper's inkling, to contradict themselves, but they are not supposed either to take their beliefs or the beliefs around them for granted; *Objectivation* (rather than objectivity), the transformation of real objects—or historical objects—into theoretical objects, is the peculiar mode of intervention of sciences. Real-life processes are externalized into things that can be measured, compared, contested, and integrated (theory) through supposedly transparent and replicable modes (methods) of transformation. This last methodological moment is key, as it is the one through which refutation can take place and the scientific character of the transformation asserted. It is these conditions—not the professional identities of the investigators or the production of grand metatheories (designed to live their lives and disappear)—that characterize scientific inquiry. These conditions can also apply to participatory endeavors, whether or not professional scientists conduct them.

Method, as a recognizable set of principles or procedures designed to explore regularities and contingencies, is, therefore, at the heart of scientific practice. Not all methods are scientific, obviously, and not all scientific methods are empowering, as we saw. In a scientific approach to participation and action, we see two embedded moments: a moment of research in action, characterized by cognitive negotiations and interactive learning, and a moment of theorization, when the agents, the action, and their contexts are externalized and reinterpreted into new theoretical and methodological integrations. In that process, different methods are combined into a fluid but coherent platform. Social facilitation, as well as investigative and analytical techniques, is mobilized to guide the exploration and interpretation of social perspectives. Figure 4-6 proposes a framework to absorb the exponential growth of techniques, methods, and approaches in social research[40] into three fundamental time dimensions of knowledge and action.

Collaborative Management

We started this chapter by questioning the link between conventional modes of knowledge extraction and local societies' continuous marginality with regard to global knowledge and power networks. Beginning with a reexamination of the participatory, action, and learning approaches intrinsic to adaptive collaborative management research, we proceeded to address three issues of interactivity, power, and science in participation. In that process, we recognized the potential and paradoxes of participatory and social-learning paradigms, challenged the notion that development sciences should shield themselves from social engagement, and pointed at the potential epistemological and ethical traps of trying to engineer social change through the consolidation of social sciences' professional identities.

Using case studies from Indonesia and Cameroon, we illustrated in different contexts how social science methods could go beyond just extracting information from local actors to serve as valid platforms for learning interactively and for negotiating meanings, powers, and representation. In that light, the scenarios that we see as the most promising lie with models of social facilitation in which scientific procedures are embedded into interactive methodologies to produce critical interactive knowledge. Within such frames, power is not "injected from the outside," as Long and Villareal (1994) fear; it is captured through plural perspectives and negotiated within a leveled interface that makes self-awareness and awareness of the other possible. In that process where new meanings are attributed to familiar objects, the actors in the interface, including the scientists who maintain their autonomy of analysis and action, appropriate knowledge differentially.

From the viewpoint of science and its peculiar way of structuring practical histories, further analysis and instrumentation of the information is not precluded, as this chapter is a proof. We can legitimately argue, however, that most actors have come out of the interface with more ability to address their own knowledge-power disabilities. This is only a small fraction of a longer process of facilitating social dialogue and negotiation in forest management, but it brings home important characteristics of social analysis applied to dynamic situations. Beside a "science for understanding" and a "science for action," both valid in their own right, we thus document the possibility of a "science in action" that does not take sides nor claim neutrality, but accepts itself as an integral part of the social interface, to be influenced and transformed in the process.

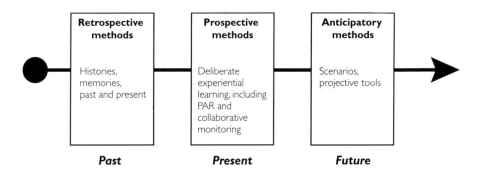

Figure 4-6. *Methods in a Time Field*

For development professionals, there is not much alternative to "getting development right." As Chambers (1997) lucidly points out, "We are all development professionals," from both North and South and from all realms of banking, law, research, academic, and donor professions, "and our decisions and actions impinge on local people and places." Our claim to science or development can only come with being more ethical in our work as scientist, professional, or development institution.

Notes

1. Epistemology refers here, simply, to the way scientific knowledge is generated, applied, and interpreted.

2. See Dissanayake 1986 in Long and Villareal 1994.

3. Selener (1997) distinguishes four major approaches of participatory research and action: (a) participatory research in community development, which he links to the influence of Paulo Freire and the theology and sociology of liberation in Latin America; (b) action research in organizations, inspired by the early work of Kurt Lewin (1946), who tried to link behavioral sciences with group research and action; (c) action research in education, which is an extension of the previous school; and (d) farmer participatory research (the farmer-first approach).

4. These included labels such as rapid appraisal (RA), rapid assessment techniques (RAT) or procedures (RAP), and rapid assessment of agricultural knowledge systems (RAAKS; Engel and Salomon 1997).

5. The MARP (méthode accélérée de recherche participative), PLA (participatory learning and action), PALM (participatory analysis and learning method), GRAAP (groupe de recherche et d'appui à l'auto-promotion paysanne), DRP (diagnòstico rural participativo), and other denominations in various languages are the offspring of PRA, which testify to its diversity and spread in various regions of the world.

6. In the mid-1970s Heinz Moser, a Swiss-German educational scientist, also tried to frame an epistemology of action research based on Habermas' Critical Theory. In a series of untranslated articles, he attacks the relational dilemma of theory and practice (science and action) and subject and object (social interface), as well as the "criterion of emancipation" (empowerment), which are at the core of this chapter (see Finger, no date).

7. These include DELTA (Development, Education, and Leadership Teams in Action), first developed in Kenya in the 1970s, and Theater for Development. Both use "listening surveys" and problem-posing art "codes" to place local experience at the core of research and extension. Like PAR, they bear the influence of Latin American thinker-activists, such as Paulo Freire and Augusto Boal (Cornwall et al. 1994).

8. Associated with this quest of alternative methodologies was the emergence of the constructivist inquiry (e.g., Guba 1990; Röling and Jiggins 1993), soft-systems and contextual investigation (e.g., Checkland 1981, 1989; Engel 1997), knowledge system thinking (e.g., Röling 1992; Engel 1997, Engel and Salomon 1997), and stakeholder analysis (e.g., Grimble and Chan 1995). A related development is the diligent search, by scientists in doubt of instrumental rationalities, for inspiration in sciences of the mind such as social psychology (Jiggins and Röling 2000).

9. On learning and adaptation, see, for instance, Röling and Jiggins 1998; Groot and Maarleveld 2000; Wollenberg et al. 2000; Buck et al. 2001; Leeuwis and Pyburn 2002; on "platforms," see Röling and Jiggins 1998; Dangbégnon 1998; Steins and Edwards 1999; Buck 1999.

10. Although in recent years social learning has held promise for an interactive design of participatory methods, the belief that they have empowering effects on less powerful groups by building their social and human competence has largely been implicitly assumed.

11. In conjunction with the devolution research conducted by CIFOR in Asia (Edmunds and Wollenberg 2000).

12. The SSM tests also took place in Brazil (Porro and Porro 1998).

13. The term for village in Indonesia is *desa*, a term that was used on Java and adopted throughout Indonesia to denote the smallest administrative unit of the government bureaucracy. The village of Baru Pelepat consists of five *dusun* (hamlets), of which three are original Minang settlements and two recent settlements established under a government resettlement project.

14. The diversity of population is historically linked to both the official placement of social groups (i.e., under the resettlement project) and the spontaneous settlement of others. The site exemplifies many other places in Indonesia where official resettlement—often ignoring preexisting social systems—has resulted in communities composed of a mix of social groups.

15. Editor's note: Orang Rimba were previously referred to as *Kubu*, a name with pejorative connotations locally.

16. In conformance with the customary regulations of the Minang group, the right to use communal land is connected to the planting of perennials, a right that holds as long as there are perennials on the land. Often this means permanent use by a family, since the integration of perennials in the system is stretched out over several generations.

17. Participatory assessments identified several village governance problems: lack of representation in community decisionmaking, particularly for natural resource management; weak customary leadership; and unclear authority for customary institutions in the presence of a formally appointed village government.

18. Because the formal and the customary legal systems coexisted, village boundaries of Baru Pelepat were unclear. The official village demarcation of the early 1980s did not match the customary territory. The community perceived that formally recognized village boundaries were necessary to halt the excessive logging in the village forest by outsiders.

19. Long (1989) suggested taking the "social actor" as the unit of analysis, rather than the "individual," because the former represents the notion of agency, which is a relational capacity rather than an individual one.

20. Röling 1994 (in Groot et al. 2002) uses the soft-systems term "human activity system" when referring to a set of actors who operate in a particular domain and are expected to or who manage to work synergistically in constructing a collective reality.

21. The choice to use the election of members of a village representative body (*Badan Perwakilan Desa*, or BPD) as a learning opportunity had been made in connection with present decentralization policy. For under this policy villages have been required by the government to set up their BPDs. However, given a lack of effective technical guidance from the government, many of them tended to elect BPD members who already held power in the community or were favored by the district government. Inexperience in developing representation mechanisms at the community level thus tended to allow village elites to capture the right to represent the community. CIFOR-led activities in Baru Pelepat on this representation issue received support from the district government because, lacking sufficient experience and resources for implementing this policy, it had welcomed any "outside expertise" it could find. The research will continue to assist community stakeholders in building their representation and decisionmaking capacities, as well as their abilities to participate in wider political processes.

22. The scenarios centered on the issue of village devolution in the context of current decentralization. Groups were asked to make a scenario about how their village would take on new responsibilities under the new conditions and how it would organize decisionmaking to include the various social groups.

23. The field team kept records of evidence of changes related to learning processes in process documentation reports.

24. In two other sites, these consultations were done in the form of one-day, open-door meetings with local actors. A sixth site was not included because of calendar issues.

25. See workshop reports (2001) by Nguiébouri et al., Malla, and Oyono and Efoua.

26. In one case, in the very first round of individual presentations when each participant was asked to recall the last time he/she had had a good laugh, a community member sternly responded that he had not laughed since 1930 (the time the community's land was unilaterally classified as a state forest reserve).

27. ONADEF (now ANAFOR), the Agency for Forest Development, is an arm of the Ministry of Forests and Environment (MINEF). It had the vision that C&I-based comanagement with local communities could be a way to resolve the reserve's crucial livelihood and conservation problems and saw in ACM an opportunity for doing that.

28. Groups or individuals represented the ethnic communities, the administration (including MINEF and the commander of the military base), the intervening agencies (including the Campo Ma'an Project), and the logging industry. The logging representative came too late, however, to participate in the exercises discussed here.

29. The village of Bosquet is the site of the first Pygmy community forest in Cameroon, which makes of Lomié a test case for understanding the changing position of social minorities in forest management (see Chapter 5).

30. Anne Marie Tiani, our team's ecologist, introduced this game, while the techniques of visualization and analysis were jointly developed during the workshop. Except in one case, the scoring of these two indicators was identical; Figure 4-3 is based on the quality of collaboration.

31. Four pebble games were used during these tests. The technique was the same, but it was applied to different types of issues: the sharing of forest benefits; the amount of resources accessible to past, present, and future generations; stakeholders' influence and decision-making powers over selected resources; and the stakeholders' rights and means over selected functions of management, which is the focus of this section.

32. This is an individual measure of handicap, while the management handicap, as we will see later, is a measure of the combined perceptions of handicap across a management function or a site.

33. The distribution of handicaps in a given management area is a zero-sum game. Since excess powers and power deficits automatically balance each other, their sum is necessarily zero. This is an artifact of the game and does not mean that management is in equilibrium. The only way to obtain an aggregate indicator of the overall management handicap is to weight the absolute values of the individual stakeholder handicap against another measure. There are several ways to do that: we are using a measure of tolerable levels (± 10), which was defined during the forums and is more meaningful in that context.

34. Recognizing the relative nature of the notion of equilibrium, we established during the workshops a comfort zone or tolerance zone of ± 10 around the point of zero handicap. This means that a stakeholder's R&M deficit was considered acceptable as long as it stayed below 10 or above -10.

35. For the R&M to sanction in Campo, for instance, we have 3 givers $\times$ 4 receivers $\times$ 10 = 120.

36. In this case, the index should be calculated from the Ottotomo community's total = 460/90 = 5.1.

37. In Lomié; in Campo the logging industry did not participate in the game. Industry workers took the same position, however, during the SSM tests in that area.

38. From our observation, the only case of systematic bias came from the sole Bagyeli Pygmy representative in Campo. Confined to a minority position in the community group, this person adopted a position of silence and escape classic to the defense strategies of Bagyeli and started talking only after the end of the forum! The specificity of that case is highlighted by the anti-Bantu discourse of the Baka in Lomié, where they had achieved equality in number and autonomy. This shows the risk of merging Bantu and Pygmy actors into an undifferentiated community group.

39. By "situated facts," we mean the mechanisms through which social actors perceive and select facts that carry special meaning to them and support their view of the world.

40. By "social research," we do not mean only social science research, but all research that bases outcomes directly on interactions with people. Opinion leaders, activists doing action research, ecologists, biologists, and geneticists—as much as social scientists such as sociologists, anthropologists, and economists—all use data that are not derived exclusively from the direct observation of the natural world but rely instead on significant human inputs. Without negating the role of disciplinary identities in science, this understanding underplays the institutionalization of disciplinary interests and the separation between science and nonscience to better highlight their shared social character.

PART II

Africa

THE AFRICAN CONTRIBUTIONS come from three sites in Cameroon (Campo Ma'an, Ottotomo, and Lomié) and two in Zimbabwe (Chivi and Mafungautsi). Chapter 4 set the stage for this set of chapters on forest management in Africa by using Indonesian and Cameroonian examples to raise questions about the theory of collaborative work with communities. The next five chapters will focus on the real and potential roles for women and minorities in managing forests. Questions include who has or should have the right to define gender roles, what roles NGOs and conservation projects should play in bringing about social changes relating to gender roles, and how social learning can be catalyzed into adaptive collaborative management of forests.

Cameroon

Campo Ma'an National Park (Chapters 4 and 6)

Campo Ma'an is a large national park in southwestern Cameroon, bordering Equatorial Guinea on the south, the Atlantic Ocean on the west, and plantations and logging concessions on the north and east. Many local communities of various ethnicities live within the park, but are often at odds with the park's management. The prestige, connections, and wealth of the park have made it a formidable adversary for local communities striving to protect their heritage and livelihoods. Many migrants have also come into the areas surrounding the park in search of employment, complicating the ethnic mix in the area.

Ottotomo (Chapter 4)

Eight villages surround the Ottotomo state forest reserve, just west of Yaoundé. Off-limits to human activities since its creation in 1930, the reserve faces growing encroachments from the local Ewondo population, who claim use rights in the forest, and from the increasing population pressure in all of the surrounding villages. The reserve's managing agency is ONADEF (now ANAFOR, the Agency for Forest Development), an arm of the Ministry of Forests and Environment. Other important stakeholders in the area include a mix of NGOs and other intervening actors, including staff of the Ministry of Agriculture and facilitators working to introduce adaptive collaborative management.

Lomié (Chapters 4 and 5)

The Lomié community forest area is in eastern Cameroon. The ACM research includes five community forests (Kongo, Koungoulou, Eschiambor, Bosquet, and Ngola), which in 2000 were among the first to be legally recognized under the 1994 Forestry Law. Important stakeholders include representatives from the logging industry and SDDL (*Soutien au Développement Durable de Lomié*), a project attached to the Dutch cooperation agency SNV, as well as the local Bantu communities and the usually marginalized, but very forest-dependent Baka Pygmies.

Zimbabwe

Chivi (Chapter 7)

Chivi is a Shona farming community in the *miombo* woodlands of central Zimbabwe. Like many Zimbabwe communities, the population was forcibly resettled there after World War II, when white soldiers were repaid for their wartime efforts with prime agricultural lands (thus displacing the existing populations). Chivi has been the site of considerable conservation research and activity by the University of Zimbabwe in Harare and by CARE International. Gender roles are clearly differentiated in Chivi.

Mafungautsi (Chapters 8 and 9)

Mafungautsi is a national forest reserve (*miombo* woodlands), surrounded by communities. The area is characterized by an ethnic mix, but the indigenous peoples are the disdained Shangwe, with a large influx—both voluntary and involuntary—of the more dominant Ndebele and Shona. The government, which does not acknowledge any rights to forest management by local communities, is experimenting with comanagement in Mafungautsi. Resource management committees (RMCs) have been set up in the surrounding communities, and the ACM project has been working to introduce adaptive collaborative management into three of these committees. As in Chivi, gender roles have been found to be clearly differentiated.

CHAPTER FIVE

From Diversity to Exclusion for Forest Minorities in Cameroon

Phil René Oyono

*T*HE FRENCH ETHNOLOGIST Laburthe-Tolra (1985) considers the Beti—
the majority human group in the rainforest of Cameroon—as the "*seigneurs
de la forêt*" (lords of the forest). One may, on that basis, say that the Pygmies are the
"gods" of that same forest: they have long lived in the area, and, accordingly, are
considered as "indigenous peoples."[1] After Le Roy (1929), Bahuchet (1992) con-
firms the similarities between the Pygmies of Central Africa and other "mobile" or
"semi mobile" peoples in Papua New Guinea, Sumatra, Sulawesi, the Molluccas,
the Philippines, Malaysia, Ecuador, Guyana, Peru, and elsewhere. Like these other
peoples, Pygmies are short in stature. It has also often been said, from the viewpoint
of physical anthropology, that they have remained at the initial stages of human
evolution. With lifestyles reminiscent of past hunter-gatherer civilizations (see Schkopp 1903; Wilkie 1988; Vansina 1990; Seltz 1993; Bailey et al. 1992), Pygmies,
alongside other human groups, would seem to have survived from an earlier era
into the modern age, as pointed out by Guillaume (1989).

Nonetheless, Pygmies have no major genetic differences from their neighboring
ethnic groups and communities; biologically, there is no "Pygmy marker gene"
(Koch 1912; Bailey et al. 1992; Cavalli-Sforza 1986). Clear social and economic
differences persist, however. After more than a century of colonial and postcolonial
"modernity," the Pygmies of Central Africa in general, and those of Cameroon in
particular, seem to live outside of mainstream society and to be the losers in a game
of national diversity. Because they are so few, in contrast to the neighboring communities, they form part of the forest minorities of the tropics. Because they remain
so dependent on forest resources, they are said to be forest peoples *par excellence*[2]
(Bahuchet 1992). It is this situation that is likely, in the short term, to determine the
future of their relationship with the forest, or in other words their ecology.

This chapter gives a brief historical overview of the Pygmies' presence in the
forests of Cameroon and thus assesses their status as indigenous people. It also

reviews the measures taken to correct the Pygmies' marginal status and describes the prevailing official rhetoric, which reveals no intention whatsoever to reverse the legal and political inferiority that the Pygmies suffer. Political inferiority is a concept applied to social actors excluded from the public sphere and from any political debate (Habermas 1995; Cohen 1999; Bigombé 2001). Using an analytical grid from the sociology of natural resources, this chapter will describe the elements of inequity, which in the context of social diversity govern the allocation of social, economic, political and, especially, natural resources. Finally, suggestions will be made about how to bring the issue of equitable access for the Pygmies into the discourse on forest policy and social sustainability.

Theorists have tried to tackle the central notion of this chapter—access—for a long time (see Commons 1968; Lukes 1986; Bromley and Cernea 1989; Nugent 1993; Nelson 1995; Ribot and Peluso 2002; and others). According to Ribot and Peluso (2002), access is defined as "bundles and webs of powers that enable actors to use and control a given resource," the "ability to benefit from things," with specific rights, structural mechanisms, and relational mechanisms. In applying this broad notion to the situation of Pygmies in Cameroon, this chapter concludes that they should be treated fairly as a part of the national community and therefore should have legitimized access to resources. If they do not have legitimate access— as is the case now—they will be pushed back to their ancestral way of life in the forest. But they will bring knowledge of modern lifestyles and technologies that they never had in the past. They already know the attraction of money, and accordingly can be used by those who would overexploit forest resources. In these circumstances, there is a danger that, if they are rejected by the national community, Pygmies could contribute extensively to the erosion of forest resources.

The Historical Context

Since antiquity, chroniclers have referred to the existence of Pygmies, as demonstrated by Trilles (1932), and historians now agree that Pygmies have a long history in the African equatorial forest (Seltz 1993). De Foy (1984), for example, notes that more than 4,500 years ago, the Egyptian general Harkhuf, heading an expeditionary force sent by Pharaoh Pepi II, discovered "Black Dwarves" around Nubia (modern-day Sudan). To the Europeans, these Black Dwarves were long believed to be simply imaginary (Schlichter 1892), one of the many myths about Africa. Monceaux (1891) speaks of "a legend of equatorial dwarves." Thorbecke (1913) characterized them as "a new race of dwarves." Seltz (1993) reveals that it was not until 1870, when the German explorer Schweinfurth discovered Pygmies in Central Africa, that Europeans were finally convinced that Pygmies truly were human beings. Today, the Pygmies are scattered throughout the immense tropical forest from the Atlantic Ocean (Kribi and Campo in Cameroon) in the west to the Great Lakes region of Uganda, Rwanda, and Burundi in the east, and from the center of Cameroon in the north to the Democratic Republic of Congo in the south.

Cameroon is thus an integral part of this vast forest amphitheater. There are three centers of concentration of Pygmies in the country, where they cohabit very closely, in satellite hamlets, with Bantu ethnic groups. The latter, known as Tall Blacks in contrast to the Pygmies, form a heterogeneous constellation of human groups found from Cameroon to southern Africa. According to Mveng (1984), the Tall Blacks (Bantu) of Cameroon joined the Pygmies in the southern Cameroon amphitheater in the second half of the nineteenth century. The three different Pygmy groups in Cameroon may be described as follows (Althabe 1965; Loung 1981):

- *The Baka Pygmies,* with a population of 35,000, live mainly in East Cameroon, although a small group resides in the Dja et Lobo division, South Province (Vallois and Marquer 1974). Within this area, they gravitate around the many protected areas, including the Dja Biosphere Reserve, the Lobéké National Park, and the Nki and Boumba-Bek Faunal Reserves. In addition to receiving the attentions of charitable organizations for more than 50 years, they are, on paper, a focus for biodiversity conservation projects set up in that part of the country.
- *The Bakola or Bagyeli Pygmies,* with a population of 4,000, are centered in Ocean and Nyong et Kellé administrative divisions, with some pockets also found in the western sectors of Mvila and Vallée du Ntem divisions. The Bakola Pygmies, like the Baka, are fairly well known, and many aspects of their strategies for exploiting the forest have been carefully documented over the past 30 years (Loung 1959, 1981; Letouzey 1967, 1975; Joiris 1986; Bahuchet 1988; Seltz 1993).
- *The Medzan Pygmies,* with a population estimated at 1,200, are found around Yoko (Barbier 1978; Vallois 1947) in the western part of the central Cameroonian plateau to the north of Yaoundé. This group is much less well known to ethnographers and anthropologists.

There have been other attempts to describe the distribution of the Pygmy groups: according to Seltz (1993), Vallois (1947) identified in Cameroon a "southeastern group" (the Baka), a "southwestern group" (the Bagyeli), also found in Equatorial Guinea (Castillo-Fiel 1949), and the "small people of Yoko region" (the Medzan). Dupré (1962) talks of the northwestern Pygmies (the Bagyeli) and the northeastern Pygmies (the Baka), without mentioning the Pygmies of Yoko at all.

The Policy Context

At the beginning of the 1990s, Cameroon embarked on a new program of democratization and of institutional/policy change. As in several other African countries (Bratton and van de Walle 1997), this tropical *perestroika* in Cameroon brought new national development strategies and new "public ideologies," such as the devolution of authority, the expansion of public participation, and the importance of good governance. Mbembé (1989) and Bayart (1993), in a political economy approach, analyze the "sociogenesis" and reproduction of the practices of material and symbolic power in Cameroon. They conclude that access to resources—and the income those resources bring—is the subject of violent competition between state managers, sections of the elite, foreign actors, and civil society. This competition occurs within the legacy of a French colonial system for administering access to resources, which was highly centralized and affirmed the state's legal and symbolic hegemony over all natural resources.

In 1994–1995, under external pressure and domestic demands for liberalization, social justice, and equity, Cameroon agreed to introduce a new forestry code better adapted to the new political climate favoring democratization. Three community oriented fundamental innovations came out of this new law for managing the forestry sector. First, the new law clearly and officially affirmed the participation of local populations in the management of forest ecosystems. Second, the new law transferred powers and management responsibilities to peripheral actors, including the local populations. This approach of sharing rights and transferring responsibilities to peripheral entities is characterized as a process of devolution (Diaw and Oyono 1998; Vabi et al. 2000; Etoungou 2003). Third, in the guise of a "decentralized forestry taxation system" (Fultan 1992; Carret 2000; Milol and Pierre 2000; Bigombé 2003), the new law defines new modes and mechanisms for the redistribution of forestry royalties, in recognition of "localism" and "autochthony." The real benefits of the Cameroonian experiment with devolution seem to be the creation of community forests and hunting zones, the empowerment of village communities to manage those areas, and the redistribution of forestry and oil royalties to the local communities in many parts of the country.

The new forestry law is one of an array of legal and administrative efforts in Cameroon to promote equitable access to forest resources. These efforts include the National Environment Management Plan (NEMP), the Forestry Sector Reform Plan (FSRP), and the Forestry and Environment Sector Program (FESP), and, before the 1994–1995 reforms, the Tropical Forestry Action Plan (TFAP). Moreover, the 1996 Constitution defines citizenship broadly and equitably:

> All persons shall have equal rights and obligations. The State shall provide all its citizens with the conditions necessary for their development; the State shall ensure the protection of minorities and preserve the rights of others and the higher interests of State (Constitution of the Republic of Cameroon).

Alongside the objectives of sustainability and social justice, the new Constitution also establishes protection of the environment as a priority. Thus, in principle, the various existing legal instruments are based on equity and justice amid social, ethnic, and national diversity.

Caught between the Tall Blacks, the State, and the International Community

During their migrations in the eighteenth and early nineteenth centuries, the Tall Blacks,[3] or the Bantu peoples of Central Africa, forged relations with the Pygmies based on simple attraction, proximity, and curiosity. As the migratory movements of the Bantu came to an end, these relations grew and were increasingly shaped by economic and social exchanges (Bahuchet and Guillaume 1982; Guillaume 1989; Bailey et al. 1992). Loung (1959) emphasizes that relations that had previously been circumstantial and contingent became more structured and cooperative, and went on to become additionally determined by heightened interdependence. In the meantime, the Pygmies had come to live in quasi-symbiosis with their Bantu neighbors.

Close examination reveals a complex system of relations, although little is known about the exact conditions in which this system emerged and the logic behind its articulations. All that is known from observation and, more sporadically, from documentation (see the synthesis in Seltz 1993) is that the Pygmies and the Bantu cohabit on the basis of a network of unequal exchanges.[4] Vallois 1948; Turnbull 1966, and De Foy 1984 note that the resultant domestic ties reproduced down the decades and even centuries are based on condescension, and mistrust. This may be explained very generally by the fact that the Bantu have never viewed the Pygmies as being social actors on the same footing as themselves.

The Pygmies—described both as "indigenous people" (Winterbottom 1992; Bahuchet 1992) and as "forest people" (Bailey et al. 1989; Dyson 1992)—differ from the national community in general and the Tall Blacks in particular in a number of characteristics: their long presence in the forest; their extreme dependence on the forest; their cultural specificities; and their political isolation, all of which translate into marginal participation in public affairs and decisionmaking. Guillaume (1989) identifies and describes the major features of the Pygmies' way of life. He considers the fundamental factor to be the "homology" or correspondence between these human societies and nature, with, in addition, the "mobility of domestic groups" and their "flexibility in their occupation of the natural space and in the common ownership of resources."

Several authors state that the Pygmies are no longer real hunter-gathers (Schebesta 1936; Schumacher 1947; Althabe 1965). Other analysts emphasize the role played by the Pygmies' acceptance of the agricultural mode of production, and the technological changes and mutations in their material and economic culture (Joiris 1986; Guillaume 1989; Seltz 1993; Nkoumbélé 1997; Bigombé and Bell 1998; Oyono 1998a). Some other researchers believe that the degradation of the Congo Basin forests will have irreversible effects on the Pygmies, more than on other populations (Bahuchet 1992; Dyson 1992; Seltz 1993). For Mimboh (1998) and Biesbrouck (1999), the Bakola and Bagyeli Pygmies are already living in "two worlds" because of current levels of deforestation. Commentators have written exhaustively about the Pygmies' "original" way of life and its metamorphoses (see Seltz 1993), from their umbilical link to the forest universe to the risk of their disappearance via the exploitation of their labor by the Tall Blacks.

During the colonial period, there was little public interest in the Pygmies of Cameroon (Althabe 1965). The Germans (1884–1916) were more concerned with pacifying the country than paying attention to such a "vagrant" human group. The French did little better during their period of colonization (Vallois 1948). Even the post-colonial state did not draw up an agenda for the Pygmy issue as such. Although the post-colonial authorities lacked a proactive interest in the Pygmies' situation, they had an "instrumental" and political interest in settling them in villages. Thus, between 1960 and 1970, the ruling political class in newly independent Cameroon planned a strategy for the "sedentarization" of the Pygmies. The task was left to the administrative and military authorities in the zones of Pygmy concentration, who enjoyed the support of the Bantu village chiefs in this respect. This sedentarization policy, partly motivated by a desire to build the nation-state, was also an attempt at sociopolitical control.[5]

The speed with which post-colonial Cameroon introduced the sedentarization of the Pygmies should be set in the context of the unsettled political climate of the newly independent state. In the early 1960s, the central state's authority was strongly contested, and the first Cameroonian head of state (Ahmadou Ahidjo) was perceived as a French agent. The country was in the grip of an armed struggle over the idea of independence and its materialization (Mbembé 1986). As in other parts of the country, the areas of Pygmy concentration—Djoum subdivision in the present-day South Province and Moloundou subdivision in East Province—suffered sporadic incursions of an armed Maoist movement. The central state secretly suspected the Pygmies of acting as guides for the guerillas; the pygmies therefore had to be brought out of the forest and settled along the rural roads, beside the Tall Blacks (Bigombé 2002). Such strategies of enclosure and forced integration were implemented in other countries of Central Africa, as noted by Guillaume:

> There have in fact been more symbolic and sporadic actions, which nevertheless show the ultimate aim of the states, than real policy concerning the Pygmies. This would seem to be true for Zaire, with the inclusion of Pygmies in the state apparatus (armed forces) and political activities … In the Central African Republic, the "Pygmy integration campaign" led primarily, in 1988, to the printing of two postage stamps showing the campaign slogan alongside the Nola Pygmies' football team (Guillaume 1989,*80*).

The Pygmy Issue: Private Replies and Public Answers

Since it is the Pygmies' way of life that is seen as a problem, for reasons that vary from one observer to another, different "solutions" have appeared over time. Catholic missionaries were first to provide sociocultural and economic support for the gradual "transplantation" of Pygmies and their settlement in villages. As a result, since the beginning of the 1950s, Pygmies have been the subject of well-intentioned initiatives such as the East-Cameroon Pygmies Project, set up and coordinated by Dutch Fathers Lambert van Heggen and Ignace Delhemmes (Loung 1998). Socioeconomic actions were developed, extended, and systematized at the beginning of the 1980s under a project entitled Support for Self-Help for the Pygmies in their Environment, which works alongside the *Loti* (solidarity, in English) Project in the Lomié region. The Little Sisters of Jesus also introduced educational programs for young Pygmies in the Bipindi/Lolodorf region in the south of the country in 1952 (Bigombé and Bell 1998).

A modernist, civilizing, and well-meaning vision of the Pygmies' future has thus been built up since colonial times (Guillaume 1989; Bigombé 2001). The post-colonial Cameroonian state promptly subscribed to this vision and easily adjusted itself to the good intentions already unleashed by the missionaries. Here and there, the Cameroonian state facilitated access for Pygmies to services (for example, by issuing them identity cards, encouraging their participation in elections—as voters, not candidates—and by recruiting them to attend schools for teachers or nurses).[6] Certainly the most publicized state initiative took place at the beginning of the 1980s, when the Ministry of Social Affairs began to talk about its project on the socioeconomic integration of the Baka Pygmies, with the technical support of a Dutch nongovernmental organization (NGO), *Stichting Nederlandse Vrijwilligers* (SNV). Sporadic actions did take place, primarily in Lomié, Yokadouma, Doumé, and Abong-Mbang regions, in farming, health care, education, housing, and civic education. As noted by Atsiga Essala (1998), the project, which had no strategic vision and suffered from a lack of planning and resources, died at the end of the decade.

The economic crisis at the end of the 1980s, the effects of which can still be felt in Cameroon's socioeconomic fabric, led to a severe shortage of state financial resources at the beginning of the 1990s. At the same time, development aid became more heterogeneous and multidirectional and relied on the NGOs as primary actors. As the Cameroonian "welfare state" crumbled, weakened by the economic recession and in fact already paying little attention to Pygmies, the Pygmy question faded for a while. It was usefully relaunched in the process of promoting North–South cooperation and the arrival, en masse, of international NGOs, on which most of the domestic NGOs had to depend.

The other parameter for the reactivation of the Pygmy issue was the rise of interest in "sustainable forest management," a topical issue with links to the treatment of forest peoples. Many international NGOs, especially in East Province, have each planned a "Support to Pygmies" section in their programs (Oyono 1998a). They include the World Conservation Union (IUCN) and its IUCN/Dja project; SNV, through its Project Support for Sustainable Development in Lomié region (SDDL); and the Catholic Relief Service (CRS). Several local NGOs have already committed themselves to developing mechanisms for the sustainable involvement of Pygmies in the management of forest ecosystems.

A Political Economy of Particularism

A range of factors, some the visible signs of a long history of discrimination and some the hidden psychological forces of their own culture, have led the Pygmies to turn their backs on the outside world and to set themselves apart from other human groups (Bailey et al. 1989; Bigombé and Bell 1998; Atsiga Essala 1999). This tendency to set themselves apart is, in itself, a fundamental factor in their marginality because it allows the rest of the national community, both public and private actors, to rationalize their own ideological representations, culture, and practices that further marginalize the Pygmy community.

The basis for these rationalizations by the public and private actors varies according to their interests, social proximity to the Pygmies, political motives, and level of "developmentist populism," an ideology and an attitude found among researchers and practitioners and consisting in exaggerating the interest in the poor

(Olivier De Sardan 1990). It should be pointed out that commercial logging is the primary economic interest in the areas in which the Pygmy families live. Pygmies derive little benefit from this activity, and it is in no way a lever in rural development in the Moloundou, Lomié, Djoum, Yoko, and Messamena regions. Bigombé (2001) notes that, because of this intrinsic and regional "backwardness," or turning away from conventional economic interests, the Pygmies have been perceived throughout the national community with the whole range of prejudices: archaism, residuality, laziness, chronic nomadism, etc. Despite the gradual introduction of social, ideological, economic, and policy programs to legitimize the specific culture and status of the Pygmies, negative social and political attitudes have developed throughout the national community with a much clearer outline than the marginalization mentioned above. This gap—between negative stereotypes and efforts to address them—has not been overcome, even in Cameroon's recent history, and its effects will be discussed in the next section.

The most exclusive form of differentiation was originally marked out and institutionalized by the Tall Blacks and the two variants of the state (the colonial and the post-colonial states).[7] Even though it has often been documented that the Pygmies have also been actors in their own marginality (Atsiga Essala 1999), it can be seen from the historical accounts cited here and from social field observations that the Tall Blacks and the national community created and compounded this "insecurity" and specificity. These two dimensions of the Pygmy question—which are basically historical and political constructs—have continually infiltrated the subconscious of Cameroon's national community.

Behind the establishment of an ideology and practices organized on the basis of the "others/selves" dichotomy, there is also the issue of control. From the outlines of the social and administrative control of the Pygmy community described above, it can be concluded that the state and the Tall Blacks have manipulated diversity and, thus, difference as a structural element in the symbolic threat represented by the Pygmies. If the Pygmies are considered to be "separate" human beings, "biological aberrations," and different "social agents," then representations and practices tend to crystallize around the "rationalization" of this situation. This quotation from a Bantu observer demonstrates that this *essentialization* and almost unanimous subscription by the national community and policy makers is based firmly on the exclusion of the "others" in access to resources:

> All these Pygmies you see in this meeting room are only here because they know that the Reserve Conservator has brought wine and cigarettes. That is what interests them. Since we have been calling for compensation for the loss of the forest, which our ancestors bequeathed to us, and for it to be managed as a reserve today, the Pygmies of this village have never shown any interest in the cause. It seems that they do not consider it to be their problem. We, the Bantu, are fighting alone. They have never been involved in the issue. When it comes down to it, it is not the forest itself that they are interested in, it is only the game. If we win our fight, I hope that they will not ask for any of the benefits (Liboire Mpouam, an inhabitant of Mintoum village, on the edge of the Dja Biosphere Reserve, Lomié region, East Cameroon, April 13, 2000).

Diversity, Identity, and Access to Resources

This section examines some key parameters of inequity in the access to resources, focusing on social, political, and economic resources. The following section then look specifically at how these parameters and local practices and policies affect the Pygmies' access to natural resources. The section tackles access to social resources, access to political resources and access to economic resources.

Access to Social Resources

In the part of the country where the Pygmies live, Catholic missionaries have been working since the 1950s to promote access to modern schooling for the children, especially in the Djoum, Bipindi, Lomié/Messok, and Moloundou regions. Charities also provide health care. Government ministries, however, have done little to encourage public demand or political will, however basic, for equal access to these social services. When charities allocate educational materials to "Pygmy schools," the event is staged to the utmost by the administrative authorities, the elite of the Cameroon People's Democratic Movement (CPDM), the governing party, and the regional representatives of the Ministry of Education, despite the fact that the inequity in access to educational resources has not been challenged for decades. Moreover, there is no official global or sectoral strategy defined or planned for even a minimum integration of Pygmies into formal education. And what is true for education is also true for primary health care and other social services, such as community development.

Two examples clearly show that Pygmies are being overlooked and underrepresented in community development efforts. The first involves Common Initiative Groups for agricultural development, and the second involves community forest management. Since a rural reform program was launched in Cameroon in 1992, villages have feverishly set up Common Initiative Groups for agricultural development, with the support of NGOs. Exclusively Pygmy villages have not been part of this organizational whirlwind. In villages with both Bantu and Pygmies, the ethnic make-up of the Common Initiative Groups does not show any Pygmy members (FIMAC 1999).

Similarly, five village communities in the Lomié area began in 2001 to manage their community forests, as part of a decentralization experiment. Each village community has set up a management committee, as required by the forestry legislation of 1994. In a recent study carried out in this area, Eoné notes that Pygmies are involved in neither management committees nor decisionmaking:

> The issue of the involvement of Baka Pygmies in community activities is really crucial in the management of community forests in these villages … No single Baka plays a role in decision-making processes … Marginalization is really brought up by this total lack of representativity (Eoné 2003, *28*).

Access to Political Resources

Theorists have identified three fundamental aspects of citizenship: (a) the juridical status of legal personhood; (b) the political principle of democracy and participation in decisionmaking; (c) a form of membership of an exclusive community (see Brubaker 1992 and Cohen 1999). It is difficult to include the Pygmies of Cameroon positively in any of these three components, even though official rhetoric gives Pygmies the status of "full citizens," with the same rights and obligations as any other ethnic group in Cameroon, and avoids political references to divisions between Pygmies and Tall Blacks. Nonetheless, the Pygmies are not truly considered as "full" citizens anywhere in the country. In the countryside, where they live alongside the Tall Blacks, the Tall Blacks depict themselves, often quite openly, as the "tutors" and "owners" of Pygmies (Guillaume 1989; Bailey et al. 1992; Dyson 1992; Atsiga Essala 1998). Public meetings on local issues show no historic proof of any capacity to integrate the Pygmies or involve them in decisionmaking.

It is true that those Pygmies who so wish do take part in elections, but they are almost completely absent from local representation[8] and local government. No Pygmy has truly played the role of social leader alongside the Tall Blacks, either at local or regional levels. Atsiga Essala (1999) notes that in many polling districts in the East Province, Bantu vote in the Pygmies' place. Quite plainly, the Pygmies' difficulties in gaining access to the most "reachable" political resources, such as voting power and local participation, lead to a form of "subcitizenship," in what Vandergeest (2003) calls the "racialization" or "ethnicization" of citizenship.

Access to Economic Resources

East Province, the demographic heartland of the Pygmies of Cameroon, is one of the poorest in the country (Mpol et al. 1994). Before the economic recession at the end of the 1980s radically reduced its room for maneuver, the Cameroonian state had many village development programs in the province. There was, for instance, the National Rural Development Fund (FONADER), the Priority Development Action Zones (ZAPI-Est), the Food Crops Development Mission (MIDEVIV), and the National Small Livestock Development Board (ONDAPB). The poorest of the poor, the Pygmies were hard hit by the rural effects of the economic recession.

For families in Lomié region, annual income fell between 1984 and 1996 from $25 to $6.50 (Oyono 1998b). Available documentation (Mpol et al. 1994; Atsiga Essala 1998) states that, even before the above-mentioned state services dissolved or crumbled, the Pygmies never had access to these services, even in districts such as Moloundou or Messamena, where they form the majority of the population. Even the most accessible services, such as the Intervention Fund for Agricultural and Community Micro-Projects (FIMAC), which was set up with the support of the World Bank in 1993, in the middle of the economic crisis, did not reach the Pygmies.[9] Sometimes, by pure administrative demagogy, a "farmers' bonus" was allocated to a Pygmy farmer (MINPAT 1985), although it never had any sustainable impact on the economic condition of the rest of the community.

Current forestry legislation in Cameroon, as noted above, provides for a forestry royalty to be paid to the communities adjacent to forest concessions being logged. Logging companies also have to pay local communities $1.50 per cubic meter of timber felled. Between 1998 and 2002, the village populations of Lomié and Messamena districts, where there are large Pygmy populations, received $210,000 and $43,000, respectively (Bigombé 2003). These sums are intended primarily to fund socioeconomic micro-projects. But nothing has been done specifically for the Pygmies (Kouna 2001). When the money was divided up and distributed between families, the Pygmies were left out.

Moreover, in Lomié, where community forests established in villages with both Tall Blacks and Pygmies are currently being logged, the Pygmies are not included in the management of the income from the logging (Efoua 2002a). Nor are they taken into account in the distribution of the expected income from the logging of Dimako council forest (Assembe 2001). The same is true for access to the oil royalties paid for the Chad-Cameroon pipeline. This huge World Bank project, running from the south of Chad to Kribi in Cameroon, is managed by the Cameroon Oil Transportation Company (COTCO) consortium. The compensation paid so far has been distributed on an inequitable basis between the Bantu and the Pygmies (Bigombé 1999), despite a "Plan for Vulnerable Indigenous Peoples" being drawn up in support of the project.

Practices and Policies on Access to Natural Resources

Access to natural resources, whether formalized or not, is the central point of reference for this chapter. This section will therefore look more specifically at access to the resources offered by the natural environment in which the Pygmies have lived for millennia. We shall first examine the practices that have developed with time and then consider the choices that have been made to produce current policies.

Different Representations and a Dominant Appropriating Attitude

For the Pygmies, their relationship with natural resources is governed by symbiotic representations, within which human beings and nature form a whole. Nkoumbélé (1997) and Oyono (2002) show that these representations are based on perceptions of the forest resources, which are both vertical (metaphysical) and horizontal (utilitarian). A simple and irreducible argument is built on this mental base, centered on the idea that, since human beings live in and with the forest, they must have free access to exercise their power over the resources available (Biesbrouck 1999). This attitude goes against the Bantu representation. The Tall Blacks consider that the forest is no longer absolutely the natural continuum of social life. Thus, in the Bantu view, access to its resources should not be governed in a *laissez-faire* way.

When commercial cacao farming was introduced at the beginning of the twentieth century, Bantu and others came to view the forest space as a good that might be appropriated by remodeling and farming (Diaw 1997; Oyono et al. 2000, 2003). Pygmies, however, have always centered their mode of economic production on taking the resources provided by nature, rather than transforming them through labor and exploitation (Biesbrouck 1999). As the Bantu attitude became dominant, access to forest resources was systematically restricted, which has led to an atomization of the commons and the emergence and establishment of an "exclusivist" argument (Diaw and Njomkap 1998; Dkamla 2003). According to this argument, the Pygmies, because of their semi-mobility and "vagrancy," cannot exercise rights of access or ownership over the "goods" in the neighboring forest. Initially restricted to those villages with both Bantu and Pygmy populations, this argument has become dominant across the countryside over time.

Conflicting Views on Forest Access

Although their livelihood is on the whole determined by hunting and gathering, the Pygmies of Cameroon—and elsewhere—have long since stopped being exclusively hunter-gatherers (Althabe 1965; Seltz 1993; Nkoumbélé 1997; Oyono 1998a). Farming has become an economic and "technological" innovation over the past 40 years (Joiris 1986; Bailey 1988), enabling Pygmies to work as farm laborers, paid in cash or kind (Oyono 1998b), or as independent farmers opening up small food-crop plots around the fields of the Tall Blacks. In contrast to the delineated, permanent land use patterns favored by the Bantu, the way in which the Pygmies appropriate forest space is seasonal, loose, unstable, and discontinuous (Biesbrouck 1999; Bigombé 2002). This logic of free, open, and easy access to resources has never been accepted or "internalized" in rural, settled Cameroon.

The Bantu's mode of appropriation of space and resources is clearly marked by rules of access. The establishment of villages, farmland, commercial farming, and now the proliferation of community forests are all examples of the institutionalization of prohibitions and restrictions in access to historically common resources. The Tall Blacks' modes for appropriating the forest's resources have finally suppressed those of the Pygmies. Bantu practices of access to the forest have covered over the traditional Pygmy structures of cohabitation, social and economic decisionmaking, and land ownership. The result is the Pygmy's exclusion from land ownership.

It is difficult to draw up an exhaustive list of practices that create inequity in the access to resources in the lands inhabited by the Bantu and the Pygmies. Some examples, however, are worth citing: Mimboh (1998) notes that the Bagyeli Pygmies of Bipindi region do not have land to cultivate or to extend their hamlets. In Lomié region, the long process of demarcation and negotiation of village boundaries required for the establishment of community forests, with participatory mapping, involved the Pygmies—who have lived in these areas for decades—only passively, as simple spectators. In many cases, they are affected by restrictions on access to land for hunting, a men's activity, for gathering, a mixed activity, and to sections of streams for fishing, a women's activity (Nkoumbélé 1997). After nearly 10 years of negotiation, the Sisters of the Catholic mission in Djoum, South Province, have only recently been able to win land for the Pygmies to build huts and a school for Pygmy children (Abilogo 2001).

Blind Policies and Marginalizing Programs

The literature demonstrates that policymakers and planners concerned about the Central African forest ecosystem are fully aware of the Pygmy issue. They agree that greater attention must be paid to the close dependence of the Pygmy community on the forest. Alongside this consensus, policymakers and planners tend to promote paradigms such as local community participation, comanagement, good governance, and environmental justice as policy and methodological "conditionalities." In the case of the management and development of protected areas in Cameroon, Loung (1981) sounded the alarm by showing that, if these biodiversity conservation programs did not take proper account of the presence of the Pygmies, they could in the long term prove less successful than expected. This type of policy warning does not seem to have received any favorable attention, although national opinion is well aware of the Pygmies' "at risk" status.

Cameroon's forestry law, although resolutely intended to achieve decentralization and devolution, and fairly progressive compared with those of other countries of the Congo Basin, makes no mention of any special concern for the Pygmies. No parliamentary representative from the regions where the Pygmies live ever raised the sensitive issue of their links with the forest ecosystems during the drafting of the forestry law in 1994, under the guidance of the World Bank. In this "revolutionary" legislation on community forest creation and management, reference is made to the overall idea of "community," which is understood by the Tall Blacks to mean their communities.

No major national program, such as the National Environment Management Program or the Forestry and Environment Sector Program, makes any reference to the Pygmies. When the Chad-Cameroon Pipeline was planned, Pygmies were given scant attention and that only on paper. The Kongo Mining Project in Lomié region, currently being set up with the U.S. company GEOVIC Ltd., will have a direct effect on the environment of the surrounding Pygmies, but does not mention them once in the 24-page agreement signed with the Republic of Cameroon (GEOVIC-Republic of Cameroon 2002). Even the participatory management programs for protected areas (Lobéké National Park, Campo-Ma'an National Park, Dja Biosphere Reserve, etc.) give no specific consideration to the Pygmies.

This "refusal" in national policies and programs to promote access for the Pygmies to forest ecosystems and the goods they contain is also found in the international programs throughout the Congo Basin. The Tropical Forestry Action Plan (TFAP) did not manage to look specifically at the relationship between deforestation and the social and historical reproduction of the Pygmies of Central Africa. On this topic, Winterbottom (1992) considered that, if major reforms were not made to the Tropical Forestry Action Plan, it might end up contributing to the destruction of indigenous peoples. More than a decade later the huge Congo Basin Forest Partnership (CBFP), aiming at improving governance and sustainable natural resource management within six Central African Countries (including Cameroon), seems equally affected by this blindness to the Pygmy issue. This initiative, supported by the United States Agency for International Development (USAID), is currently in its planning stage. The CBFP initiative is focusing on 12 "forest landscapes" covering protected areas. The experts working on it have so far omitted any explicit reference to the issue of the Pygmy populations.

The Pygmies are well aware that they are being disregarded. For example, in a pebble-game exercise intended to reveal the perceptions of stakeholders about rights to manage forests, Pygmies felt themselves "disempowered" and without recognized rights. To the question, "Who has the rights to define limits in the forest?" they excluded themselves and clearly referred to the Bantu and to the state (Diaw 1997; Oyono and Diaw 1998). Diaw and Kusumanto demonstrate this trend in their presentation of the scoring of stakeholders' perception of rights to define forest limits and to manage conflicts (see Chapter 4 in this volume). In the following quotation, an old Pygmy summarizes the situation his community is facing in Bipindi region, south Cameroon:

> We Pygmies are also human beings. I think that is why you come towards us. We were living in this forest before these villagers claiming now to be our patrons joined us. The forest was our life: we were in the forest and the forest in us. They asked us to come and live here. The result is bad. You have vehicles, clothes, and money. You have good houses. In your sense, you are rich. Do you want us to be like you now that we are living in villages like the Bantu? Unfortunately, nobody is taking care of us. We are abandoned. We have nothing since we are living here in this village. Our children are suffering: I don't know what will be their future. The whole life, our life, is changing. We have no weapon to face this change. The modern school is not good for our children, because when they go there Bantu children treat them very badly, and they give up. Logging companies don't care about us. We see them every day going into the forest of our ancestors and our spirits, to log timber. Maybe one day we will go back to the forest, our forest, but to do what and to live how? The Government is ignoring us. How can we get out of this *limlim têh* [deadlock]? (quoted in Nguiébouri 2001, 8).

A Potential Vicious Circle: Pygmies and Symbiosis with Nature

Throughout this chapter, we have attempted to show that policies acknowledging "national diversity" have in fact generated lines of differentiation, which have been historically and symbolically prejudicial to the Pygmies of Cameroon. This prejudice is indicated by the obvious inequity in access to various resources, first and foremost natural resources, and by the exclusion of Pygmies from the management of these resources. The result has been a "racialization" or "ethnicization" of access to resources. Although the process of sedentarization, or the settling of the Pygmies around the villages of the Tall Blacks, has given the Pygmies some economic autonomy, with their subscription to farming activities, it has intentionally undermined their traditional lifestyle based on a symbiotic relationship with the natural environment. The sad results of ethnicization and sedentarization have made the national community, NGOs, and other private agencies more aware of the Pygmy issue, but have not changed their plight. The Pygmies today are rejected by the nation, denied equitable access to resources, and unprotected by any mechanisms for integration, as emphasized by Guillaume (1989).

Our prediction is that the Pygmies will reinstall themselves in nature and become one with it again. In other words, whereas sedentarization and the merging of the Pygmies into the proclaimed "national diversity" were intended to dissolve their links with the world of the forest, in fact those links will be strengthened. The rejection of Pygmies by the larger society will reestablish their ties and dependence on the forest, but the "ecological cost" of their "reinstallation" now will be augmented by chainsaws and links to the global market. The Pygmies' new technological capability represents a huge risk for forest resources.

The Pygmies' reinstallation in the forest, which can already be felt, will be very different from their traditional lifestyle. In the 40 years of sedentarization, the Pygmies have acquired new tools and new practices for exploiting nature as capital (see Althabe 1965 and Joiris 1986). A decisive threshold has thus been crossed. The withdrawal of Pygmies back into the forest will have a direct spatial effect, given their proximity to the resources from which they have in recent years been prevented access. In that sense, a strict exclusion regime will, in turn, become more attractive to the government and may lead this time to the Pygmies' finally taking a firm stand.[10] A number of worrying scenarios already exist.

First, it is commonly agreed that, over the past 40 years, Cameroon's Pygmies, using modern guns, have become significant actors in poaching and hunting large mammals, such as elephants, gorillas, and panthers. They hunt for themselves, but also for the Bantu elite who provide their guns. This practice is prevalent around the protected areas (Dja Biosphere Reserve, Campo-Ma'an National Park, Lobéké National Park, etc.), which have significant populations of protected mammals (Carroll 1986; Akono 1995). In fact, if the Pygmies are driven back into the forest, their reliance on this double socioeconomic function of hunting is likely to increase and, in the short term, will make them a central vector in the increasing scarcity of certain protected wildlife species.

Second, the Pygmy community's possible return to the forest will have direct effects on the extent of industrial logging. More extensive than intensive in Cameroon, commercial logging requires knowledge of the forest. As it is practiced now, this logging is both anarchic and illegal: evidence shows, for example, that high-value species are selectively depleted; unsaleable timber is felled but abandoned in the forest; and access roads are extremely destructive of flora (Van Soest 1995; Bikié et al. 2000). The Pygmies, who are increasingly employed by the logging companies because of their familiarity with forest geography, are in fact a prime factor in the worsening of these indicators. Their current exclusion will only accelerate this process, since, once they are pushed back into semi-isolation, they will be even less open to environmental information campaigns on deforestation and its consequences.

Third, the Pygmies' own increased commercial and subsistence use of the forests will be detrimental in various ways. There is the effect of poaching on the flora, characterized by the abusive and illegal use of prime species such as *bubinga* (*Guibourtia tessmanii*) and *moabi* (*Baillonella toxisperma*). There is also the intensive and extensive removal, for commercial purposes, of bark and roots for many different uses. Everywhere, the Pygmies are likely to increase this type of exploitation of the forest, supporting the small economic operators attracted by the subsector, as the urban and European demand for such forest products is increasing, according to Ndoye et al. (1998).

Thus the withdrawal of the Pygmy community, initially into itself, and then into the forest, will create a whole series of imbalances. These developments may perhaps not take the same form as they did 40 or 50 years ago, when cohabitation between the Pygmies and the natural environment was comparatively balanced. Now the Pygmies are in the midst of a systematic confrontation between "sociodiversity" and "biodiversity." The "romantic" and "developmentist populism" solutions thought up here and there by the state and the NGOs to establish equitable access to resources have fallen short of expectations (Guillaume 1989; Bigombé 2002). All in all, it is extremely difficult to see a future balance between the Pygmies' ecology and the forests of Central Africa. The following three-stage scenario attempts to illustrate the situation. The first stage represents past, relatively balanced conditions, in which the Pygmies coexisted with the forest's biodiversity. In the second stage, this balance is upset by policies and practices such as sedentarization and other prohibitive or restrictive regimes. The rupture of the historical balance, together with the rejection of the Pygmies by the nation and its citizens, opens the way for the third stage, which is a violent reconstitution of the initial relations between the Pygmies and forest resources.

Conclusions

Need it be said, now that the relationship between human beings and the natural environment has become so important, that the Pygmies are the first inhabitants of the equatorial forest? From a historical perspective, the forest basically belongs to them. But where they once freely reigned, they are now marginalized. Lack of official concern about the Pygmy issue, socioeconomic exploitation by the Tall Blacks and others, exploitation for mystical and magical aims by these same Tall Blacks (Koch 1968), and religious "instrumentalization" by missionaries have all structured the process of marginalization. Many different actors have—voluntarily or not—helped to obstruct the Pygmies' access to resources.

Surely the best form of devolution for forest management in Africa would transfer and maintain the rights and functional obligations of the peoples who depend on the forest. Because, after 30 or more years of interventionism, NGOs have not achieved an equitable "bottom-up" approach to devolution, a "top-down" approach could be a productive alternative in dealing with the problem. National forestry and social policies that transfer specific rights to the forest populations could, in the end, help to redress inequities and lead to environmental justice. Other policies could be facilitated by the political advocacy of NGOs. For example, pilot sites among the Pygmies could be developed to test the rights to create community forests and the security of access to forestry royalties, both key features of the 1994 forestry reforms. Once the lessons have been learned and corrections made, the pilot efforts could be replicated, monitored, and evaluated. Success in establishing equitable access to natural resources and the related benefits could augur the deconstruction of restrictive regimes of access to other resources.

All this is no doubt easier said than done, given Cameroon's current policy environment. One thing is nevertheless sure: tackling the issues associated with access to natural resources can go far toward solving other access issues. Natural resource sociology—focusing on improving natural resource management and promoting social equity—also requires the determination of broader policy issues, such as the assignment of appropriate rights and powers, the role for local decision-making, the definition of equity, and the development of social indicators of diversity. In such a process of research, social action, policy formulation, and advocacy, particularism should strongly be avoided, as recommended by Bailey:

> It should be clear that, for the purpose of designing programs for development or conservation, Pygmies cannot be considered in isolation from indigenous forest farmers. Central African farmers and Pygmies exist together, are interdependent, and should be considered as an integrated economic and social system. This is a system that is generally not recognized by African governments and is only minimally integrated into the formal politics and economy of the national societies. Yet for the people themselves, the system facilitates the spread of risk in an uncertain, seasonal environment and offers support to people vulnerable to unpredictable changes brought by outside agents" (Bailey et al. 1992).

Acknowledgments

I am grateful to the European Union for funds provided to CIFOR-Cameroon during the research work on Adadptive Collaborative Management of forests. Some of the lessons drawn from the study of decentralization in Cameroon—funded by CARPE/USAID, with the support of WRI—are also included in this paper. I would like to thank Samuel Efoua, Samuel Assembe, and Joachim Nguié-bouri of CIFOR-Cameroon for field information. My thanks also go to Guy Patrice Dkamla of CARPE/USAID-Cameroon.

Endnotes

1. This is an ambiguous notion, if anyone born in the country in question is to be "indigenous." It is, moreover, suggested that the term "indigenous populations" should be used instead of "indigenous peoples" when the human group concerned does not demand the right to self-determination (Bahuchet 1982).

2. Winterbottom (1992) also gives a list of criteria to define a forest population. The most significant criteria are "long presence in the forest," "dependence on the forest ecosystem" for physical and cultural well-being, and "sociocultural isolation."

3. The expression "Tall Blacks" for the Bantu is used in contrast to the "small stature" said to be characteristic of the Pygmies (De Foy 1984). From the strict viewpoint of physical anthropology, however, there is no proof that the Pygmies are exclusively "people of small stature," compared with the Bantu. In many cases, the physical measurements are the same. Seltz (1993) speaks of "a heterogeneous racial image."

4. These relations are often one of the most important variables in explaining the Pygmies' socioeconomic situation (see Bailey et al. 1992). Such relations have created social, economic, and technological dependence, as found by Althabe (1965), Seltz (1993), and Nkoumbélé (1997). According to Bigombé (2001), the nature of the domination of the Pygmies by the Bantu is similar to that of the former relationship between the colonizer and the colonized. He notes similarities with slavery. From more recent observations of the management of community forests in Lomié region, Efoua notes that:

> The Bantu always ask the Pygmies to leave the forest forever and to come to live near them, in the village. Once they are established in the village, they are used as almost free labor, working for old clothes and cassava tubers. The Bantu do not recognize any rights for the Pygmies in the villages where community forests exist (Efoua 2002b, 4).

5. Once the southeast was freed of the Maoist guerrilla movement in 1966 (as described later in the text), this control had no political purpose. Nonetheless, because Pygmies were considered to be both inferior beings and beings with supernatural powers, popular superstitions supported efforts to subjugate—and even to murder and cannibalize—the Pygmies for political and social advantage. Mimboh (1998) reports the popular sentiment that some Pygmies might use the magical powers they are said to possess to win elections or be appointed minister. At the beginning of 2003, international radio stations—Radio France Internationale, Africa Numero 1, Voice of America, Deutsche Welle, and the British Broadcasting Corporation—reported a complaint, verified by evidence from the UN Peacekeeping Mission in the Democratic Republic of Congo, that troops of the Congolese rebel leader Jean-Pierre Mbemba had eaten Pygmies in the belief that doing so would make them invulnerable.

6. The economic crisis and the rationale of reducing state expenditure ended this temporary operation.

7. In social and political action, the ideological construction of "others" (Calhoun 1994; Somers 1994; Berezin 1999)—which is a collective, as well as an individual, stance—is based on a eulogy of difference. Hence, the Pygmies of Cameroon and, by extension, of Central Africa have been pejoratively labeled by their national community as "others," different from the "selves" (Turnbull 1966).

8. In a somewhat eccentric case, a Pygmy was elected (perhaps appointed) as a municipal councilor on Lomié rural council a few years ago (Bigombé 2001). The councilor himself recognized that he had never spoken during discussions in meetings of the municipal council. There may be other such cases in other regions.

9. Of all the ethnic groups in Cameroon, only the Pygmies have been forgotten in the distribution of this largesse from the World Bank. The initiative was designed and planned as "rural credit for small farmers," but was captured and taken over by the middle-class urban elite and the political classes, who set up fictitious community groups in their villages. Once they got their credit, the group disappeared, without doing anything. The result may have been the same in any case, since the reason given for "forgetting" the Pygmies was that they could not establish agricultural projects and repay the credit.

10. In African wisdom, people say, "If you prevent someone from having access to something, it will in turn take on a greater attraction for him." That is what happened in the Book of Genesis to Adam and Eve with the forbidden fruit.

CHAPTER SIX

Women in Campo-Ma'an National Park

Uncertainties and Adaptations in Cameroon

Anne Marie Tiani, George Akwah, and
Joachim Nguiébouri

*L*OCAL COMMUNITIES GENERALLY perceive forest management as a public affair. And yet, as shown by Kanji and Menon-Sen (2001) concerning "spending priorities" or Brown and Lapuyade (2001) about the "separate spheres of action for men and women" in the household, the public domain and investment fall within the competence of men, since women are responsible for "private," domestic business. Vabi and others (2001) assert that men and women not only play different roles in society, with distinct levels of resource control, but also have different needs and interests. Because of their deciding role in household food security, women are most affected by disruptions in the availability of and access to resources.

Drawing inspiration from Bisilliat's assertion (1992) that "information about women is necessarily information concerning men," this chapter aims to show how international and national environmental policies may affect local communities by affecting the women of two villages adjacent to the Campo-Ma'an National Park in Cameroon. Following descriptions of the two villages, the stakeholders involved in resource management, and the economic livelihoods and strategies of adaptation they pursue, we analyze the situation of women in light of constraints imposed on them by the proximity of the national park. Some lines of reflection and action are proposed for ameliorating the differences of opinion between the park's managers and local communities, for effective conservation and sustainable local development.

Site and Context

The Campo-Ma'an forest, located in the southwestern part of Cameroon, borders Equatorial Guinea and is endowed with an almost unique wealth of flora and fauna.

The modern history of the area began with the creation in 1932 of the game reserve. In 1945, the first logging company—*La forestière de Campo,* now part of the Bollore Group—settled in the region. This company is the main economic actor in the whole southern part of the area. Later on, large agroindustrial plantations developed in the northern part: HEVECAM, a rubber company created in 1975, and SOCAPALM, a palm oil company installed in 1980. These economic drivers led to extraordinary demographic and urban development in the area formerly occupied by small lineages of 10–50 people spread throughout the forest. In 1999, the Cameroonian government upgraded this forest to a Technical Operational Unit (TOU) of 771,000 ha, with the aim of preserving and developing the forest's economic, ecological, scientific, and cultural values in an integrated manner. The Campo-Ma'an TOU has a land allocation plan comprising protected areas and forest (260,830 ha), forest management units for timber production (235,485 ha), a state maritime estate (320 ha), and a multiple-use agroforestry area (275,033 ha) mainly devoted to human activities. The TOU is under the control of the Campo-Ma'an Project (PCM). This project has three institutional components: the Ministry of Environment and Forestry (MINEF), which is responsible for conservation and deploys forest police known as ecoguards in the field; the SNV (Netherlands Development Organization), which is responsible for local development; and Tropenbos International, which is responsible for research and drafting of the park's management plan.[1]

In 2000, the protected areas and protected forest became a national park, in compensation for foreseeable environmental damages along the Chad-Cameroon pipeline. This 1,090-km long pipeline opens onto Kribi, some 80 km to the north of Campo-Ma'an. All the communities situated along this pipeline have been compensated (Ravignan 2000). However, the Campo-Ma'an communities whose forests were confiscated were not taken into account.

In September 2001, the Campo-Ma'an Project requested the support of the Center for International Forestry Research (CIFOR) in studying the park's impact on the socioeconomic activities of communities adjoining the park, in order to make management proposals that effectively integrate community perspectives. The research reported in this chapter was conducted within this framework.

Among the more than 116 villages and hamlets surrounding the national park, 9 were selected for study. Major criteria of selection included proximity to the park, pre-existing rights in the land occupied by the park, and dependency on the resources of the park. This chapter concerns two of these nine villages: Bifa and Ebianemeyong.

Among Cameroon's forest populations, discussions about the management of the forest are first and foremost the business of men. If the creation of the national park affects all the surrounding communities because it takes away a great part of "their forest," it should in principle affect men especially and directly, since it is their hunting territory that has been drastically reduced. Women's economic territory, in contrast, is close to the villages and should probably not be very much affected by the park's limits. Thus, it was not surprising that the men mobilized themselves for meetings related to the management of the national park. In seven out of the nine villages, few women attended the meetings. Those who were present seemed to be observers more than actors.

We were surprised therefore by the strong mobilization of women in Bifa and Oveng for the community meetings we organized: In attendance were 17 women and 35 men in Bifa and 18 women and 30 men in Oveng, as compared with 1 woman and 17 men in Mabiogo and no women and 23 men in Messama I, II, and III. Bifa was especially noteworthy. From the first meeting, many women were present. From the determination they displayed, it was evident that they had something particular that they were eager to share, or at least, to express. But what? The aggressiveness with which they reacted when we introduced the topic of the national park for discussion showed that they were strongly affected by its presence. But how? This is what we tried to understand during our stay in these two villages.

We initiated community discussion sessions that revealed the content, frequency, and quality of the relations and interactions between the various stakeholders, between the social groups, especially men and women, and between each social group and its environment. Games borrowed from participatory rural appraisal and social science methods developed by CIFOR helped facilitate discussions and the exchange of ideas within separate male and female subgroups (Colfer et al. 1999a; Colfer et al. 1999b).

Below, we first discuss the two villages, Bifa and Ebianemeyong. We then compare our findings from the two communities and draw some conclusions about and policy implications for people's interactions with protected areas.

Stakeholders and Livelihoods in Bifa

Bifa is a village of 306 inhabitants, situated to the northwest of the Campo-Ma'an TOU. It is jammed between the national park and HEVECAM, an agro-industrial complex made up of vast rubber plantations, factories, and workers' camps with approximately 18,216 inhabitants (ERE Développement 2001). The local communities are Bulu, an ethnic group that settled in Bifa around 1860 and is part of the large Fang-Beti ethnic complex, formed by Fang, Fon, Mvae, Ntumu, Zaman, and Bulu ethnic groups (Diaw and Njomkap 1998). They have or keep preferential and complex relationships with their neighbors and brothers of Nzingui.[2]

As in neighboring villages, the people of Bifa have experienced external influences over the years, which have gradually modified their way of life. The most recent event was the creation in 1975 of HEVECAM. HEVECAM's rubber plantation took away part of the village land and caused great changes within the local communities, including:

- exacerbation of inter- and intracommunity conflict for the remaining land;
- competition between Bifa and Nzingui, the neighboring village, for the remaining resources;
- destruction of large areas of forest and reduction in resources and incomes;
- influx of strangers into the area in search of jobs; and
- increased poaching and illegal occupation of land by plantation laborers and their families.

HEVECAM also represented an opportunity for local communities, not only because it provided jobs but, above all, because it created an important and easily accessible market.

Given this context, we felt that a full understanding of the roles of the various stakeholders, from the perspective of local populations, was critical. We held a series of community meetings to determine community members' views of other forest resource users, stakeholders' interactions with each other, and local perceptions of concepts such as equity, well-being, and sustainable management. The results, summarized in Table 6-1, represent consensual community perceptions.

Economic Activities, Resources, and Gender in Bifa

Like some 80% of the populations who inhabit the forest zones of Equatorial Africa (Bahuchet 1993), the men and women of Bifa carry out all of the following traditional activities: agriculture, hunting, gathering and harvesting of non-timber forest products, fishing, small-scale poultry keeping, and breeding of small ruminants. This way of life is found elsewhere among the peoples of the great Amazonian forest (cf. Pinton 1992). Even though agriculture remains the major activity, people do not hesitate to proclaim that their main source of income is hunting and other activities are subsidiary. Each of these economic activities is described in the following pages.

Table 6-1. *Community Perceptions of Stakeholders in the Management of Forests in Bifa*

Stakeholder	Position and impact on well-being	Impact on resources
Local population	(+) Is an eligible party	(−) Exerts pressure on resources, especially on wildlife
HEVECAM	(+) Provides jobs for Bifa youth (+) Provides market for food crops and game	(−) Allows illegal occupation of land by workers (−) Allows poaching by workers, mostly nonnative (−) Exerts enormous pressure on resources because of high population density
Zingui (neighboring village)	(+) Carries out common socioeconomic activities with the people of Bifa (community forestry)	(−) Has dormant conflicts with Bifa over hunting territory or grounds and other forest resources, which speeds up degradation
Campo-Ma'an Project	(−) Deprives the local population of traditional hunting and fishing grounds and cultural sites (−) Is indifferent to local concerns; is located far away (head office in Kribi) (−) Has conflicts with local populations caused by harassment of ecoguards	(+) Preserves the forest by reducing poaching

Note: Perceptions were derived from a sample of 35 men and 17 women interviewed in Bifa in September 2001. (+) indicates positive impact; (−) indicates negative impact.

Agriculture. The main agricultural products are food crops. In order of importance, both of consumption and cash income, women prioritize food crops as follows: cucumbers, cassava, plantains, groundnuts, corn, yams, potatoes, and peppers. Food crop farming is principally for household consumption. Any surplus is then sold in the HEVECAM markets located some 10 km from Bifa.

The main crops grown by men are plantains and fruit trees. These have a high market value. Cocoa crops are of little importance in Bifa. Since the selling price of cocoa fell in 1987, and since the government has adopted a hands-off attitude regarding subsidies for agricultural inputs, most cocoa farmers have abandoned their farms. However, they did not destroy them, hoping for better days ahead. They have now turned to caring for the fruit trees found on the farms.

Hunting. There are 68 hunters in the village. The most common hunting method is trapping. There are between 86 and 130 traps per hunter with three animals caught, or approximately 12.4 kg of bushmeat, per hunter per week. The hunting area is 580 km^2, and the maximum distance of penetration is 30 km from the village (Ngueguim and Ohanda 2001). Camps are set up when the hunting area falls within a 5-km radius. The game is then smoked on the spot, and the hunters' family members carry it from the camp to the village.

In Bifa, hunting is mainly a commercial activity. According to Ngueguim and Ohanda (2001), 78.6% of the captured game is sold elsewhere, for example, in Nzingui or in HEVECAM. Of the game harvested with guns, 87.5% is sold, while only 60.6% of the game caught in traps is sold. This difference is explained by the fact that a hunter with a gun has to buy ammunition and pay "rent" for the gun.

The most commonly trapped animals are hares (*Cephalophus monticola*), antelopes (*C. dorsalis, C. sylvicultor*), porcupines (*Thrionomys swindarianus*), monkeys, moles (*Cricetomys emini*), hedgehogs (*Atherusus africanus*), and civet cats (*Viverra civetta*). Occasionally, hunters catch pangolins (*Manis spp.*), reptiles, and buffaloes, and exceptionally, black tigers, and panthers.

Women hunt by setting traps around farms. They do this in order to protect their crops, and the activity provides some meat for the family. The animals caught in this way are mainly rodents, hares, and sometimes antelopes and black snakes. Women also help the men smoke the game and, together with children, transport the smoked meat from the camp to the village. They are also responsible for selling these products.

Fishing. Fishing is a minor activity in Bifa. The men fish using nets on the Nyete River and in big streams flowing some 8 km away from the village, where they set up camps. These camps serve as places for processing and storing hunting and fishery products. Unlike the men, who sell part of their catch, women catch fish and freshwater crabs exclusively for food. In the dry season, women also do dam-fishing in small streams found near villages.

Gathering. Non-timber forest products are also secondary resources for Bifa's residents. The men are involved in exploiting a very limited number of such products: palm or raffia wine; fuelwood; and tree barks. Women gather fruits and seeds: *andok* (*Irvingia gabonensis*); *Fan; Adzap* (*Baillonelle toxisperma*); nuts (*coula edulis*); *Mvout* (*tricoscypha spp.*); *Djansang* (*Ricinodendron heudelotii*); mushrooms; caterpillars; tree barks; fuelwood; and broad leaves for wrapping food stuffs. The women maintain that fruit-bearing by wild trees has become low and irregular, although the women cannot explain why. For instance, wild mangoes (*Irvingia gabonensis*) normally have a biennial cycle. For nearly a decade now, they have borne fruit only every four to five years.

Small-Scale Animal Husbandry. Women are responsible for poultry farming, while men rear sheep and goats. These products are meant neither for daily consumption nor for sale. They are kept as a sort of social security, to take care of unforeseen contingencies (receiving guests) or for exceptional festive occasions (weddings, dowries, and funerals). However, some of these products are sold within the village during periods of scarcity or to cope with cash needs, such as school fees or health expenses.

Market Activities. In Bifa women usually sell agricultural products, smoked game, and fish. Two times a month, they carry all the family's production destined for sale to the HEVECAM market located about 10 km from the village. These market days correspond to the period when the HEVECAM laborers receive their fortnight's pay.[3] It is for this reason that hunting, though mainly undertaken by men, seems to be beneficial to women who do the selling.

The high demand for consumer products in Bifa's large neighboring towns (HEVECAM and, to a lesser extent, Nzingui) stimulates the population to produce for sale. This is seen in the case of hunting, where only 33% of the total game biomass caught is consumed or sold in the village, even though bushmeat is the only regular source of animal protein in the diet. The staple foodstuffs are plantains and cassava, which are the most important sources of energy.

Gender Differences. Studying time allocation in Ntumu ethnic group[4] in Campo-Ma'an region, Annaud and Carrière (2000) found that both men and women spend the same amount of time on livelihood activities (about 4.5 hours per day). The daily trekking for livelihood activities concerns both men and women and takes about 2.5 hours per day. The gender difference appears when considering the management of "free time."[5]

An exercise intended to measure the relative importance of each activity, according to time invested and income generated, was done with 35 men and 8 women separately. We asked villagers to distribute 100 pebbles, representing either the total time worked or income derived, among as many plates as there were activities listed (see Figure 6-1). This figure shows that hunting is the most profitable activity in Bifa. Hunting occupies 20% of the men's time, and it accounts for 40% of their cash revenue. Women also spend about 22% of their time on game-related activities, which account for 58% of their revenue. The cash income women get from these activities is twice what they get from agriculture (58% of their time and 24% of their revenue). Their fishery products are all for home consumption. Women maintained that they kept account of the family's money and were involved in deciding how to spend it.

With their dual role as product sellers and family bread-winners, the women play a central role in their households and in their community. They collect all the

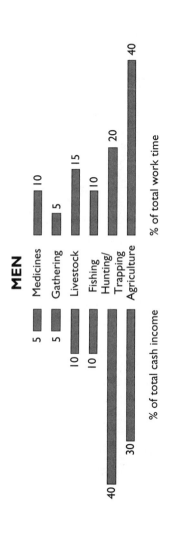

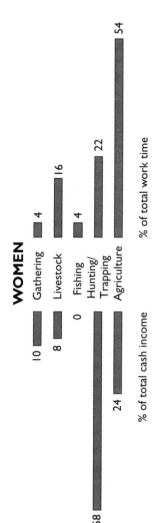

Figure 6-1. *Relative Distribution of Time Spent and Cash Income Generated by Activity in Bifa*

Note: Perceptions were derived from a sample of 35 men and 8 women interviewed separately in Bifa in September 2001.

resources and redistribute them for purchases, sales, gifts, and various social exchanges. Women also control the income from the sale of the products.

The National Park and Bifa's Population: Constraints and Adaptations

Local populations gradually adapted their way of life to cope with the changes induced by external factors over the years. In the process, women initially got the lion's share by positioning themselves as the salespeople of the family products, notably game caught by the men.

The creation of the national park led to new disturbances, which have disrupted the very basis of village economic life and put in question the achievements of all the local communities, especially women. The women accuse ecoguards, who have been present in the area since the creation of the park in 2000, of failure to demarcate the park clearly and to spell out the rules and regulations governing hunting, in a bid to seize any game found with the women in the market or in the village.

Although the national park has been demarcated on the map, it has no real existence on the ground because its boundaries have not been marked out. Beyond the areas with natural boundaries, none of the partners in the park's management seems to know the real boundaries. Therefore, it is not easy to prove whether the traps set by villagers in a particular place are in or out of the national park. Likewise, it is not easy to prove that the game sold on the local market or carried by a villager was captured in or out of the park.

The women complain of being harassed by the ecoguards, who do not hesitate to "enter into kitchens to examine the contents of pots" or to "seize our game anywhere and anytime. Be it for a smoked porcupine, hare, or antelope being sold in the HEVECAM market, or two moles carried by a man returning from visiting his traps, or in kitchens, the ecoguards will search pots."[6]

Ignorance by the local communities of the rules and regulations governing hunting exacerbates misunderstandings between the parties. The population accuses the ecoguards of keeping the information close to their chests and deliberately leaving the people in ignorance so as to better dominate them.

Without clear and convincing arguments, ecoguards will find it difficult to make the population understand the validity of their control and conservation actions. This argument is supported by the fact that the population does not know what becomes of the game seized by ecoguards.

A stranger who arrives in any village around Campo-Ma'an National Park is immediately suspected by the villagers of being an ecoguard in disguise. The women will say, when offering food, that the whole village is now made up of vegetarians because hunting is forbidden. Any signs of the existence of game in the village are quickly hidden away from newcomers (Nasi et al. 2001; Tiani et al. 2001). A few days later, once the stranger has been accepted by the community and has proved that he or she is not opposed to game consumption, it becomes clear that bushmeat is eaten on a daily basis, fresh or smoked. Hiding meat from strangers is a reaction stemming from the strained relationship between the population and the ecoguards.

Local men have put in place a new system of selling game and smoked fish that eliminates the women's role. Customers no longer wait for these products in the

HEVECAM market. Rather "the buyer comes to the seller" (Dounias 1993). How- ever, these buyers do not come to get the game from the village, as noted by Dounias, but go directly to the camps set up deep in the forest. Women, who derived their income from the sale of these products, are thus sidelined. Since the sale of game was the main source of income for Bifa's women, they have become increasingly poor, unable to work out adaptation strategies in time like the men. The women see their incomes dwindling while the problem of poaching still ex- ists. This has had a negative impact on the equilibrium between men and women.

Beyond these constraints, villagers' discussions show signs of hope for the fu- ture—hope of seeing animal life reconstituted, as abundant as in the past, and their livelihoods improved thanks to the alternative solutions that usually come with the creation of protected areas.

Stakeholders and Livelihoods in Ebianemeyong

Ebianemeyong is a village of 103 inhabitants situated in an enclave on the south- east edge of the national park. It is made up of many hamlets, one of which is Oveng where the chief's compound is found. The people of this village all belong to the Mvae ethnic group[7] and to different clans, settled along the road intended to link Campo in the west to Ma'an in the east. Based on the people's comments and discussions with resource persons of the village, it was possible to determine the stakeholders and their characteristics (see Table 6-2).

Economic Activities, Resources, and Gender in Ebianemeyong

Like the people of Bifa, the population of Ebianemeyong are traditional farmers, who make their living by practicing agriculture, hunting, collecting nontimber forest products, and fishing. Differences among these activities are related to the scale of the activities and the resources generated.

Men are involved in cocoa farming, but are moving increasingly into food crop cultivation, especially cucumbers and plantains for both consumption and sale. They hunt by setting lines of traps in the forest, near or around farms. Some men hunt with guns. Generally, as noted by Annaud and Carrière (2000), men have two houses: one in the village and another that is used seasonally as a trapping and fishing camp. The most commonly captured animals are antelopes, hares, hedge- hogs, porcupines, monkeys, civet cats, and pangolins. Men also collect forest prod- ucts, such as palm wine, fuelwood, tree barks, and creepers. They fish with nets on the big rivers, such as the Ntem, Djo'oh, and Biwome. Men also domesticate fowl, pigs, and some ruminants, which are used as food during visits by important guests, at funerals, or for dowry elements.

Women are engaged principally in food-crop farming and increasingly in the cultivation of fruit trees. Agricultural products are diversified and include cassava, plantains, cocoyam, groundnuts, cucumbers, corn, potatoes, green amaranthus, okra, and yams, among others. The orchards behind the houses contain pear, mango, guava, palm, plum (*Dacriodes spp.*), citrus fruit, and bush mango trees (*Spondias cytherea*).

Table 6-2. *Community Perceptions of Actors and Stakeholders in the Management of Forests in Ebianemeyong*

Stakeholder	Characteristics
Local population	Beneficiaries. Found in an enclave in the national park. Totally dependent on the forest.
Campo-Ma'an Project	Distant institutional managers. Headquarters in Kribi.
World Bank	Threat to the population.
Ecoguards	Protection of the national park. Lukewarm relations with the population.
HFC, also known as *La Forestière de Campo*	Logging company considered by the population as a necessary evil, because it destroys the forest while providing nearly all the socioeconomic needs of the people and the council.
Nkoelon (neighboring village)	Has a clan relation with the population of the area. It is the nearest village to the west coast of the national park. Some disagreement with the people of the area as to the control of the Nkoelon gate.[a]
Poachers	Absent, but very present through their activity and its impact on the ecosystem and the daily lives of the population.

Note: Perceptions were derived from a sample of 30 men and 18 women interviewed in Ebianemeyong in September 2001.

[a]A control gate erected by the Nkoelon villagers at the western entrance of the national park has been operational for some 20 years. From time to time, these villagers clash with their neighbors and brothers from Ebianemeyong on the origin of the game they transport. In fact, it was difficult to prove whether the game being transported was captured in the park, given that the eastern entrance of the park is open. This problem has today been solved with the setting up of a control post and a radio station at the eastern entrance of the park.

In the dry season, women fish for crabs, silurids, and small fish with dams on small streams found near the village. Like in Bifa, they also set traps around farms and collect forest products, such as different types of fruits, broad leaves, mushrooms, and caterpillars. Some women are also involved in handicrafts, particularly basket-making.

As in Bifa, we assessed the relative importance of each activity by gender, in terms of time spent and income generated. Villagers were asked to distribute 100 pebbles representing the total time worked or income earned into as many plates as there were activities (see Figure 6-2). Both men and women reported devoting more than half of their time to agricultural activities. In Ebianemeyong, women's activities are more diversified than men's, for whom the collection of fruits and other nontimber forest products are minor activities. Women's cash incomes essentially come from agriculture and to a lesser extent from non-timber forest products. Fishery and hunting products are used for household consumption. Men have additional sources of income because their hunting and fishery products are also sold.

Finally, women work mainly to feed the family, while men's activities are geared toward earning an income. This relationship establishes an equilibrium in the household and shows the division of responsibilities between the sexes: women feed the family, and men take care of actions that require a certain amount of money, particularly the education of children and health care. Apart from cocoa, all products are

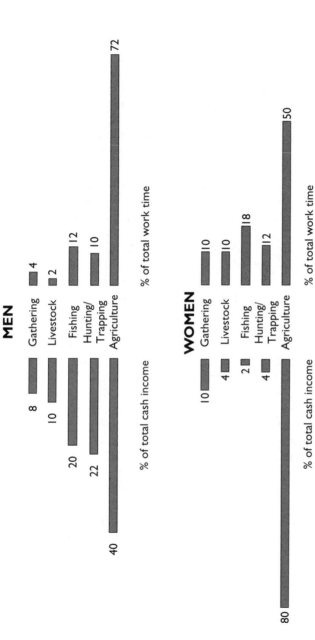

Figure 6-2. *Relative Distribution of Time Spent and Cash Income Generated by Activity in Ebianemeyong*

Note: Perceptions were derived from a sample of 30 men and 19 women interviewed separately in Ebianemeyong in September 2001.

primarily for home consumption. Only the surplus is marketed. Activities typically undertaken by men are financially profitable to them. Women's activities, in contrast, are more focused on meeting the household's subsistence needs; only agriculture and to a lesser extent harvesting bring in cash income.

The Road and Poaching: Controversial Topics in Ebianemeyong

The road linking Campo to Ma'an was started in 1941 by the French Administration and was built in fits and starts. The section connecting Ma'an to Nyabizan was constructed in 1941 by the French Administration; the Nyabizan to Ebianemeyong section was added in 1959 by the Cameroonian government; and finally, the Ebianemeyong-Campo section was completed in 1998 by the most important forest company in the region, La Forestière de Campo (HFC).

The route connecting Campo to Ma'an is not continuous. It is interrupted for 2 km between Nyabizan and Ebianemeyong, so that for populations of Ebianemeyong Campo remains the unique opening to the exterior and the point of access to the market, the hospital, etc. In 2000, with the establishment of Campo-Ma'an National Park, use and maintenance of the Ebianemeyong-Campo route were suspended at the request of the World Bank, which financially supports the Campo Ma'an Project. The people of Ebianemeyong were thus cut off from the rest of the world, without access to either Ma'an or Campo. The closing down and repeated reopening of this road are the only indicators local people have that negotiations are taking place between the more powerful belligerents. For these reasons, women said that the above-described economic activities do not really represent the current situation, but rather the situation before the closure of the road. They said that there is no way for women to earn money at the moment.

At first glance, a road that crosses a national park would appear to be a threat for wildlife. The Campo-Ma'an road, however, has had a very limited impact on poaching, based on testimonies collected from people we visited in various villages and from forest service officers. This information coincides with that of Nasi and others (2001). By reconstituting the various penetration routes used by hunters into the national park (Figure 6-3), we found that, indeed, poachers cross the entire forest. They settle in camps for several weeks and freely indulge in their hunting and fishing activities. Game and fish are smoked or dried in steam rooms and transported right to points of sale, without using the road. The road is considered unsafe, moreover, because the ecoguards use it regularly in their surveillance and monitoring of the park.

The closing down of the Campo-Ebianemeyong route to HFC timber trucks affected the entire population in the southeast part of the park. Women were especially affected, because they derive their financial income from the sale of agricultural products, which they convey to the Campo market, about 50 km from the village. These products, unlike game and smoked fish, are of low market value and do not attract customers to the village. Therefore, the road closure affects the income of women more than it affects that of men.

Besides the enclavement, the distant conflict between the Campo-Ma'an Project and the World Bank, on the one hand, and HFC on the other, psychologically

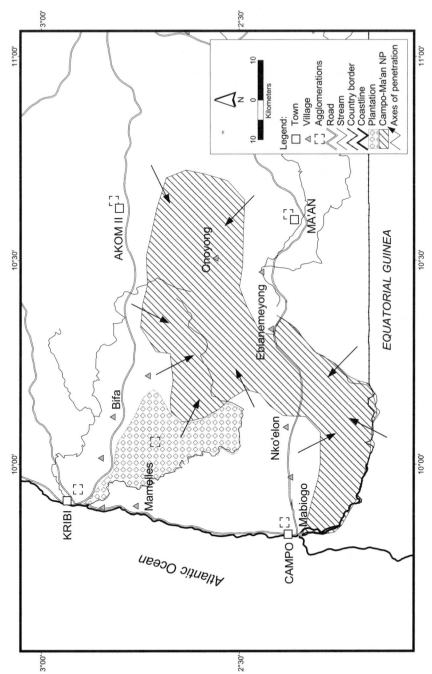

Figure 6-3. *Map of the Area and Axes of Penetration by Poachers in the Campo-Ma'an National Park*

afflicts the local communities and is tantamount to "disempowering" them. Villagers participate passively in these events initiated and manipulated from outside, which nevertheless have profound effects on their lives. Their fate is decided elsewhere, without taking their opinions into account. People feel that they are less important than the animals that are being protected at their expense.

Under conditions of rapid change and increasing uncertainties, the weaker community members tend to develop a feeling of insecurity. The women of Ebianemeyong expressed this feeling in the following words: "Why are we hemmed in here as in a prison? That makes two days that you have been here. Have you heard the least sound of an engine? Why do you come to mock us? How do we evacuate or convey our sick ones to hospitals? The closest is in Campo. How are we going to survive?" (Tiani et al. 2001).

Comparative Analysis

Apart from the reduction of living space, which is a common problem to all communities living adjacent to the national park, the problems affecting women in Bifa and Ebianemeyong are specific to their own time and space. They concern conflicting relationships with the ecoguards in Bifa and the enclavement of Ebianemeyong.

The conflicts between the ecoguards and Bifa women are due to the authorities' failure to demarcate the national park and the communities' ignorance of the laws governing the possession and marketing of game. Meanwhile, the women of Ebianemeyong are victims of the misunderstanding (or disagreement) between two powerful external actors, the World Bank and a forestry company, HFC. The two villages present problems of various natures, with various causes, but the outcome for women is similar, namely their impoverishment and loss of status. Below, we emphasize several patterns and issues in the Campo Ma'an situation.

Specificity in Activities and Sources of Income

The comparative study of people's activities and sources of income in the two villages highlights the following differences: Bifa villagers devote more time to hunting or hunting-related activities (20% for men and 22% for women, compared with 10% for men and 12% for women in Ebianemeyong), while Ebianemeyong villagers practice more fishing (12% for men and 18% for women, compared with 10% for men and 4 % for women in Bifa). Whereas the women of Bifa derive their greatest income from the sale of hunting products, the women of Ebianemeyong depend almost exclusively on the marketing of agricultural products. This means they are especially vulnerable to changes in the agricultural sector.

In Bifa, male activities are diversified, with agriculture and hunting the dominant activities (40% and 20% of their time, respectively). In Ebianemeyong, men are essentially farmers, devoting 72% of their time to agricultural activities. However, these male patterns involve opportunistic choices, inasmuch as the chosen activities of each village constitute its regular source of income. When all is said and done, the people of Bifa are farmer-hunters, while the people of Ebianemeyong are farmers. These data confirm the findings of Dounias (1993) and Annaud and Carrière (2000), which showed that the Mvae have a complex agrarian system, rich and adapted to the environment.

The fact that almost all peoples of the Cameroonian forest are subsistence farmers conceals specific community adaptations in their strategies for survival. Each of the many external interventions has induced different dynamics within the respective communities. In Bifa, the establishment of industrial plantations introduced a market economy within the local community. Women forged a place in the socio-economic circuit by constituting themselves as the agents for marketing household products, a strategic position that conferred on them some prerogatives and authority over household, indeed, community business. Since the creation of the national park, however, Bifa men have been trying to break away from the control women have over income by creating new marketing arrangements, benefiting in spite of themselves from the conflicts between the community and the ecoguards. This will alter the balance of power within the households and the community to the detriment of the women.

Vulnerability of Women in Periods of Uncertainty

In the face of changes, the women of Bifa and Ebianemeyong are experiencing more difficulties than the men in adapting to new circumstances. This is not an isolated case. Studies conducted in various areas have demonstrated the vulnerability of women in the face of externally induced upheavals (Brown and Lapuyade 2001; Pokam and Sunderlin 1999). This vulnerability is linked to several factors: the competing demands on women's time, the circumscribed scale of women's activities, the concentration or uniqueness of their income sources, and the low market value of products derived from their activities.

One reason women have less room to maneuver in times of uncertainty is the fact that they are daily overloaded with work. In addition to production activities (agriculture, hunting, fishing, harvesting, breeding, transportation, etc.), there are those of reproduction. These include all the various tasks such as fetching water, feeding the family, raising children, and managing the home. Already fully occupied with their roles as mothers and educators, women hardly have the time to organize themselves to adopt reasoned and common strategies in the face of adversity. Annaud and Carrière (2000) assert that for the Mvae of Campo, the time men and women allocate for all production activities is 4 hours and 3.6 hours per day, respectively. The time devoted to the food budget is 25% for men and 68% for women. These authors furthermore indicate that men allocate 60% of their waking time to discretionary activities and leisure, compared with only 28% for women.

Although the men and women in the Campo-Ma'an region seem to be undertaking the same activities, the scale of these activities is quite different. Men cultivate cash crops (cocoa, plantains, fruit trees, oil palm, etc.), which fetch them a comparatively good return. They fish with nets in the great waterways (Djo'oh,

Biwome, Ntem). They hunt with guns and with snare cable traps in primary and secondary forests, as well as in fallows. They also raise sheep and goats. Women, in contrast, grow food crops largely for household consumption, set a few traps around their fields, fish in the dams of small waterways, and breed poultry. Consequently, the income of women is relatively low.

Women's income stems from hunting and agriculture in Bifa and agriculture alone in Ebianemeyong. In both cases, women are very dependent on agriculture, where the yield is subject to climatic fluctuations, soil fertility, and plant diseases and the overall return depends on changing market prices and accessibility. In Bifa, hunting has become an illegal activity, and selling bushmeat in the market has become very unstable because of the presence of game guards.

The vulnerability of women is also related to the types of resources that they produce, resources that are in turn linked to the nutritive role that is theirs within the society. Women cultivate, fish, or hunt primarily to feed their families. Only the surplus is available for trade or sale. These food products have less value than forest products that are hunted or harvested for sale. Because of their lower market value, women's products have to be transported to the market to sell them. The enclave's distance from the markets and poor transportation exacerbate the instability of women's income.

In short, the smallest disruption of the agricultural sector directly affects women's capacity to feed their families and deprives them of their main source of income. Even where women have succeeded in positioning themselves in the orbit of male activities to the point of earning substantial income, as in the case of Bifa, they are the first to be eliminated as soon as the least disruption occurs.

Recommendations and Conclusions

Since the Stockholm Conference in 1972, the spirit and the letter of international policy on the environment are unequivocal as regards their concern for both ecosystem conservation and the sustainable development of local populations dependent on these ecosystems for their survival. Principle 4 of the Rio Declaration (1992) established the inseparable nature of environmental protection and development. Since then, international decisions imposing restrictions on access to resources in inhabited areas are always accompanied by compensation and development alternatives. Thus the Campo Ma'an Project (in charge of the management of the Campo-Ma'an TOU), created in 1999, includes a component responsible for ecodevelopment. The role of the component is to promote local development and alleviate poverty, by supporting communities' efforts to establish economic alternatives to hunting in the national park. Yet even before the national park has been effectively demarcated, workers responsible for monitoring and protecting the park were already in office, while development efforts have remained timid. Compensation and other alternatives have been lacking. All these give the impression of a discriminatory application of texts.

The January 6, 2000, Decree that created Campo-Ma'an National Park stipulates in Articles 3, 4, and 5 that the management plan shall determine the users' rights of surrounding populations; that boundaries and conditions for managing buffer zones shall be established; and that indemnification shall be provided for people whose land has been expropriated or who have been occupying an internal portion of the park. But what is meant by the occupation of an internal portion of the national park? We know that in forest communities, the territory of a village or a hamlet spreads well beyond cultivated land (Bahuchet 1993). As shown by Joiris (1997) and Diaw and Njomkap (1998), traditional customary rights concern both agricultural land and village forest land on which many other subsistence activities are undertaken, such as hunting, fishing, and harvesting, and on which ancient villages or *bilik* are also situated. These customary rights conflict with modern rights, which specify the state as the sole owner of all forests found on the national estate.

The interpretation of texts thus depends on the interests or priorities of each user. Between designing conservation or development strategy and its implementation, there is a shift (whether intentional or not), depending on the interests of intermediaries, linked to the interpretation of texts and regulations. From there, information is hidden, biased or re-oriented, the most powerful parties confiscate authority, and conflict takes hold.

Solutions Proposed by Women: "Living Together with the National Park"

After hours of discussions, women in Bifa and Ebianemeyong recognized that the best strategy was to accept the national park as part of their existence and make a sustainable profit from it, if possible. They proposed three areas of focus to reduce the conflicts that set them against the workers in charge of protecting the park.

First, they recognized the need to collaborate with the other stakeholders to conserve the national park. They considered indicators of successful collaboration to include the frequency of encounters, the communities' contribution to the park's protection, and the sharing of information. Second, they wanted to improve transparency in the management of the national park. Suggested indicators include the demarcation of boundaries and the dissemination of rules and laws applying to hunting and the possession of game. Third, women agreed on the need to promote negotiation with park authorities with the objective of adopting Dounias' (1993) recommendation of demarcating the buffer zone and legalizing hunting for subsistence purposes. The latter should be oriented toward fauna of secondary forests, which have adapted to living with human beings.

In Ebianemeyong, women offered their services to check or control timber trucks at park entrances and exits and to report to the authorities the presence of poachers in the forest. They also offered to provide accommodation for the ecoguards. Their requests in return were for improvements to the communication infrastructure, efforts to create jobs for young people, assistance in developing alternative economic activities, extension of rural electrification, improvements in the water supply, establishment of schools and health centers, and promotion of ecotourism.

These proposals do not constitute a final solution to the problems arising from the national park. Nonetheless, they do have some advantages:

- They constitute a basis for discussion between the parties in conflict.
- They emanate from the persons concerned and are not imposed by other authorities.
- They may result in securing the national park at a lower cost.
- They could, if they are considered, give more authority and assurance to the local populations, more confidence in themselves as partners in forest management.
- They could contribute overall to linking conservation to development.

Recommendations of This Study

This study shows that the current mode of management of the Campo-Ma'an National Park may weaken women's stature because it affects their sources of income. The main constraints met by women in the two villages result from a shocking lack of collaboration among the stakeholders involved in forest management, both at the international and local levels.

Collaboration is a mutually beneficial relationship. To local communities it could bring training, access to useful information, development, and empowerment. In return, institutional administrators of the park would be able to reach their goals, namely, more efficient and less expensive protection and conservation of the national park. Collaboration would also make it possible to prevent and resolve conflicts among actors in a more lasting manner.

Avenues for collaboration already exist. The national policy of decentralization of forest management has been pursued for a decade now. The political will to enlist neighboring communities in managing Campo-Ma'an National Park has been explicitly included in the TOU master plan. Local populations have expressed their will to collaborate with other partners in managing the park. International and national partner organizations, such as CIFOR, are ready to provide methodological support to the process. Everything is in place to institute a frank dialogue between the parties for shared benefits. To go beyond the intent to collaborate, however, requires a formal or legal framework.

Our findings highlight the importance of pursuing concrete actions at different levels. At the policy level, these measures are needed:

- Policies need to be reformulated concerning protected area management to legalize and validate local communities' participation.
- Conservation and development initiatives must be harmonized and synchronized, so that local populations are not pushed into poverty and misery, which are problems that must be addressed.
- Negotiations must be undertaken with all stakeholders to obtain their support for conservation and development efforts. Stakeholders span all levels, including the ministry in charge of forestry and its decentralized services, officials in charge

of agricultural development, the territorial administration, the economic operators in the area, and the local communities and their councils.

- Women must be ensured an active and significant presence in decisionmaking bodies in connection with the park's management.

At the local level, the following measures are important:

- The park's management plan must be negotiated with local populations, so that they can identify with it and participate in its implementation.
- The mobilization capacity of women should be used to secure the park. The park's administrators should consider women's proposals to this end.
- The on-going demarcation of the area should also be negotiated, so that boundaries are known, accepted, and respected by all parties. Similarly, some compromise should be negotiated regarding village hunting and the conditions for possessing and marketing game near the park to avoid misunderstanding and conflict.
- A balance has to be found between satisfying the vital and immediate needs of local communities and global environmental needs.

The culmination of these efforts should be that communities cease to consider themselves as "hostages of animals"[8] in Campo Ma'an.

Notes

The authors acknowledge the communities of Mabiogo, Messama I, Messama 2, Messama 3, Bindem, Ebianemeyong and Bifa for their hospitality, especially the women of these last two villages for the interest and the commitment they expressed during CIFOR field work. Thanks to Abessolo Evina Isaie and the late Medjoto Martin for their precious support, and thanks to Carol Colfer for her maternal encouragement.

1. Tropenbos' management responsibilities for Campo Ma'an were terminated in late 2002, with future management arrangements in the hands of the Worldwide Fund for Nature (Nguiébouri 2001, *210*).

2. Nzingui is a neighboring village to Bifa. The population, which is estimated at 1,062 inhabitants, is also of the Bulu group.

3. Laborers are paid every 15 days.

4. Ntumu, Mvae, and Bulu ethnic groups are close culturally, geographically, and linguistically.

5. Compulsory personal needs, such as eating, resting, and bathing, take 3.4 hours per day for men and 2.8 hours per day for women. Social needs require two hours per day for men and 35 minutes per day for women, according to Congels and Pasquet (2000) cited by Annaud and Carrière (2000).

6. Declaration of Bifa women's meeting in the community hall, September 24, 2001.

7. Like the Bulu ethnic group, the Mvae people belong to the large Fang-Beti ethnic group.

8. Declaration of an Ebianemeyong woman, September 2001.

CHAPTER SEVEN

Women, Decisionmaking, and Resource Management in Zimbabwe

Nontokozo Nemarundwe

INSTITUTIONS AND ORGANIZATIONS play a fundamental role in determining the access to and management of natural resources, such as land, water, and woodlands. Studies of common-pool resource management have shown that institutions have a strong, positive influence on sustainable use, if users are involved in decisionmaking, rule formulation, monitoring, and enforcement (Ostrom 1990; Murphree 1991). In this chapter, the term "institutions" refers to rules, norms, and enduring practices that guide people's behavior in a given society and make their interactions more predictable (North 1990; Uphoff 1986; Havnevik and Harsmar 1999; Leach and others 1997). Human structures that oversee the implementation and enforcement of these institutions are referred to here as "organizations."

As the state shifts formal responsibility and rights over woodlands, water, and other natural resources to local users' organizations (as representatives of the communities), these organizations control access to and facilitate management of the resource (Meinzen-Dick and Zwarteveen 2001). It therefore becomes important to examine who within these organizations performs what tasks, who controls resource use, decisionmaking, and related benefits. These organizations and their internal dynamics[1] will be analyzed to determine their bearing on the position of men and women in relation to the institutions governing access to natural resources. In the context of communal lands,[2] the role of institutions that determine who has access to and control over which natural resources is often influenced by formal and informal relations between women and men and other social interactions in the community.

This study draws on examples from woodland resource and water use to highlight the importance of gender in the institutions governing use of natural resources. Focus is on the internal dynamics of these institutions and organizations at the local level, particularly the roles of women in the management of natural resources. The chapter further examines the tactics and strategies women develop to influence natural resource management institutions and organizations for their gain. In this context, it is argued that while women may not always be represented in traditional or modern formal institutions and organizations, they use a variety of avenues to exert their views (cf. Chapter 8). Findings of the study presented in this chapter concur with those of Goebel (1998) in the resettlement areas and Fortmann and Nabane's (1992) study on social forestry in the Mhondoro communal area in Zimbabwe. While a study undertaken by May (1979) found that women use hidden strategies, such as withholding sex from men in order to be heard, this study finds that women's strategies for influencing decisionmaking are becoming more visible with the emergence of nongovernmental organizations (NGOs).

There is a growing concern about improving women's role in and influence over resource management institutions and organizations and creating situations where their voices can be heard in the decisionmaking process (Jiggins 1997; Rocheleau 1991). The need to focus on decisionmaking processes, social organizations, and formal and informal rules in ensuring both women's and men's access to and control over natural resources is also highlighted (Rocheleau and Edmunds 1997). Literature on gender in development in general, and natural resource management in particular, often underestimates the role of women, because emphasis is often on formal institutions, rather than informal decisionmaking processes. This chapter shows that women employ informal strategies to ensure representation of their views regarding access to natural resources and that they have more bargaining power when they operate as groups rather than as individuals. Regarding access to woodland resources, women lack strong group organizations, so they have less bargaining power than men. With regard to garden plots, however, women tend to operate in groups, and their influence is much higher.

The specific study site is the Romwe Catchment, which is located on the southern side of Chivi District about 7 km from the main Masvingo-Beitbridge road in Zimbabwe. The mountain area, where Romwe Catchment is located, receives rainfall of about 800–1000 mm per year—higher than is typical of southern Zimbabwe, where the average is 400–600 mm per year. The environmental variations in Chivi District often lead to competition over the most productive areas, and in such circumstances, women may be disadvantaged. Mazarire (1999) suggests that precolonial Chivi society acknowledged the importance of women in the environment in their day-to-day cosmology, their myths, totems, and even bodily decorations.[3] As a result of the colonial land policies, the Romwe Catchment has a mix of Shona and Ndebele ethnic groups, who were relocated into this area in the early 1950s. The relocation history has influenced the formation of institutions and organizations governing access to natural resources, as well as influencing which leadership structures are more respected and viewed as legitimate by the local community and neighboring villages.

Conceptual Framework

The study of institutions is not a recent phenomenon in sociological research, but dates back to early institutional theory in sociology spearheaded by scholars such as

Weber (1957; 1962). While this literature highlights issues of social differentiation, especially based on class, it is generally silent on the role of women in forming and maintaining institutions and the influence of institutions on women's access to productive and reproductive resources. Similarly, recent analyses of institutional arrangements for common-pool resources [4] (North 1990; Ostrom 1990) has often centered on understanding what institutions and organizations are in place, how they are constituted, and what role they play in resource management, without recognizing how the internal dynamics of the local community are influenced by gender, age, and other intrahousehold characteristics. The analysis of formal decisionmaking arenas at the community level often takes a similar form. This chapter investigates the role of women within institutions for common-pool resources and examines the internal dynamics of organizations that influence the form and nature of resource access (North 1990; Ostrom 1990; Scott 1995). Investigating the internal dynamics of organizations entails analyzing the methods for making decisions, the nature and structure of membership and the processes for changing leadership. The different interests and levels of power among members lead to different patterns of representation and cooperation at the community level.

Women's access to many natural resource management institutions has not increased, even under the pressure of decentralization policies, because the institutions have been so dominated by men, especially in patrilineal societies (Meinzen-Dick and others 1997). Attention should therefore be paid to gender issues, especially women's involvement in local formal and informal institutions that govern access to and exploitation of natural resources. Gender analysis often complicates the analysis of institutions governing access to natural resources, given the unequal power relationships between women and men in most societies (Rocheleau and Edmunds 1997). These power relations are, however, not static and are influenced by complex webs of connection between various natural resource user groups and the resources that sustain them.

In the gender and natural resource management literature, women have often been viewed as being excluded from formal decisionmaking processes, although they have widespread knowledge of natural resources in their locality and they may be the only users. A few studies recognize that women often devise strategies to ensure their voices are heard. Ranger (1999, 8) highlights strategies that women use to exert their influence over men.[5] Similarly, Schroeder's (1993) work in the Gambia shows how male landowners manipulated customary law to compel women to plant trees in their rented gardens, thereby attempting to phase out women's crops and displace the women from the garden site they had fenced, watered, and fertilized. Women resisted the attempts by male landowners to convert women's lucrative vegetable gardens to orchards and woodlots through both formal, legal means and subtle acts of sabotage.

With the recognition of such internal differentiation within communities, this study adopts the idea of gendered spaces and places, a conceptual framework for the analysis of women's role in natural resource management institutions (Fortmann 1995; Leach 1994; Rocheleau and Edmunds 1997). This framework emphasizes the separation of women's and men's *activity* and *authority* in geographic space. Women's spaces in the landscape are not easy to identify (Rocheleau and Edmunds

1997, *1355*). These are often the "in-between" spaces not valued by men but still quite useful to women. Firewood, medicinal plants, and wild foods (including fruits, insects, mushrooms, and wild vegetables) are found in these spaces and are often critical to women's efforts to meet their personal, household, and community responsibilities (Rocheleau 1991). The notion of gendered spaces can focus on both the household and community levels to illuminate the differences among women's interests, rights, and responsibilities; here the framework used to analyze data at the community level.

Recent studies emphasize that women are not a homogenous group (Bradley 1991; Nabane 1997; Nabane and Matzke 1997; Fortmann and Nabane 1992). For instance, Bradley (1991) finds that among the Luhya in Kakamega in Kenya, older widows have significantly greater decisionmaking power than do younger widows, with regard to the planting, care, and disposal of woodlot and fencerow trees. Nabane (1997) finds that in the Zambezi Valley in Zimbabwe, Korekore women have better access to benefits from a community-based wildlife management program, the Communal Areas Management Programme for Indigenous Resources (CAMP-FIRE), as compared with women from the small Vadoma ethnic group. In Gokwe, Sithole and Kozanayi (2001; see also Chapter 8, this volume) find that wives of powerful leaders play a pivotal role in the running of local organizations. Differences among women with regard to access to space and resources have implications for their participation in decisionmaking in organizations. The gendered space framework therefore also provides a platform for investigating internal differentiation among women themselves.

Methodology

The research used qualitative methods of analysis, as they were found to be more appropriate for collecting data on decisionmaking processes and institutional relationships. These methods included participatory rural appraisal (PRA) tools and techniques, attendance at village meetings, residence in a village for extended periods, development of extensive social networks in the community, participation in community social gatherings (such as funerals, weddings, and church services), and documentation of case histories of women in leadership.

In the PRA workshops, women's and men's groups worked separately on similar tasks designed to describe their use of the woodlands and their perceptions of institutions regulating access to natural resources. As Goebel (1998) has highlighted, PRA data may be fraught with problems. These problems include a falsely homogenized view of village perspectives and a static and misleading picture of social practices and beliefs. Despite such weaknesses, this study found that PRA generates useful data on general trends and perceptions. The data need to be cross-checked and verified though alternative research methods, such as individual and key informant interviews, observations, and informal interaction with community members, as was done for this study.

I lived in the village of Romwe for a period of about 18 months with frequent two-week visits to the field between July 1998 and August 2001. My research assistant, based in the field on a full-time basis for three years, gathered additional information using participant observation. During the research process, I also developed extensive social relationships with individuals from wide-ranging socio-economic backgrounds.[6] Making friends with people from both rich and poor households,[7] I was able, with the help of Ndebele and Shona speakers, to interact with a variety of informants.[8] These social interactions revealed complex webs of power relationships in the community and helped to explain why certain decisions were made and how they were implemented.

While based in the community, I attended both formal meetings and informal gatherings. During formal village meetings, the process and deliberations were documented (e.g., who said what and whose ideas were adopted), and male and female attendance was recorded. Among youths, it is often the boys who attend village meetings; rarely do girls participate. Yet, girls make crucial decisions about the environment, such as where to collect firewood since they are the main collectors.

Informal gatherings included spending time with groups of women while they watered vegetables at the garden projects, socializing at private homes, as well as attending church services, funerals, and other open events where I could engage in discussions with community members. I also had to make contributions, both in monetary and labor terms during funerals. If a household member dies, other households in the village are expected to make contributions toward funeral expenses averaging about Z$20 (US$0.36 at the official exchange rate of US$1 = Z$55). I contributed the same amount as other households in the study area, so as not to raise people's expectations. Another in-kind contribution I made was transporting people to the business center to buy funeral items or to the Ngundu clinic, often at night. This service brought me close to the community members, and my interaction with them greatly improved over the period spent in the village. Thus I could better observe interactions between women and men and obtain some information otherwise not available.

For example, in early interviews women professed Christianity and thus denied using any medicinal plants. At a funeral, I took a walk into the bush with the women (there are no toilets in most households), and during the process saw women showing each other some plants that are used by pregnant women to relax muscles of the birth canal and other plants that are used to enhance sexual pleasure. If it were not for this close relationship, I might never have had access to such information. Thus data collection is a complex process that requires both formal and informal dimensions to generate a reliable picture.

Case histories of women in leadership positions were also collected during the course of the research. In addition to the various methods used for data collection on the ground, a review of the research literature, as well as Rural District Council (RDC) and NGO documents, provided information on the broader context of community-based natural resource management in Zimbabwe and the study site.

Access to Natural Resource Spaces and Institutions

In reflecting on how gender influences institutions and organizations for community-based natural resource management, it is important to visualize the natural resource spaces. In this study, this visualization was done by mapping resource areas and products to capture the complexity of woodland resource spaces. Often, the conception of natural resources such as woodlands is of discrete areas of trees located away from human settlement. In the Zimbabwean communal lands, including Chivi District, this is not the case. The miombo woodlands, characteristic of the savannah region, comprise different tree areas in and around human settlement, pastures, arable lands, and hill areas. Resource areas highlighted by the community members in Romwe include the mountains, the river banks, and the edges of arable land and grazing areas, where people collect a wide variety of woodland products, such as poles, firewood, thatch grass, wild fruits, mushrooms, herbal medicines, and fiber (Table 7-1). Access to these resource areas is gendered in the sense that women and men have distinct uses of these areas and differential control over their resources. These patterns of resource use have implications for women's roles in natural resource management institutions and organizations.

Based on their roles and responsibilities, women and men have different mobility patterns, which reflect gender differences in resource use and other divisions of labor, e.g., in agriculture and cattle herding. The differential use of particular resource areas and products found in those areas leads to the association of certain spaces as gendered spaces. Women's spaces are not always easily identifiable and may overlap with men's. For example, women gather mushrooms, fruits, medicinal plants, and other nontimber forest products in areas, such as the Mapande Range, that are predominantly used by men for grazing and gathering poles and fiber.

This overlapping gender identification of spaces has implications for woodland-related management activities, such as participation in the establishment and monitoring of management institutions. Although women use the woodlands in the Romwe Catchment, they are not actually represented in formal woodland management structures and are therefore less likely (formally) to influence the institutions that determine access to these woodlands. Women therefore often devise a variety of strategies to ensure that their interests and views are considered in resource management decisions and in the formulation and enforcement of resource use institutions (see Chapter 8).

In Romwe, the hills emerge as the most important of the overlapping resource areas, because relatively large quantities of a number of products are gathered here. People also depend on a variety of areas for collection of the same products, indicating the interconnected nature of the resource areas. For instance, women collect firewood from the Mapande Range, Barura, Romwe, Siyawarayira, Magegesa, and Chamanyoka Hills. Similarly, both women and men collect different resources from the same areas. For instance, in the Mapande Range, women collect mushrooms and firewood, while men collect fiber and poles and hunt small game.[9] Institutions governing access to these resource areas are either area- or product-specific or can be justified by sacredness or civil contracts.[10] For example, while specific rules govern mushroom harvesting in Mapande Hill, other rules govern access to the whole mountain range. In this case, rules governing access to mushrooms in Mapande specifically affect women, while the other general rules, such as prohibition to burn the area, apply to both women and men.

Observations indicate that men are more likely to break established rules, particularly those on felling trees and encroaching into grazing areas. Women's chores,

Table 7-1. Resource Areas and Gender Access to Various Products

Resource area	Products	Major users	Who controls access	Rules governing access and management practices	Comments on adherence
Mapande range	Firewood, fiber, poles, mushrooms, sacred pools	Women, men, boys	*Sabukus*,¹¹ chiefs, headman, Forestry Commission (FC)	Do not make bad comments about mushrooms. Do not cut wet wood. Do not start bush fires. Respect sacred wells.	Women no longer respect traditional beliefs, such as not making bad comments about mushrooms, because of the influence of Christianity. Men cut wet wood for poles, and the local leadership does not apprehend them. FC could enforce, but the officer is based about 60 km away and is responsible for the whole district. People adhere to the rule on forest fires.
Barura mountain	Grazing, poles, wood, fiber, mushrooms	Everyone	*Sabukus*, everyone in the village, FC	Do not cut live trees. Do not start bush fires. Collect dead wood only.	Rules are not always followed, especially those relating to tree cutting, because people do not have alternative sources of poles. Men sometimes cut wet wood for sale.
Romwe mountain	Grazing, firewood, sacred pools, poles	Women, men, boys	Ancestral spirits	Do not bathe or do laundry in the sacred pool. Do not cut wet wood.	Men cut wet poles because traditional rites are no longer held in Romwe. Women generally follow rules, as they collect only dry wood. Headman appears to have lost interest in enforcing the rules. Only elderly women past menopause were traditionally involved in managing the sacred pool; now no one has that responsibility.
Chana chaRomwe	Grazing, firewood	Women, girls, boys	*Sabukus*	Do not cut live trees. Do not start bush fires.	Rules adhered to because the hill is close to homesteads, thus making it easier to see offenders.
Siyawaraira mountain	Grazing, poles, mushrooms, firewood	Women, men	*Sabukus*, chiefs, headman	Do not cut live trees. Do not make bad comments about mushrooms.	Men cut live trees for poles, although they do not cut from one spot.

Chamanyoka hill	Fiber, poles, firewood	Everyone	*Sabukus*, everyone should police	Do not cut live trees. Do not start bush fires.	Rules here said to be adhered to because hill is close to homesteads, thus making it easier to see offenders.
Magegesa hill	Firewood, poles, grazing	Women, men, boys	*Sabukus*, chiefs, FC	Do not cut live trees. Do not burn grass.	Men do not follow rules on tree-cutting.
Grazing areas	Grazing, firewood, fruits	Women, men, boys	*Sabukus*, FC, Agritex	Do not burn grass. Do not cut live trees. Do not harvest unripe fruit. Do not sell fruit.	Naughty children sometimes harvest unripe fruit.
Fields	Winter grazing, fruits	Everyone	Field owner	Do not allow outsiders to use the fields during the cropping season.	These rules are strongly enforced. There are a few cases of children stealing fruits in the field during the cropping season.
Mawunga stream	Firewood, water, grazing	Women, men	*Sabukus*, Department of Natural Resources (DNR)	Do not cultivate or graze along stream banks.	Neither women nor men follow this rule because they claim they do not have enough land in the area. Also DNR is not actively enforcing the rules, and *sabukus* say they are constrained by kinship ties.
Homestead areas	Mainly exotic fruit trees	Everyone in household	Household head and spouse	Do not steal fruit from other people's homes.	Rules are followed. Community members tend to respect other people's spaces as well.

such as collecting firewood for household use, can be met without cutting live trees, whereas men's chores, such as house construction or collecting firewood for sale, can be significantly more tedious if fresh wood is not cut. Thus, men are forced by the nature of their household responsibilities to cut fresh wood. Men, for example, are said to prefer live poles to dry poles for construction and to seek fresh wood rather than dead wood as firewood to minimize the time needed to fill the ox- or donkey-drawn carts they use to collect firewood for sale. Women collect head loads of firewood and thus find it easier to limit their harvests to dead wood.

Men also play a larger role in securing land for the family. Because settlement and arable land is in short supply in the area, men have been found guilty of encroaching into areas designated for grazing and woodland product collection. One example is that of a farmer who had a conflict with the traditional leadership over the privatization of the common-pool land resource in May 1999. In one of the villages, an area above the Barura Dam (part of the dam catchment) was designated as a grazing area. Some local young men, however, claimed that the local *kraal* head[11] had demarcated and reserved the area for his sons to use as crop fields. One of these young men decided to clear that land for his crop field. He vowed to continue converting the grazing area for his own crop field even if the *kraal* head tried to stop him.

This case is similar to that reported by Nemarundwe (2001) in which a young man established a well and a vegetable garden in a grazing area in Tamwa village in 1998. The *kraal* head did not take action about encroachment into grazing areas. It is usually young men who encroach into the grazing areas, which may be a result of the livestock ownership patterns. Generally, elderly men own cattle, while young men have few or no livestock. The *kraal* head's rules are sometimes broken because people feel they can afford to pay the normally small fines. The majority of interviewees first said that rules are respected and adhered to, but with further probing they revealed that certain rules are broken, as shown in Table 7-1.

Natural Resource Management Organizations

This section describes the natural resource management organizations found in the study area. The study analyzes the form and nature of membership in these organizations, as well as which actors are influential in decisionmaking regarding access to natural resources. Three main categories of organizations found in Romwe are explored, namely, the traditional authority, the local government administrative structures, and the project-related structures, such as the garden committees. The gender composition of membership was found to vary from one organization to another (Table 7-2).

It is common to find a woman being elected as treasurer in these organizations, because that position is viewed as one that has to be given to a trustworthy person. Women are viewed as trustworthy with regard to management of money. The variation in the organizations' membership can be attributed to a number of factors, including the organization's role in resource management, the composition of its membership, the personalities involved, and the nature of kinship ties and other

Table 7-2. Membership in Organizations in Romwe, by Gender

Organizations	Total members	No. of women	No. of men	Comments
Traditional authority[a]	6	0	6	Membership is all male. When asking people about the possibility of having a female *kraal* head, the first reaction was always surprise. They have never heard of a female *kraal* head.[b]
Village Development Committee (VIDCO)	7	1	6	The woman on this committee was elected because of the quota that specifies that there should be a woman representative in the VIDCO.
Ward Development Committee (WADCO)	7	1	6	The role of the WADCO is not directly felt in Romwe. Only the councilor is known and recognized. Asking villagers who was in the WADCO revealed that people did not even know the names of their representatives.
Chidiso Garden Committee	4	3	1	This committee has been in place since the inception of the garden project in 1991. At inception, there were 4 women and 2 men. One man, the original secretary, resigned in 1993, citing harassment by the vice- chairwoman and was replaced by the vice-chairwoman's sister-in-law. One woman who was a committee member also resigned the same year over a disagreement with the vice-chairwoman on how the project should operate. The original vice-secretary of the committee, a community mobilizer, is no longer active as she concentrates on the Barura Garden Project. It is rumored that the vice-chairwoman has strong personal relations with the chairman of this garden committee, who is also a respected *kraal* head in the area. Thus she has power to harass other members regardless of whether they are women or men.
Barura Garden Committee	7	6	1	The garden began operating fully in 1999. Membership in the garden is predominantly made up of women, as is the committee. The majority of the members are also plot holders in the Chidiso Garden.
Barura Dam Committee	7	1	6	The very influential *kraal* head in the area, who also holds the position of chairman in the Chidiso garden project, is the chairman of the Barura Dam Committee. The Barura Dam was constructed in 1985 to harness water for livestock, hence the predominantly male committee. The chairman of this committee claims that the dam project was his brainchild, so when it came to electing the committee, villagers agreed that he should be chairman. He has remained chairman of this committee. The Barura Garden Project Committee, made up predominantly of women, reports to the Barura Dam Committee.

[a]Traditional authority covers the *kraal* heads, headman, and the chief. The chief is at the top of the hierarchy, followed by the headman, with the *kraal* head as the lowest rung in the hierarchy. In the case of Romwe, there are three *kraal* heads (Tamwa, Sihambe, and Dobhani); two chiefs (Madzivire and Nemauzhe); and one headman (Chikanda).

[b]These findings concur with what Lindgren (2002) found in Matebeleland South where conflict arose in the community because a woman was appointed chief after her father (the chief) died.

social networks affecting decisionmaking in each organization. The following section further elaborates on the form and nature of the three broad categories of organizations: traditional authority, local government (modern institutions), and project-related organizations.

Traditional Authority

The use of the term "traditional" is problematic in the literature because it has been applied to practices and institutions that are evolving, not static, and subject to modification by negotiation between various natural resource user groups and state institutions (Berry 1993; Havnevik and Harsmar 1999). The property-rights regimes often labeled as traditional constitute complex outcomes of cultural and environmental change, and thus should be understood as dynamic phenomena (Ranger 1993). In this chapter, "traditional authority" refers to precolonial and native colonial authorities, such as the chiefs and *sabukus,* as well as the religious beliefs, norms, and current rules associated with them, which are not codified. Most of the so-called traditional authorities were remolded to extend colonial rule over the African population through a system of indirect rule. For example, the term *sabuku* literally means the keeper of the book, i.e., records of taxes extracted from the African population by the colonial governments through these leaders. Although traditional institutions did not have legal recognition at the local level until the enactment of the 1998 Traditional Leaders' Act, traditional authorities are respected for having mythical powers and applauded for their spiritual powers. A general tendency has been to perceive traditional authorities as though they are static, yet they have responded to both environmental and socioeconomic changes over time (Matondi 2001).

Traditional authority positions are inherited, and there is no history of women being, for example, *sabukus.* The mechanism of transferring power from an old *sabuku* to a new one is complicated by the fact that this transfer is often done after the death of the old *sabuku,* leaving no opportunity for a formal hand-over ceremony. Thus the new leader often acts on the basis of a mixture of experience and familiarity with what the old leader used to do. Similarly, traditional rules and regulations are not applied in a systematic manner. The majority of the rules are not formal rules but more "habits of the heart and minds," fitting within the broader social relations found in the community (Matondi 2001). Frameworks for understanding rule formulation processes tend to fragment rules (e.g., rules for water, grazing, and trees), yet at the local level rules are viewed in a more holistic manner within the broader way of life in the community. The traditional regulatory mechanisms include explicit and implicit rules and taboos.

There is some degree of social control over the traditional institutions and leadership, if the community does not approve of their decisions, through reference to the power of the ancestral spirits, as well as other forms of social denunciation. At a superficial level, traditional institutions may be seen as oppressing women, because women lack formal representation within their supporting structures, but women often develop strategies for manipulating the system. The case in Box 7-1 illustrates some of the strategies used by women to ensure that their interests are considered within traditional institutions for natural resource use and management.

Box 7-1. Challenging the Legacy

Women's Response to Traditional Authority Decisions

Historically, traditional authorities have been respected and rarely challenged. With recent support from external NGOs, however, women now sometimes challenge certain decisions made by the traditional authority, especially the *sabuku*. A case in point was the conflict over water in Barura Dam that occurred in January 2001. In December 2000, there was a dry spell in the area, and water levels in Barura Dam dropped. An influential *sabuku*, who is also the chairman in the Chidiso Garden Committee, unilaterally decided to close off the dam to reserve water for livestock. This restricted access to water by an innovative female farmer, as well as the rest of the Barura Garden Project, whose membership is predominantly women. There were claims that the influential *sabuku* held a meeting with a few of his right-hand men from his village to discuss the water level in the dam. At the beginning of the meeting he told them that no one should oppose his stance. He then selected three men to seal the spillway and instructed some women to cook food for the three men. This was done. The affected innovator and members of the Barura Garden Project Committee protested and requested the *sabuku* to revoke his decision, but the *kraal* head just ignored them. The group decided to send representatives to the CARE International offices in Masvingo to report the matter, as their crops were wilting because of lack of water. CARE officers came to resolve the issue and asked the *sabuku* to open up the spillway and repay the bag of cement he had forcibly taken from the Barura Garden Project for sealing the spillway. The *sabuku* was angry that people had disobeyed him and sold him out to outsiders, but at the same time he opened up the dam because he did not want external development agents to see him as a "bad" leader. He threatened community members that it would be their fault if their cattle died of thirst, as the water in the dam dried up. The women in the garden project and the female innovator were very happy that at last this influential *sabuku* had found his match, as he has always made unilateral decisions without being challenged.

Source: Key informants in Romwe (including said *sabuku* himself), February 2001.

The case also illustrates that, where women cannot employ internal strategies, they now rely on external development facilitators for support to ensure that their views are considered. In this case, the women sought and received support from an external agent. An alternative option would have been to seek assistance from the traditional village court (*dare*), but there was no immediate assurance that their case would be treated fairly as the traditional village court is made up only of men. Women participate only as witnesses in the cases tried by the village court.

Local Government/Modern Institutions

The term "modern institutions" is used here to refer to natural resource management institutions of the state, such as those instituted by the village development committees (VIDCOs) and ward development committees (WADCOs). In theory, VIDCOs and WADCOs are applauded as truly democratic institutions that allow broad participation, because individuals are elected into their positions. VIDCOs and WADCOs were created under the prime minister's directive in 1984, ostensibly to democratize the process of planning for local development (Murombedzi 1994). VIDCOs are demographically defined administrative units that, in principle, are based on a system of popular representation. A VIDCO consists of seven elected committee members and represents about 100 households. Two of these seven positions are reserved for the ruling ZANU PF party's women and youth leagues. A VIDCO is presided over by a chair and is tasked to design local development plans. About six or seven VIDCOs constitute a ward, which is presided over by an elected councilor, who represents the ward at district level.

The fact that one position in the VIDCO is reserved for women and another for the youth emanated from the numerous attempts by the government to make legal and institutional moves to promote gender equality. The government was influenced by the socialist ideology of revolutionary struggle and the role women played in the revolutionary war. This early commitment to gender equality affected the formation of new local institutions. Often ruling party, women's league, and youth wing representatives were voted in or co-opted into the VIDCO. Although women have a specified seat in the VIDCO and WADCO, it is important to note that numbers do not always mean effective representation.

The community votes by show of hands for members of the VIDCO. Thus, voting lacks any privacy, and people can be influenced by peer pressure to vote for someone. In the Romwe Catchment, the VIDCO was found to be inactive. During the two and a half years of the research period, the VIDCO held only two meetings where the process of decisionmaking could be observed. The first meeting was held to replace the late VIDCO chair, but this meeting was poorly attended. The second meeting was called to compile a list of beneficiaries of the food-for-work program. Thus, while there are women representatives on these committees, the committees' limited activity makes it difficult to assess women's contribution to decisionmaking processes.

The RDC expects the VIDCO to enforce rules on woodland use at the local level. In practice, however, enforcement is lax, and the current VIDCO chair says there is no incentive to enforce rules because there is no compensation. Members sacrifice time they could have spent in fields and doing other household chores to attend RDC workshops outside the area. The current VIDCO chair has threatened the councilor with resignation if the members' demand for allowances is not addressed. In talking to the chair's wife during his absence, I observed that the wife might have pushed him to resign. She pointed out that she did not support the idea of her husband continuing as chair of the VIDCO. All the meetings that he had to attend without pay at the RDC held up their household work. Their main house was blown away by heavy winds in November 2000, but as of August 2001 the house had not been repaired. Furthermore, before her husband became chair of the VIDCO, she used to go to Museva business center, about 5 km away, to sell vegetables and tomatoes. Now she has to do a lot more work at home and can no longer go to Museva. Thus, she has very little income of her own.

Project-Related Organizations

A number of organizations set up by NGOs or other government departments were found in the study site. These structures—for example, the catchment management, dam management, and garden committees in Chidiso and Barura—theoretically are established by democratic elections, as is the case of the VIDCOs. Participants are free to choose people they want to be in committees, and there are no conditions. NGO facilitators set the timeframe, usually two to three years, for the operation of these organizations. In the majority of cases, election of members is done at community meetings. Similar to the VIDCO elections, voting during the project committee elections is by show of hands. While this method is considered transparent, the downside is that some people may vote in a particular way out of fear of being accused of voting for the wrong candidate.

Committee members often attend training workshops and meetings at hotels or venues outside the village—they thus have more exposure and sometimes view themselves as superior to other people in the community. In some cases, committee members are seen as being "too forward" as a result of their exposure. This is especially so if they are women without husbands. Their "forwardness" is believed to be a result of lack of control over them by a man (see Box 7-2).

Women in the NGO-facilitated committees are very assertive. This may be partly due to the fact that women tend to predominate on those committees, and committee members receive formal training on leadership skills and attend confidence-building courses.

Where Are the Women?

Recent commentators on community-based natural resource management initiatives have questioned the representativeness of those institutions, especially at the local level (Narayan 2000; Mukamuri 2000). This literature argues that the more powerful members of a society may create institutions in order to regularize and entrench mutually beneficial relationships. Thus institutions do not necessarily serve the needs and interests of all, but mainly preserve the power of influential persons. Women and other poor groups are often at the periphery or excluded from natural resource management institutions. As a result, women have restricted access to land and other key natural resources. In these situations, women sometimes devise strategies (both intentionally and unintentionally) to ensure they gain access to such key resources. Some of these strategies are highlighted in Box 7-3, which describes how women participate in community projects as a way of gaining direct access to land through ownership of plots in the gardens.

Box 7-3 demonstrates that decisionmaking processes are complex and cannot be easily traced, especially where decisions are made outside formal arenas. For instance, while the chairman of the Chidiso Garden is well respected in the area, he is influenced by the vice-chairwoman, who may be his lover. I observed that women use the garden projects to gain access to land that they can claim as their own. Although all communal lands in Zimbabwe officially belong to the state,[12] at the local level, from the perspective of custom and tradition, men own land. Women gain user rights to land through their husbands or other male relatives. Membership in the Chidiso and Barura Garden Projects gives women an opportunity to own some piece of land registered in their own right. Thus women have become

Box 7-2. Women Leaders and Innovators

The Community Mobilizer

One of the female leaders and innovators in Romwe is the community mobilizer for CARE International. She is a widow; her late husband was a soldier in the Zimbabwe National Army. She is also a daughter-in-law to the influential *kraal* head. She is in her mid-40s, and her household is considered to be rich. She owns a car, has positions on a number of committees, and has attended many training workshops. She has primary school education and can therefore read and write. Her views are respected in community decisionmaking arenas, but other community members do not copy some of her innovations. She often stands up in village meetings and speaks out in a polite way. Being polite is a strategy that she has used well to ensure that people listen to her. People respect her for being cool and to the point.

As a result of being innovative, she has diverted water from Barura Dam spillway for her own use and has a flourishing 2-acre (approximately 1 hectare) plot of vegetables, sugarcane, citrus fruits, and maize, which she crops all year round (outside the Barura Garden). Other farmers once raised concerns that she was using community water from the dam for her own private use, and she argued that she was only using water spilling on its own from the dam. She uses her open truck to ferry her produce to the nearby markets.

She also works with a variety of organizations on trials, e.g., maize seed trials with Seedco, drip irrigation with the Intermediate Technology Development Group (ITDG), wheat trials and indigenous tree planting with the Institute of Environmental Studies (IES). Being a progressive farmer, she interacts with many external agents and thus has been accused of being a prostitute, especially by men in the community. This perception was confirmed by one *sabuku*, who called on the field-based male IES research assistant and said to him "My son, you have to be careful with some women that you work with. We know them. They are not good. They have had love affairs with many people, and they may disturb your work." One married woman noted, "This woman is such a hard-working person. It is her jealous relatives who accuse her of being a prostitute."

In other circles, the female leader in Romwe is also accused of manipulating people to her own advantage, e.g., making Barura Project participants buy her sugarcane for planting. She is believed to have made false claims that people from outside the area were coming to buy her sugarcane the following day. Thus if the project did not buy on that particular day, then she would sell to outsiders, and project members would have to buy elsewhere. This sent panic among the Barura Project members, and they decided to buy the sugarcane. Given that she has clients who come from outside the village to buy her garden produce, it was probably not a false claim that the sugarcane would be bought by other people if the project did not buy on time.

"official owners" of land portions, despite men's traditional rights to own the fields, homesteads, and bushland near their homestead area.

A boom in women's gardening projects is contributing to a gradual erosion of total male land-holding privileges. These findings concur with Schroeder's findings in the Gambia, where women became landowners through the introduction of market gardening (1997, *489*). In Romwe, women's economic status is restricted by lack of access to vegetable markets and the resulting low profits from vegetable crops. This low profitability may explain why men are less active than women in the gardening projects. Men tend to prefer projects that generate relatively high incomes. This shift to women's becoming "official" landowners through vegetable gardening underscores the contention that rural property systems in Africa are often quite dynamic (Bassett 1993; Shipton and Goheen 1992). Because women

Box 7-3. The Chidiso and Barura Garden Projects

Established in 1991, Chidiso was the first garden project in the area. The project was externally funded during the early stages. The garden is known as "Chidiso chaMwari" (God's will) because, according to community members, no one from the community applied for it, not even the councilor. The villagers say the project came from God because they just saw a group of whites coming to drill a borehole well for them. Membership is predominantly female, and the members own plots in their own right. Project members point out that they have benefited from the garden and collector well since they now have a reliable source of water. They also get cash income from the sale of vegetables, as well as supplementing their dietary needs. Farmers participating in the garden are from the three catchment villages. The project has a committee that was put in place in 1991, which has not changed since it was established and is no longer active.

Formal decisionmaking in the committee is dominated by two members: the chair, who is the most influential *kraal* head in the catchment, and the vice-chair, who is a woman. It is rumored that the chair and vice-chair of the project have a strong personal relationship, thus decisionmaking is unlikely to be transparent. The vice-chair also belongs to a powerful family in the catchment. Members of this family call themselves *Zvidzazvopo* (the powerful ones) and claim that no substantive decisions can be made without their involvement (see Nemarundwe 2001). The majority of the decisions are made outside formal arenas, and the two leaders have taken the opportunity to push their ideas and interests forward. Reacting to this setup, some original members who are not happy with decisionmaking processes are now subletting their cultivation beds to tenants because they do not want to give up their claims to the land in the garden, yet they refuse to accept the garden decisionmaking process. At the time of the study, 80% of the plot users were tenants, without the right to vote. The majority of original members had moved to Barura Garden, where the use of a flood irrigation system makes watering a lot easier than in Chidiso, where they have to pump water from the borehole well and carry it about 500 m.

The Barura Garden Project was established in 1998 with support from CARE International. Membership in the project is drawn from beyond Romwe Catchment villages because water used in the garden is from Barura Dam, which also serves other villages outside the Romwe Catchment area. Like Chidiso Garden, Barura membership is predominantly women, and the same applies to the project committee. Influential female personalities in the area such as the "community mobilizer," one of the wives of the late councilor, and a widow from Tamwa village are all members of the committee. Thus women's views are often taken into account, much more than among the Chidiso garden managers.

hold the predominant membership in community gardening projects, they also have an opportunity to gain leadership experience that they are highly unlikely to get in joint projects with men.

A recent strategy used by women in Romwe to gain access to traditionally male decisionmaking domains is to penetrate activities that men have dominated, such as tree-planting. Women, for example, are participating in action research by planting tree species traditionally used to meet male needs. The action research on trees involves experimenting with planting indigenous tree species. The project started off with 48 participants (12 men and 36 women) in December 2000. Tree species planted were *mukamba* (*Afzelia Quanzensis*), *muuzhe* (*Brachystegia spiciformis*), and *mupfura* (*Sclerocaryia birrea*). These tree species are used predominantly by men in the area for poles and carving. It is interesting to note that women are enthusiastically planting these tree species traditionally believed to fulfill male chores. The selection of species was project driven and not community initiated, but participants have developed a lot of interest in the project. This group has no formal representative structure, group members readily share information with each other. Participatory monitoring and evaluation sessions are conducted every six months.

Women indicated that one of the major reasons they wanted to participate in the tree research group, besides gaining knowledge, was to claim ownership of the land on which they plant their trees. Another was to secure unrestricted access to the three tree species that the group is planting. The participants, especially women, indicated that once their trees are big, they would be able to use them without restrictions. The common response when asked about their visions for the trees they have planted was, "If my trees grow, I will have my own trees that I can use the way I want without anyone refusing me access to the trees." Asked what she would do if her husbands divorced her and she had to move elsewhere and leave her trees,[13] one woman said, "I will never leave this homestead, if my husband wants me to leave, he would rather kill me here. If it were exotic fruit trees I would leave because those trees grow much faster and I could go and plant other trees wherever I go. But now, these indigenous trees take a lot of work and if you see them grow big, you should be a very patient person." This shows the determination of the women in the action research regarding ownership of the trees they have planted, to ensure that the trees mature, while retaining control over their use.

Discussion

Power relations in local natural resource management institutions and organizations are gendered and also interwoven with other kinds of social relations, such as kinship ties. This is highlighted in the case of the vice-chair of Chidiso Garden Project, who draws her power from social networks and kinship ties as detailed in Box 7-3. To understand how final decisions are made, one needs an understanding of the broader community context of social interactions and networks that influence resource management institutions. This calls for a research methodology that allows for extensive observations in formal and informal arenas, as well as longitudinal studies that involve an extended stay in the community and participation in community activities.

Similarly, women's roles in natural resource management decisionmaking processes cannot be easily understood apart from other social processes, in which they are embedded. For instance, marital status is important, because most women in key decisionmaking positions are assertive and are predominantly widows, elderly single parents, or de jure household heads (e.g., the CARE International community mobilizer, the Chidiso Garden Committee vice-chair). The need to separate women's and men's spaces has been overstated in the gender and community-based natural resource management literature. Findings from this study show that women and men share the same spaces (woodlands, for example) but use different resources. These findings concur with Goebel's findings from a study carried out in the resettlement areas in Zimbabwe. There is a need to take these findings into consideration when using the gendered spaces and places framework for the analysis of natural resource management. The sharing of resource spaces by men and women to obtain different products is especially applicable to the study of woodlands products, where a variety of resources may be found in the same spaces and places.

The power relations between women and men are constantly shifting as they are negotiated and renegotiated in response to changing natural resource availability, and needs and interests of the actors involved. The challenge facing researchers is how to capture such continuous changes and their influence on decisions about who has access to what natural resources, which in turn has implications for the sustainable use of natural resources. Women's bargaining power appears to be lower for woodland resources, where access is more individualized and where overlap occurs in the areas and products used by women and men. The women's bargaining power is much higher in the garden projects where they operate in groups and where resource use areas are more discrete.

Decisionmaking arenas are not always public or formal. Informal platforms for decisionmaking also exist, and they are often very influential. For instance, when meetings are called, the agenda is not always formally announced but is made known through the informal social and information networks in the village. By the date of the meeting almost everyone will know what will be discussed in the meeting, and some decisions are even made before the formal meeting. The majority of women in Romwe do not seem to speak up at public meetings. They largely use informal means to control powerful male figures in the community. An example is the alleged love affair involving the powerful and respected *kraal* head in the area, which sometimes constrains his power to make certain decisions (cf. similar situation in Mafungautsi, Chapter 8). Nonetheless, conclusions should not be drawn on the role of women outside the public arena without in-depth analysis. In the Zambezi Valley (Nabane 1997), I found that there was minimal communication on the agenda of meetings prior to the meeting, highlighting the need in those circumstances for women's participation in public decisionmaking processes, in order to have any influence.

Women in Romwe are actively involved in decisionmaking in smaller group projects, such as the garden project and the action research on trees, rather than in larger community arenas such as woodland management through formal administrative and traditional authority structures. The analysis shows that the membership in smaller projects is predominantly women. This partly explains the number of women in these project committees. Despite the fact that gardens are traditionally viewed as women's domain, the problem of marketing and limited profits may make these projects less attractive to men. A similar conclusion may be drawn for the action research on trees project, where the benefits are realized only over the long term, if at all.

This study finds that participating in projects with membership structures that are predominantly female gives women access to natural resources that they would not otherwise have. For instance, membership in gardening projects has given rise to new forms of property ownership by women. These findings contradict what Meinzen-Dick and Zwarteveen (2001) found in South Asia, where, with the exception of female-headed farms, women often continued to be perceived as helpers of their husbands, and the household was seen as a unit of common interests. In the case of southern Zimbabwe, women now own individual plots of land.

Conclusions

Gender relations in decisionmaking processes within institutions that manage natural resources are complex. Sometimes outcomes are resource specific (e.g., woodland use versus gardening projects); sometimes they are influenced by social relations in the form of kinship ties, age, and the issue being discussed. For women, marital status seems to be a key to determining their chances of being elected to committees, as well as being listened to. For instance, in Chivi, widows and elderly single women tend to be given more positions in management structures, compared with married or younger single women. Widows and elderly single women tend to stand up and speak out in public meetings. These findings concur with what Bradley (1991) found among the Luhya tribe in Kenya, where older widows had greater decisionmaking power with regard to tree-planting and disposal of woodlots than younger females.

While the cases in this chapter are anecdotal, they are examples of how informal and less recognized ways of participating in organizations and institutions for natural resource management may be a platform for increasing women's participation in broader decisionmaking about the use and management of resources. The informal strategies women use to gain access to natural resources do not guarantee secure rights to natural resources. These strategies are very dynamic and dependent on patterns of social interaction that women may not control. Thus, the key issue facing women—and in turn the development projects seeking to support them— is not whether to choose between statutory (e.g., WADCOs and VIDCOs) or traditional (e.g., *kraal* heads and headmen) institutions to secure their rights and access to resources, but to maximize their claims under either or both depending on which benefits them the most at a given time.

The study investigated natural resource management institutions through a framework of gender spaces and places. It reveals that the dynamics of natural resource management institutions can only be understood fully if attention is broadened to consider formal and informal institutions. The evidence suggests that lack of formal representation of women in resource management institutions does not mean that women have no influence over what happens within the organization. Women devise strategies to ensure that existing institutions address their interests and needs regarding access to natural resources. Given that informal strategies for resource access do not provide women secure rights to natural resources, it is essential that there be awareness raising on the importance of involving both women and men in formal decisionmaking processes. Simply mandating a quota system that sets aside positions of authority for women is not sufficient. While the quota system may be a good starting point to facilitate women's participation in formal decisionmaking, if not accompanied by awareness raising and leadership training, this approach may not yield the expected results.

Notes

This study was undertaken through joint funding from the Center for International Forestry Research's Stakeholders and Biodiversity in the Forests of the Future Project, funded by the Swiss

International Development Agency, and the Department for International Development (UK) Micro-catchment Management and Common Property Resources Project R7304. Funding was also received from the European Union–funded Management of Miombo Woodlands Project, which enabled me to participate in the Adaptive Collaborative Management Program (ACM) writing workshop in Bogor, Indonesia, where this chapter was further developed. However, funders need accept no responsibility for any information provided or views expressed.

1. The nature of membership, roles, responsibilities, and jurisdiction over such structures influence the roles of institutions in determining access to and control over resources by women and men in a given society.

2. In the context of Zimbabwe, communal lands are the black African smallholders' farming areas, formerly called reserves and tribal trust lands during the colonial period. These were originally created through land alienation by the colonial regime using the Land Apportionment Act of 1930.

3. Mazarire (1999) quotes Theodore Bent who describes an example of bodily decorations as the "breast and furrow pattern" believed to signify women as the chief agriculturalists in the area. Bent wrote, "At Mlala, too, we were first introduced to the women who have their stomachs decorated with many long lines or cicatrices … executed with surprising regularity and resembling the furrows on a ploughed field" (Mazarire 1999, 46–47).

4. In the context of natural resource use and management, common pool resources are those that generate finite quantities of resource units so that one person's use subtracts from the quantity of resource units available to others (Ostrom 1990). Most common pool resources are sufficiently large that multiple actors are able to use the resource system simultaneously and efforts to exclude potential users are costly. Examples of common pool resources include woodlands, ground water, grazing lands, etc.

5. Ranger (1999, 21) narrates the story of Princess Koswa of the Varozvi Mambos (chieftainship): "They [Koswa and Nyahuwi] came into the country occupied by one Ganganyama, said to have been a giant of a man, and very hairy. Since fighting seemed dangerous, the two women seduced the rustic gentleman and persuaded him to take them both to his couch … Ganganyama readily acceded to their request after being fortified for the event with liberal quantities of beer, [and] he ventured forth to do justice to the situation. However, the two girls complained that their hirsute lover's rank growth of hair so tickled them that they could not enjoy themselves. He granted them permission to shave him and so the poor man's fate was sealed. With one quick cut of the knife as he lay under the ministrations of his two 'amorous barbers' he was slain."

6. For instance, I was asked to pray at one respondent's home because she was excited that I had paid her a visit, while other development workers, according to her, only visited certain "rich" households, November 2000.

7. This categorization is based on a wealth-ranking exercise done in the village based on community members' criteria and indicators for defining socioeconomic status. Four wealth groups were identified, namely, the rich (*vanowanisisa*), average (*vanowana*), poor (*vanoshaya*), and poorest (*vanoshayisisa*).

8. Although I am Ndebele, my husband is Shona and comes from Masvingo province. This status accorded me respect and a listening ear from both groups. The Ndebele viewed me as one of them because I am Ndebele, while the Shona did the same on the grounds that I am married to a Shona.

9. Officially people do not acknowledge that they hunt small game. They tell you "people steal small game, but you should not write that down as it is like taking yourself to the police station after committing a crime."

10. See also Nhira and Fortmann (1993) for the classification of resource controls in Zimbabwe's communal areas.

11. A *kraal* head (or *sabuku*) in this area is a traditional village head who is responsible for land allocation, resolution of disputes, and other community issues that need to be dealt with in the village. This is not always his role (see Matondi 2001).

12. The Communal Lands Act of 1982 vests control of land in the state, with administrative powers conferred to the RDCs who are given authority to enact bylaws and to devolve authority to WADCOs and VIDCOs (Bruce 1990; Mandondo 2001).

13. In the case of divorce, women lose access to tree products in the homestead area (Luckert and others 2000). This pattern was also reported by Goebel (1999) in Hwedza district and Fortmann and Nabane (1992), who found that divorced women lost access to homestead trees, even if they had stayed in the village and planted and tended the trees. Goebel (1999) notes that, while the gender ideology does not restrict women from planting trees, the social construction of ownership is male.

CHAPTER EIGHT

Becoming Men in Our Dresses!

Women's Involvement in a Joint Forestry Management Project in Zimbabwe

Bevlyne Sithole

THE CONCEPT OF SOCIAL INCLUSION refers to the involvement of the populace in the structures and institutions of society, so that a shared sense of public good can be created and debated (Cleavey 2000). Debates about inclusion center on pragmatic and political concerns. The World Commission on Environment and Development (1987, 65) argues that the first step in pursuing sustainability is the creation of a political system that secures effective citizen participation in decisionmaking. Although policy recommendations generally use terms such as "collaboration," "people-centered," or "involving all stakeholders," little explicit and systematic thought has been given to how these objectives could be achieved, especially as regards the participation of women and other disadvantaged groups in decisionmaking (Meinzen-Dick 1997, 3). Nabane and Matzke (1997), for example, described how women continue to be sidelined in decisionmaking for community management of wildlife, despite efforts to target and empower them. In this chapter, I explore why existing efforts at inclusion have had limited success and, in particular, why women seem reluctant to take up opportunities to participate in structures of decisionmaking.

In general, women continue to be projected as being politically weak, without a voice, on the periphery of decisionmaking, or bypassed altogether. Development literature to a large extent continues to characterize governance of natural resources as a male-dominated arena, where involvement by women is minimal or absent. This view remains dominant among development practitioners and still underlines most of their advocacy for inclusive democratization. The literature has largely focused, however, on women's participation in the public domain. Leach (1994), for example, emphasizes the need to differentiate between decisionmaking in the public domain, which is regarded as the male sphere, and the private domain, which is regarded as women's sphere. While there is wide acceptance of the dominance of one gender in one sphere or the other, what is less appreciated is the extent to which one sphere influences the other. I argue here for an approach that recognizes participation of different groups as an outcome of social processes, which, though occurring in different spheres, do not exist independent of each other. Furthermore, I explore the extent to which nonpublic decisionmaking has an impact on or feeds into formal decisionmaking systems.

Scott (1985), in his work, *Weapons of the Weak,* challenges our conception of marginalized groups and suggests that these groups can define their own room to maneuver and can use various weapons to challenge those in power and be involved in decisionmaking. Scott's work raises three critical points in relation to women: First, that we reconsider our stereotypes of women in decisionmaking and establish to what extent they are indeed silent or weak. Second, that we look more closely at the means available to women that allow them to operate within formal decisionmaking systems. Third, that we reevaluate the notion of power, especially as regards how women view it.

The research was undertaken in Batanai Resource Management Committee, where the Center for International Forestry Research (CIFOR) has initiated a social-learning project to facilitate more effective collaboration among the stakeholders involved in the Mafungautsi Joint Forest Management Project. Participatory Rural Appraisal (PRA) techniques, especially institutional and kinship mapping, participant observation, and key interviews, were the main methods used. Names of respondents and villages where fieldwork was undertaken have been changed for purposes of confidentiality.

Overview of the Mafungautsi Project

Mafungautsi is a state forest reserve found in Gokwe District in Mashonaland, West Province (Figure 8-1). The miombo forest, the dominant vegetation type for south central Africa, is about 82,000 ha and was reserved in 1954. The forest is zoned into a core area reserved for wildlife; a buffer zone where some grazing is permitted; and a transition zone where grazing, beekeeping, and harvesting of products such as thatch grass and wild foods are permitted (Matose 1997). Timber concessions were discontinued in the early 1990s. Many ethnically diverse villages surround the forest; the dominant groups are the Shangwe, Ndebele, and Shona peoples. There is extensive literature on local use of the state forest in Mafungautsi, as well as detailed analyses of the relationships between the forestry authorities and the communities (Vermuelen 1994; Matose 1994, 1997).

The Forestry Commission has been involved in a pilot project to develop a joint management model for some of its forest reserves since 1992. Villages bordering the forest are now organized into resource management committees (RMCs) to represent local people in the joint management of the forest. There are 14 RMCs around the forest, some more active than others. The RMCs represent different areas, and these areas vary in size. RMCs cover a number of village development committees (VIDCOs). A VIDCO is an administrative boundary demarcating an area of authority under an elected committee. VIDCOs vary in size of population and in area. Old RMCs covered three or four VIDCOs. Because of

Figure 8-1. *Map of Mafungautsi, Zimbabwe*

administrative problems, some of the RMCs were reconstituted to represent small-er areas. The forestry authority views smaller institutions as more responsive to local people's needs. There are also subcommittees for special projects. At the time of research, only two were operating.

The RMCs are composed of "democratically elected" individuals from all the villages in the RMC. However, respondents feel that the electoral process is often tampered with and strongly influenced by the forestry authorities. Many respon-dents described how the first RMC had been hand-picked by forestry officials without the participation of the local people. Some respondents stated that they were unaware of the existence or purpose of some committees. The gist of their comments was, "Some committees you never see being constituted. You just wake up and find them there. You don't know who selected them, when they were created, and often you don't know how they came to exist." Some respondents further suggested that forestry officials hardly visit remote villages and tend to know people from nearer villages and hence favor them in the elections. Forestry officials are said to also "just make a beeline for the homesteads of these individuals, and they make the decisions together and then call this collaborating with the community." In general, people from one village dominate the RMC structures.

The main functions of the RMC are to issue permits for the collection of thatch grass and to monitor use of the forest. The committees rarely hold meetings or consult with the villagers they represent. Meetings are mostly held when requested by forestry officials or when facilitated by external organizations working on issues related to the project. Often villagers are not informed in advance about the agenda of the meetings. Previous members of a dissolved committee received payouts of Z$500 (US$14)[1] each, overall, and got a yearly allocation of thatch grass from the forests for their participation. Members on the current committees are entitled to allocations of free thatch grass every year. They also reward themselves with areas where they can harvest thatch grass free of charge as a token of appreciation (by the communities they purport to represent) for their services. The people do not support them and accuse committee members of hoarding resources at the expense of other members. Lately, members of the committees have been clamoring to be paid in cash at the end of every grass season. Some have even asked for "sitting allowances" whenever they attend workshops on behalf of the communities they represent. These "gifts" alienate the committee from the people and confirm local people's perceptions that these committee members are quasi employees of the forestry authority rather than their representatives.

Forestry officials rarely communicate with the community, choosing instead to communicate through the members of the various committees. The RMCs feel bound to the forestry authority and depend on the officials for direction. One respondent stated: "They created us, they should tell us and guide us so we know what to do." At village meetings, local people expressed ignorance of the role and functions of the RMC (e.g., "What is the RMC? We don't even know what it is"). According to the RMC committee, local people pretend ignorance because it suits their purposes to do so. We found that some people in the village were genuinely unaware of the RMC, while others who resented the RMC for one reason or another pretended to be ignorant of its existence to thereby contest its legitimacy.

In the joint forest management project local communities regard power as the ability to sanction use of forestry products. Power is the control over revenue derived from the forest, the ability to impose conditions on the use of forest areas, and the authority to include individuals on committees about which other people are unaware. Local people have none of this power and feel that there is no joint forestry management between them and the forestry authority. Under the terms of the collaboration, local people can use some of the products and services, but they have to obtain and pay for permits from their RMC. However, the procedures for obtaining these permits are fraught with problems and therefore people ignore them. One individual explained why he decided to harvest products without permits:

> I sometimes pay for a permit to the RMC to get grass from the forest. I pay Z$30.00 (US$0.09) and I can harvest Z$400–500 (US$12–14) worth of grass. Sometimes I do not go to get the permit. I just go into the forest—as you can see, I live on the forest boundary. The RMC members live far away, and when I do not have time to visit them I just go into the forest and get

what I want. Also some of the RMC members are too full of themselves, and they take their time processing the permits as if to make you feel their authority. Once I got caught, and they took all the grass.

Local people, through their RMCs, derive revenue from permits that are sold in each RMC. However, respondents feel that the forestry authority controls the revenue and dictates how it is used. The following statements echo these sentiments:

> The money generated by the RMC is controlled by the forestry authority. When the forestry authority comes here, they ask after the beekeeping projects or the nursery projects. The forestry authority wants us to preserve their forest.

> The money from the RMC is not our money. That is why people poach. They say the resources belong to the forestry authority, and they are not being used.

> The forestry authority says the money is yours, but tells people do this and do that in that way and this way. Where is our ownership in that?

Local people do not control the revenue or decide on its use. Instead they sanction the use of the revenue to support the forestry authority's "pet projects" rather than their own projects. The forestry authority's insistence on supporting its own forestry-related projects instead of community-led choices reinforces the argument that the state still pursues its own sectoral interests at the expense of community or joint interests outlined in the project.

Women's Participation in Project Meetings

Participation in the RMC meetings and subprojects varies by village. Of the three villages of Batanai RMC, the people of Mrembwe, which is the village closest to the district center where the forestry authority offices are located, can participate more actively than people from the more distant villages. Though the three villages in the RMC have over 400 households, less than 25% participate in meetings. For one of the big meetings recorded in the study, participation by women was less than 25% of total attendance. In other parts of the country, women's attendance is often higher than that of men. Researchers generally describe participation by women in this area as very low, compared with other communal areas.[2]

Women's participation in meetings of the RMC and its subcommittees tends to vary according to ethnicity. Some respondents describe women's attendance as being constrained by cultural conditions, although interpretations of the so-called constraints vary widely. Some respondents described the constraints as women being traditional and respectful of cultural norms and values, which clearly define roles and expectations of the two genders. Many researchers working in the area tend to have an opposite view, suggesting instead that the cultural constraints refer to husbands who forbid their wives to attend or participate, public censure of women who try to participate, or women who accept that they should not challenge the status quo. Shangwe women are less likely or willing to attend public meetings than the Shona or the Ndebele. In general, Shona women are said to participate more than women from the other ethnic groups. But even among the Shona women we found that participation is restricted to a small group of influential women who are married to powerful men, are thought to be witches, or hold important positions in other organizations sometimes unrelated to the joint forestry management project. These women are extraordinary. There is some ambiguity when they speak, since they sometimes seem to represent other women, sometimes not. Indeed, there are indications that other women sit around them, whisper and chat to them, and give all appearances of giving them a mandate to speak. But sometimes, even as they speak, it is obvious that they do not speak for all women. Wives of migrants also tend to feel outside the project. One headmaster of a school in the area stated:

> There are a lot of people among the locals that can spearhead development in this area. Unfortunately such people are not allowed to speak at meetings or even attend meetings. If you suggest something at meetings no matter how good, people discount it on account of your origin. If you persist and try to be involved, they threaten you with eviction or witchcraft (cited in Sithole 2004).

The truth of this statement was demonstrated at a village RMC meeting where local village leaders[3] silenced a migrant persistently:

> Each time the migrant contributed to the discussion some of the village leaders silenced him saying, "He is new to the area, he does not know about this area." When the facilitator persisted in trying to get the migrant involved, the local leaders and some of the women became abusive and shouted obscenities at the migrant. In the end the migrant was silenced (Sithole 2004).

Respondents state that they deliberately discriminate against migrants in elections because they are new to the area and people don't know their characters. One respondent who identified himself as a migrant said that, "In this area if you are an outsider you should not dream that you can ever be elected to office" (Sithole 2004). Early migrants are more accepted than recent migrants, and in some areas legal migrants are viewed differently than illegal migrants. In general, however, local people resent the election of migrants into democratic organizations, as these often replace or challenge existing traditional structures. For their part, migrants prefer to stay away from governance structures, as they often find themselves pitted against their benefactors. Some reports describe women as being unwilling to participate and even when they attend meetings being unwilling to voice their opinions. One field researcher observed:

> Women sit in their "designated" place away from the men and sometimes make it difficult for the facilitator to adequately bring them into the discussion. The few women who spoke at the meetings were usually the Shona women who held public office as committee members or village community or health workers. In general local people, including women, believe they should not participate in public debates as this demeans them to their audience.

In the general RMC meetings, one or two women usually speak. It is important to note that throughout the research the same women kept reappearing as members of committees or as leaders in other organizations unrelated to the joint forest management project.

In one meeting, researchers observed that men sometimes speak for the women:

> In one meeting we tried to divide the group along gender lines to try and increase women's participation. Three village leaders stood up and went to sit with the women. We could not chase them away. When we asked some questions, the three men would answer for the women and sometimes the women themselves looked towards the men for answers. If addressed directly, some women would profess ignorance or refer the questions to the leaders.

Women tended to participate publicly in women-only groups or organizations and rarely participated in the presence of men. In their own activities and women-only committees, women are elected to key positions.

Women's Participation in Decisionmaking

Are women really on the periphery of decisionmaking? Do they get bypassed by decisionmaking? Most respondents challenged the conception that women had no voice. They described a wide variety of ways that women can participate in decisionmaking. Most of them are informal, however. Furthermore, although men recognize these informal avenues of decisionmaking, they are not bound by decisions coming from them. Some respondents described these as "subtle," "hidden from view," "bedroom tactics," and "politics of the cooking pot." All these terms suggest that women have a role in decisionmaking, although in arenas less frequented by researchers. According to most respondents, women have various methods to ensure that their views are considered. Some of these situations are described below, based on respondents' comments:

> Some women will sit you down when you go home and say "Here, my husband, you have lost direction. How could you support such a decision?" They may even refuse to cook if they are angry enough with you. I tell you some men wake up the next day and they start talking to other men and the decision is reversed. That's power, I tell you. It is not true they are not involved, they are, you know, you just don't see it.

> Women are the parliament, we are the cabinet, and the two work together. That is government, you know that.

> What do you think the songs are about? They are about us and the decisions we make. Women are cunning; they sit down and before you know it there is a song and it's about you. When they don't like something, you get to hear about it. [*Editor's note:* Song-making is a more routine part of life in Zimbabwe than in many places. When Colfer visited the Romwe site discussed in Chapter 7 of this volume, a group of women spontaneously composed a song about her visit and sang it to her together.]

Men like to be led, but they don't want to let their friends see that they are being led, that is why we pretend to be ignorant, not to understand the questions and refer to the husbands. It's not because they are clever, it's just tricks, you can call it politics. Often when they are really in trouble they ask anyway but that is different; it is not like telling, it is responding.

When some decisions are taken and we see who is pushing the decision, we just talk to the wife and make her understand that her husband is foolish. She goes home and takes her husband to task and shouts at him for making her a laughing stock. That often works, and the husbands then go back to the beer hall or when he is with his friends changes his story and supports us.

The easiest way for us to participate would be to go and participate when decisions are being taken, but you know that is not our way. We have to find other ways of doing this that do not offend or embarrass our husbands. After all, after the meeting you must go home and it is with your husband that you must live.

Most of the respondents felt that current ways of communication allow them to influence decisions made in public places. However, many women insisted that they prefer to "respond" rather than participate at an equal level when asked to comment on a particular issue. Some respondents noted that in public gatherings challenges to decisions or contributions to decisionmaking were still indirect or subtle, through the use of song, mummers, expressions, disruptive laughter, leaving in protest en masse, or role-plays, with such acts of disapproval generally carried out by groups rather than individuals. Some exceptional women have directly challenged decisions made in public gatherings, but these are women with complex identities, who draw on multiple sources of power. Some male respondents indicated that it was acceptable for these women to participate, as they earned their right to do so by virtue of status or profession. Consequently, these few women get called upon to represent all women, although they do not in fact speak for all women. In one case a woman started calling herself director of a project, even though she was just the chair of a committee. In fact, she has given herself executive powers and does not consult with the rest of the committee.

An analysis of single-gender focus group discussions indicated that there is a high level of advance discussion between the genders before public meetings and review of ideas and arguments afterward, to the extent that the timing of gender disaggregation is crucial in determining whether one captures different views or not, as described below:

If meetings occurred concurrently, men interfered in the women's meetings and generally censored women's responses. Some men actually appear to lose interest [in their own group] and listen to women's discussion. Some speak for the women and start their answers with "she wants to say this," which forces women to concur. If in meetings one could keep the two groups separate, there was little exchange of ideas between the genders, but if meetings occurred on consecutive days then we found that there was a remarkable similarity of ideas and responses from both genders, suggesting that issues had been discussed in other arenas that are not readily accessible to us (Sithole 2004).

Generally, women have no desire to participate in public meetings, challenge established norms, or undermine existing channels of communication where they are represented through kinship and other social networks. They have already spoken, why should they speak again? Most women noted that they are informed and consulted before decisions are made. They often have a chance to let their views be known and are satisfied that this is so. However, I do not want to give the impression that women are always successful and influence all decisions.

Women's Positions in Decisionmaking Structures

Three committees operate under the Mafungautsi Joint Forestry Management Project: the RMC and two functioning subcommittees representing interest groups involved in beekeeping and thatch grass collection. There are also nursery groups in each village, and each has its own committee of elected individuals. On each of these committees, the forestry authority has requested that women be elected to formal positions. However, in general, respondents acknowledge a general lack of interest and enthusiasm among women to hold public office and participate in the elections. Women are most likely to propose men to positions and decline to be nominated. Some plead illiteracy. Others want to consult with husbands before accepting, while yet others fear responsibility. Some have no time for consistent involvement in meetings. Others are too shy or not senior enough in the community, while yet others fear being bewitched if they are part of making less popular decisions. They stated that mere presence in a meeting did not guarantee participation. Some women expressed no real desire to participate and indicated their fear of being labeled "too forward."[4]

Many female and male respondents felt that women should not participate in public debates over issues. Even in women-only groups, some women felt that the presence of men "would lend weight to the group."[5] Women who had been elected to some village-level committees were overwhelmed by the male presence in the committees and felt they were not contributing much. They felt that the forestry authority was forcing them to participate. Furthermore, since the committees rarely met, they did not feel part of the organizations. A number of women noted that all the unwillingness or nervousness would be dispelled if women did not feel so culturally restricted. On being quizzed about this point, many respondents noted that there are some women who are above these restrictions and can go against the norms without suffering public censure. So some respondents would say, "Ah, those can do it, who can touch or question what they do?" But the women elected to the positions also expressed reservations about participating in male-dominated decisionmaking systems.

On most of these committees, the chair and the treasurer make decisions, and the rest of the committee are passengers who do not really have a say when decisions are made. Some researchers jokingly refer to women as napping participants because they don't show much interest in what is being discussed. Women particularly are ignored when decisions are made and only come to know of some meetings when they have already taken place in a beer hall or at the business center, sometimes at hours when women are unable to attend. Other people in the committee indicated that women in these committees rarely if ever make a contribution, preferring to listen or contribute only if asked to do so directly. In general, women expressed fear of public censure if they got into office. Many practical issues undermine women's active involvement and put their character into question. For example, meetings can involve periods of time away from home or attending meetings with participants, most of whom are men. This can lead to accusations of illicit relations, and, lately, being labeled a member of the opposition political party by any disgruntled person. The social costs for women seem much higher than those for men.

One of the RMC subcommittees, the thatch collection group, is led by a woman, the same woman who is also a committee member on the RMC. The thatch collection group is an interest-based group initiated without the involvement of the forestry authority. The group comprises eight women from the village of Mrembwe, who collect thatch grass and broom grass for resale in Gokwe where the grass fetches high prices (see Chapter 9). All the women in the group have close family ties. Even though we find no formal organizational structure for this group, it functions as a formal entity, and even the forestry authority regards it as one. Within the group, there is a gradation of authority comparable to that within family structures. The wife of the traditional leader of Mrembwe assumes the leadership roles, even though there was never an election or indication from others that she should assume that position. Her husband is the only male member of the group. The decision to include her husband was a unilateral one. When asked to explain his presence in a female-only group, the leader replied that her husband's presence was important because it "lends weight to the group" as local people tend to discount women's activities as *mahumbwe,* or "child's play." Nonetheless, she did not explain why only her husband was a member and none of the other husbands in the group. Other women agreed that the presence of the husband was necessary for the group to be taken seriously.

The beekeeping group is one of the interest-based groups set up to facilitate beekeeping among the members of the RMC. The group has 24 members including one woman, the same woman on the RMC and the leader of the thatch collection group. Beekeeping is an approved activity financed by the revenue generated from the RMC. Not many respondents were present when the group was formed or when the committee was elected. Members of the RMC state that there is little interest in beekeeping in the area, but the forestry authority insisted that revenue from the project be used to develop the bee component. Some of the existing beekeepers did not join the group. Some members in the group indicated that they were not genuinely interested in beekeeping. There are no other women in the committee and in general women showed no interest in this subproject. As in the case of the thatch grass collection group, the presence of the woman was at the behest of her husband to give the appearance of equal gender opportunities in the interest groups. Women regard the project as focusing on male needs rather than female interests. Views from both male and female respondents suggest that none of the revenue has been used to target women's choices of projects:

> These bee projects are useless. They are for men. Women don't get involved with bees, therefore there are no projects for women. There is nothing to benefit us.

Women from Mrembwe village tried to do a poultry project using revenue generated from the RMC and they were told by the forestry authority that they have lost focus, what they planned to do had nothing to do with forestry. "You must involve yourselves in beekeeping."

There are also three nursery groups, all registered with the forestry authority as part of the joint forestry management project. In the village of Chanetsa all the committee members in this group are men. In the Vizho nursery group the majority of members are women, who also hold two key positions (vice-chair and treasurer). But the position of the chair was given to a man, again to lend weight to the group. Of the two women, the vice-chair is a very influential woman and is a keen advocate for more and direct involvement by women in decisionmaking. She thinks that women do not participate because they are restricted by their cultural roles and that to participate it is necessary to compromise these roles. She illustrates this opinion with her own experiences, where her family and productivity have suffered because she took her participation seriously. She states that for one of the projects she is involved in:

> I strived hard for this project to be a success, sometimes I used my own money, because our resources were inadequate. I made many sacrifices, even my marriage; it was a wonder I did not get divorced. To save my marriage I have co-opted my husband in the project so that he has a clear idea of what I am doing.

As in the thatch collection group, where the inclusion of the husband is publicly explained as an act of gender balancing, in reality women do so to circumvent cultural restrictions on their participation. Other respondents described this woman as influential but efficient in exploiting her situation for her personal gains. Some respondents say the woman has become "a man in a dress" because her actions show the difficulties of accommodating cultural restrictions. To participate effectively women believe that they must shed their domestic mantles and assume new ones. In Mrembwe, the chair of the nursery group is a woman. This is the same woman who is on the RMC committee, leader of the thatch grass collection group and the only female member of the beekeeping group.

Women Who Are Called Powerful

Respondents identified one woman, Mai Ngirandeas, as being very influential and powerful in the project. She is the only female in all the committees for the joint forest management project. Besides being elected to these positions, she has other attributes that make her exceptional among other women:

- She is the wife of a powerful traditional leader.
- She holds positions in other organizations unrelated to the project.
- She is suspected of being a witch.
- She is suspected of illicit relations with powerful individuals.
- She has kinship ties to people who hold other positions within the committees for the project.

We found that in addition to the positions that Mai Ngirandeas holds on the RMC, she also holds more positions than her husband in political, social, and development organizations. Her husband is a very powerful village leader in the traditional hierarchy. As his wife, she is also a defender of traditional social roles, hence her acquiescence in the positions she holds. This position allows her much power. She participates more than any other woman in the structures of the RMC. Because she has been elected to many committees, she is able to draw on different allegiances and networks and has become very influential in the project as a whole.

One of the most useful attributes that Mai Ngirandeas commands is her label as a feared witch. The belief in witchcraft is very strong in the area, especially among some of the Shangwe, who are regarded as more "backward" and animistic compared with the Shona and the Ndebele. Fear of witchcraft is widespread in Zimbabwe, but particularly pronounced in some areas, where it influences to a great degree how people behave toward each other. Even though particular groups are thought to have more witchcraft than others, cases included individuals from every ethnic group. Respondents recount various incidents of witchcraft in the area as proof that their fear is justified.

The subject of witchcraft is very difficult and sensitive to discuss. Accusations that someone is a witch are not easily verifiable. However, the mere accusation or suspicion that someone is a witch allows that person to benefit from the notoriety associated with the label. The case of a former treasurer illustrates the status of witches in the area. In the old committee the treasurer was accused of embezzling money for his own use. The treasurer purchased a cow with money from the RMC. He alone used the cow, for various purposes. When approached to explain the purchase, the treasurer argued that he had merely invested the money and is willing to let the cow be sold to recover the money. The treasurer still uses the cow. No one in the committee or the members has dared take the cow from the treasurer because everyone is afraid of being bewitched. An individual who is believed to be a witch commands much respect and fear and can behave unchallenged on most issues.

Another important attribute of Mai Ngirandeas is her extensive network of kinship relations both locally and with people of power residing outside the village. Locally, most of her relations hold key positions in traditional, political, and development organizations. Her two children are the most sought after research assistants, and they work with research and development organizations. She has "connections everywhere and knows everything going on." Many respondents believe that, even though all her well-connected kin are male, she can still influence their decisions on other committees. Some jokingly referred to her as being the real power behind her powerful husband. Because of the influence she has, some suggest that she is the most powerful woman in the community. But this power does not derive from the positions she holds on different committees.

Some respondents have suggested that she may have illicit relations with some powerful people. However, others note that these are just slurs from jealous and envious people. Like the claims of witchcraft, such accusations are not easy to verify. However, as with witchcraft, however false the accusation, the individual still benefits from the perceived relationship with powerful people.

Mai Ngirandeas also has an important role in the women's wing of the ruling political party. Under current political circumstances in Zimbabwe, members of this party wield absolute power and have their actions sanctioned by political leaders. This political role adds to the power this woman can wield. However, respondents observed that this type of power is not long lasting and will wane again as it did after independence in 1980.

Discussion

Inclusive democratized institutions imply organizations that have balanced representation, which are responsive and accountable to their constituency. Bratton (1994) indicates that in many places the state has yet to engage local people in genuinely participatory partnerships. Murphree (1991) observes how often attempts at democratizing power end up entrenching existing differentials between elites and other groups rather than incorporating new interests or groups. That we have institutions people profess ignorance about, institutions aligned to the state rather than the constituency, and institutions inaccurately purporting to include women suggests that our efforts at democratization have failed. Perhaps we need to rethink how truly democratic systems can be set up in these diverse cultural and unstable political settings.

We also note that even when presented with the opportunities to participate some women, especially from some ethnic groups, participate reluctantly or grudgingly. How do we interpret this reluctance to participate? One possible indication is that women are satisfied with existing channels of communication. We have argued that our preoccupation with inclusive decisionmaking has forced us to emphasize the importance of public over informal systems of governance, although the two do not operate in isolation from one another. We therefore suggest that this limited involvement and grudging participation indicate that Western notions of participation or involvement, which assume participation as an open and public action, have blinded us to the real dynamics of participation in small groups and especially the participation of women in particular cultural settings.

Instead we should focus our attention on strengthening existing pathways of communication. If we acknowledge the existence and indeed the value of parallel systems of decisionmaking, then the challenge is to find ways of making the intersections between the systems more effective and more publicly acknowledged. Thus women's informal and hidden systems of negotiations and bargaining that feed into the formal public decisionmaking systems can be enhanced and recognized as a necessary caveat to the systems in place. However, the challenge is to achieve this enhancement and recognition without compromising the current balance between the systems.

Even though the forestry authority insists that women should be elected to the project's committees, their participation is very low. Schmidt (1992), who writes on the position of Shona women in Zimbabwe, argues that despite the public indications of subordination, many researchers concur that women are not weak or silent. She argues that, while women may not have power, they have influence. Men have power to define and enforce rules by which society is governed. The power allows them to monopolize structures of governance. Influence, in contrast, is not institutionalized. It represents the strategies of those without formal power to limit the power of others that impinges upon their lives. To the extent that women are excluded from or choose not to participate in structures of decision-making, they must rely on influence.

In Shona society women wield a significant amount of influence, which permits them to exercise a significant amount of informal power. May (1979) also notes that though men might hold important public roles and are shown deference by women (who thus help perpetuate the myth of male dominance), in many cases the men are superficially dominant and wield relatively little power. She adds that in these situations, men wield powerless authority. Furthermore, the assumption that power derives from being a member of a committee ignores the importance of power derived from other attributes. It is clear from the case of Mai Ngirandeas that being powerful is an outcome of a diverse portfolio of positions and extensive local networks through kin and political affiliations. Her presence in the various committees as a noneffective or participating member reinforces the argument that to wield power one does not need to hold public positions. It seems evident that this woman wields a different type of power, power that is played out in private and draws on unconventional sources. This type of power appears to be far more desirable for women in a strong cultural setting. It allows women to exercise a veto over decisionmaking in ways that are not confrontational to established patriarchal systems and maintains the illusion of male-dominated spheres and women's subordination. What is less understood or appreciated by development practitioners and researchers is the extent to which most marginalized groups and women prefer influence to power or the extent to which our current efforts to empower women in public have undermined or bolstered existing pathways of decisionmaking.

Conclusions

The notion of inclusive democratization as it is currently conceptualized and applied by development practitioners, especially as regards the participation of marginalized groups like women, has not and will not lead to real inclusion in existing decisionmaking systems. It projects inclusion as something that is dependent on participation in formal public structures. However, we have presented data showing that women are unwilling, uncomfortable, and unlikely to participate effectively if they are forced into this current mold. There are isolated cases of exceptional women being elected repeatedly to various committees, but even they hold relatively minor positions and make no significant inroads toward making women less silent in the presence of men. Even within women-only groups, we see a prevailing trend to co-opt men to maintain this status quo and the illusion of male dominance in public decisionmaking structures.

Women can be labeled as being powerful, but theirs is not the kind of power that we see in public. Theirs is a power that draws on multiple and unconventional sources and is exercised in private arenas rarely accessible to researchers. Their power derives from extensive networks beyond the project and special attributes as wives, witches, mistresses, and politicians, rather than as members of some committee. Thus, in so-called public decisionmaking systems, women remain weak and silent, preferring to maintain and perpetuate the illusion of male dominance. The unwillingness or reluctance of women to "become men in a dress" by participating in formal, public decisionmaking forums challenges Western notions of inclusive democracy, which the women find inappropriate in certain social settings. Thus we suggest rethinking gender advocacy to focus on getting women involved in innovative ways that are more acceptable to them and strengthening their preferred ways of participation in decisionmaking so that women's views from whatever sphere have more impact on public decisionmaking.

Notes

This research was undertaken with funding from the Center for International Forestry Research (CIFOR) under the Local People, Devolution, and Adaptive Collaborative Management of Forests (ACM) Program. Some of the data used in this chapter were also presented in a longer report by Sithole and Kozanayi (2001).

1. The exchange rate at the time of the study was Z$35 to US$1. The exchange rate was in constant flux at this time.

2. "Communal areas" are those areas where black Africans live.

3. "Village leaders" are *kraal* heads or traditional leaders who inherit their titles to rule a small group of households. The number of households (as many as 20 or more) is highly variable across the country.

4. Being "forward" is a label suggesting loose morality, low stature, or bad breeding.

5. Lend weight: "*Kupa chiremera.*"

CHAPTER NINE

Learning Amongst Ourselves
Adaptive Forest Management through Social Learning in Zimbabwe

Tendayi Mutimukuru, Richard Nyirenda, and Frank Matose

F OREST MANAGEMENT IN ZIMBABWE has been governed by highly restrictive government policies that offered inadequate incentives for the participation of local people (McNamara 1993). Control and management of resources were vested in regulatory departments such as the Forestry Commission (FC), a government department responsible for a wide range of activities: forest administration; conservation of timber; state forest management; afforestation; woodland management; regulation and control of timber products; and forest support services of research, education, and extension. It is also responsible for the development of forest policy in consultation with the Ministry of Environment and Tourism.

Throughout its history, the FC has taken a strongly top-down approach to manage forests, and this approach, of late, has proved to be unsustainable, partly because of the government's dwindling supply of money. Recently, the FC has begun to realize that increased community participation in forest management is essential if sustainable management is to be achieved. As the title of this chapter suggests,[1] local forest users are also increasingly recognizing that they must learn amongst themselves in order to improve their lives and their resource management practices. This new openness to community participation was the main reason a pilot resource-sharing project was initiated in Mafungautsi State Forest in 1994. The project was intended to give communities an opportunity to participate in forest management practices (while simultaneously benefiting from them) and to generate some lessons for the country as a whole. Mafungautsi was turned into a protected state forest in 1954 and has since been managed by the Forestry Commission. Unlike the other state forests, Mafungautsi is entirely surrounded by communal areas and has two main stakeholders—the local communities and the Forestry Commission.

One of the main problems faced in this pilot project is that adaptiveness in forest management by stakeholders has remained generally low (Nyirenda 2001), despite efforts to increase it. Even though stakeholders have been learning individually, opportunities for shared learning have not been realized. At the local level, groups of stakeholders were highly polarized on the basis of ethnicity, gender, age, length of residence in the area, and social status. Sharing of information across groups was greatly limited. No deliberate effort has been made to consolidate lessons so that all the stakeholders could learn and come up with better and more adaptive management strategies at the local level.

This chapter examines how social learning affects adaptiveness in forest management by local-level stakeholders and how social learning can be enhanced. The local-level stakeholders include councilors, headmen, village heads, chiefs, resource management committees, social clubs, and political leaders. They also include various resource user groups formed at the researchers' initiative as a platform for diversified groups (in terms of gender, ethnicity, status, and length of residence in the study area) of stakeholders to share information and learn together. We demonstrate that social learning enhances adaptiveness of diversified stakeholder groups in forest management.

The next section briefly considers the definition, preconditions, and processes involved in social learning. This is followed by a description of the research area and the methodology for enhancing social learning. After the research results are presented and discussed, we offer some conclusions regarding the relationship between social learning and adaptiveness in forest management and the challenges of facilitating social learning.

The Concept of Social Learning

Social learning falls under the general experiential learning cycle that has been described by Kolb as a series of steps of active experimentation, abstract conceptualization, reflective observation, and concrete experience (Kolb 1984). Social learning has been defined by some researchers as an approach and philosophy that focuses on participatory processes of social change (Woodhill and Röling 1998). It involves critical self-reflection, the development of multi-layered democratic processes, the reflective capabilities of individuals and societies, and the capacity for social movements to change political and economic frameworks for the better (Woodhill and Röling 1998). Others define social learning as a dynamic process that involves continuous sense-making of the world through perspectives based on concrete, experience-modified knowledge, beliefs, and values, and a dynamic process of reflection and action by stakeholders through the experiences encountered by involvement with other people and the physical environment (Maarleveld and Dangbégnon 1998). It is much more than memorizing facts and acquiring intellectual understanding and includes the ability to act as well as understand and attribute meanings (Wilson and Morren 1990). Social learning involves tapping the capacities of different stakeholders, learning collectively, and sharing their perceptions on various problems before agreeing on what course of action to take (CIFOR undated).

Social learning depends on all sorts of preconditions, which have to be created and strategically negotiated in advance. It involves gaining understanding about other stakeholders' perceptions, goals, and interests and differences within stakeholder groups. It is also based upon consensus-building through cooperation by group members. Interaction among stakeholders provides opportunities for alternative and diversified ways of getting things done and is most fruitful when people are able to be nonjudgmental, entering into dialogue without dismissing views of others because they are different from their own. Rather, to enhance social learning, people should try to identify the assumptions made by others and learn from them. In Mafungautsi, groups of stakeholders were, at the onset of the research, mainly characterized by their gender, ethnicity, age, or length of residence in the area, and there was little sharing of information across these groups. The researchers therefore decided to adopt a user-group approach, which resulted in the formation of new groups that were more diversified in gender, ethnicity, age, and status. The researchers then facilitated the resource user groups as they underwent a series of steps for social learning. Those steps included shared problem definition; shared sense of mutual interdependency; shared social construction of the hard and soft systems in question; shared perception of the causes of the problem including agreed ways of looking at intractable social impasses; reflective learning; shared perspectives on the nature of solutions (in terms of hard and soft changes); and collective resource mobilization and establishment of leadership for action (Maarleveld and others 1997).

Drawing from the literature reviewed, we defined social learning in Mafungautsi as a dynamic process of reflection and action by a group of stakeholders, who are continuously interacting, communicating, and sharing their experiences and coming up with lessons to influence future decisionmaking processes. Stakeholders began the social-learning process by identifying problems and then each stakeholder's interest in being involved in the process. With multiple interests, a sense of interdependency had to be cultivated among these stakeholders, so that they would work together to solve their problems. Stakeholders then sought various ways of solving their problems, and through negotiations and discussions of their experiences, came up with solutions that benefited all. Leadership structures were then put in place to manage the implementation of the desired solution. Throughout the whole social learning effort, stakeholders went through reflection and learning processes.

The Research Area

Mafungautsi Forest is located in Gokwe South District in Midlands Province, Zimbabwe. The forest has an area of 82,100 ha, which makes up 17% of the district, while 73% of the district is covered by communal areas and the remaining 10% by small-scale commercial farms. Gokwe South District receives a total annual rainfall of around 800 mm (which falls between November and March) and suffers from mid-season dry spells and high temperatures. The region is most suited for animal production (Katerere and others 1993).

The vegetation of Mafungautsi is predominantly miombo woodland and dominant tree species are *Brachystaegia* and *Julbernadia* species (Nyirenda 2001). The dominant soils in Mafungautsi Forest are the Kalahari sands and only a few patches can be found with sodic and heavy clay soils. The forest is a catchment area for four major rivers in Zimbabwe, namely, the Sengwa, Mbumbusi, Ngondoma, and Lutope. Conservation of the watershed was one of the main reasons it was protected as a state forest in 1954. Mafungautsi Forest is a source of several resources, including pastures, thatching grass, broom grass, medicinal plants, honey, mushrooms, firewood, construction timber, game meat, Mopane worms,[2] indigenous fruits, and herbs.

The initiation of a resource-sharing project in 1994 brought some changes in forest management in Mafungautsi (see Chapter 8, this volume). The main aim of the project was to enable surrounding communities to take an active role in the management of the forest resource. Fifteen Resource Management Committees (RMCs) were set up in various communities surrounding the forest, and their main role was to monitor and control harvesting of the resources, to which communities and villagers in each site were now allowed access. The RMCs also initiated beekeeping projects in the communities to reduce the number of people cutting trees in the forest for purposes of harvesting honey.

The research reported here was conducted in 3 of the 15 RMCs that surround the forest, namely, Gababe, Batanai, and Ndarire. These RMCs consist of 10, 3, and 5 villages, respectively (according to the classification by the national extension agent, AGRITEX, a village consists of 100 households).[3] Most stakeholders around Mafungautsi belong to two ethnic groups: Ndebele and Shona. A minority belongs to the following ethnic groups: Tonga, Kalanga, Chewa (immigrants from Malawi), and Shangwe. Except for the Shangwe, the rest of the ethnic groups are immigrants to the area. Although they were the original people in the area, the Shangwe are looked down upon by the other ethnic groups. Shangwes are hunters who derive most of their livelihood from the forest, and thus are labeled as backward and normally blamed for being poachers in the forest. Because of their background, Shangwe people are embarrassed to come out in the open and normally hide under other ethnic groups and call themselves Ndebele.

A Method for Enhancing Social Learning in Mafungautsi

Facilitators identified and emphasized the following steps for social learning: shared problem definition, shared sense of mutual dependency, shared construction of the hard and soft system in question, and reflective learning. Indicators were also developed for tracing social learning; these are changes in attitudes, perceptions, and management strategies, as well as enhanced collaboration among stakeholders.

The core social-learning activities in the facilitation process were training workshops, participatory action research with resource users, and feedback meetings. The workshop entitled Training for Transformation (TT) marked the initial phase of enhancing social learning among stakeholders. The workshop was conducted in a participatory way, with extensive use of visualization techniques, games, stories, and short plays in explaining the various principles of TT. The workshop focused on six TT principles, and some of these will be described in this section.

One key TT principle is that no education is neutral; it can either domesticate or liberate you. Domestication is when people are taught in a way that strengthens and maintains the existing situation. An example of such education is the teaching of societal norms and beliefs. Parents, for instance, teach their children about gender roles and differences. As children grow, the parents and the other older children inculcate these gender roles, and thus the inequalities that such gender differences create become morally supported and accepted. The following story was narrated to demonstrate how education could domesticate people:

Once upon a time there was a mother who used to remove thighs of a chicken before cooking it. She would later cook the thighs separately. Her daughter grew up seeing this, and when she got married she also continued cooking chickens her mother's way. Now when she had a daughter, the daughter asked why she was cooking chicken that way. The only answer the mother gave was, "I saw my mother cooking chicken that way, but I have no idea why she did that." When they visited their granny in the rural areas, the granddaughter asked her why she cooked chicken that way. The grandmother said, "I just saw my mother doing it, but I have no idea why she cooked it that way." This continued for quite some generations, and finally they found the reason why the great grandmother used to cook that way. It was discovered that her pots were so small that the whole chicken could not fit. She therefore devised a plan to cook the thighs separately from the whole chicken.

Education that liberates a person was said to enable people to be critical, creative, active, and responsible. Workshop participants were encouraged to open up to this type of education and to be critical and analytical in whatever they do and to learn by asking questions. The facilitator highlighted the importance of looking critically at existing community norms (including those related to gender, ethnicity, age, and other diversity-related factors) and why they were put in place. Otherwise, people would be domesticated in trying to follow those norms.

Another principle that was emphasized in the workshop was that which promotes dialogue among stakeholders to achieve their goals and objectives. It was emphasized that no one knows everything and everybody knows something. It is therefore crucial for stakeholders in natural resource management to engage continuously in dialogue and to negotiate, discuss, debate, and come up with joint ways of solving the problems they face.

Participants were later made to play a game, called the game of squares, which illustrated the importance of collaboration and sharing of information by resource users. In this game, workshop participants were divided into groups of five. The facilitator started by explaining the objective of the game: each member of the group was to make a square of equal size to the rest of the group. Each member was given an envelope with assorted pieces for making the square. Group members were not allowed speak to each other or to take a piece from anyone. They were, however, allowed to give away their pieces to others, if they wanted to do so. After explaining the rules, the facilitator gave a signal for the groups to start the game. A plenary session was later organized at the end of the game to discuss the lessons learned by the participants.

The major lesson highlighted during the plenary was that collaboration, openness, communication, and sharing of knowledge, skills, and resources are essential elements for development to take place. It was also highlighted that communities are made up of diversified groups of people with different interests and behaviors. However, each person, even when displaying behavior that is difficult for others to accept, is still important and can contribute significantly to the development process. The example of the 11 players on a soccer team was given. Each player performs a different function, and if one of these players is missing it is difficult for the team to win. Community members were said to be like members of a soccer team, where every member is valuable if they are to achieve the overall goal of developing their area. All people—men, women, the rich, the poor, and all ethnic groups—are valuable in the development process, and all of their views need to be taken seriously.

Other attempts to facilitate social learning included a Criteria and Indicators workshop, where both local and district level stakeholders came together to share knowledge on criteria and indicators of sustainable forest management. In addition, several meetings were organized in which stakeholders discussed critical issues concerning the forest and their livelihoods. Discussions and sharing of knowledge were facilitated through the resource user groups (for thatch, honey, and broom grass), which were formed during the participatory action research process through facilitation by the researchers. Resource users were asked to join a group dealing with a resource that interested them, and membership in the groups was not fixed.

The resulting user groups were diversified in gender, ethnicity, age, and social status, and they were flexible enough to allow those interested in any resource to come and join. Some resource users actually joined all the three user groups when meetings were conducted at different times. Those who had an interest in only one resource joined meetings conducted for that resource only. The resource user groups met regularly to discuss concerns, to share knowledge, and to jointly come up with action plans to solve the problems concerning their resource. The groups offered a lot of opportunities for the stakeholders to learn together, adapt their management strategies, and better their lives. The learning process in Mafungautsi is perceived as an on-going process, with diversified stakeholders continuing to learn.

For tracing social learning a checklist of questions was developed to guide discussions among those who attended the workshop and those who did not:

- *Questions for those who attended the workshop*
  - Did you tell anyone about the workshop?
  - What did you tell him/her/them? Describe.
  - According to you, what is the most important thing you learned from the workshop? From whom did you learn (i.e., facilitators, other participants, ACM researchers)?
  - How have you used, in your everyday life, the important things that you learned?

- *Questions for those who were told about the workshop*
  - Who told you about the workshop?
  - What exactly did they tell you?
  - What did you learn from the message?

- Did the lesson make you change something in your life (i.e., the way you do some things, perceptions, etc.)? What exactly changed?
- Did you also tell someone about the workshop?
- Who else did you tell?

Data collection techniques that were used in the research included focus group discussions, participant observation, group discussions, personal interviews, and informal discussions and interactions with community members.

Findings

This section presents and discusses the research findings, which are related to the three resources—broom grass, thatch grass, and honey—that communities were allowed to harvest after the initiation of the resource-sharing project. Broom grass, *Aristida junciformis,* is an annual grass that is used for making brooms to sweep houses. Traditionally, women collect the grass from the forest. The grass matures soon after the rainy season, and resource users are allowed to harvest the grass after paying for a permit. Harvesting is controlled and monitored by the RMC. Thatch grass, *Hyparrhenia* spp., is used for thatching roofs and is also collected mainly by women. Honey can be collected in the forest from natural beehives, which are normally found in old hollow trees. Traditionally, only men harvested honey in the case study area. Now, however, harvesting honey from the forest is illegal, because in most cases it involves cutting down the tree. Community members are therefore encouraged to keep beehives in the villages near their homesteads.

The context studies conducted at the outset of this research revealed that social-learning processes had been taking place in the case study area, although these were not systematically planned. An example of where communities were involved in social-learning processes is that of an experiment that was organized in Machije Valley.

The Machije Experiment

In an effort to enhance learning together about sustainable methods of harvesting broom grass, community members in Batanai RMC, with an initiative from the Forestry Commission, decided to conduct an experiment in Machije Wetland (the area where Batanai RMC harvests broom grass) at the inception of the resource-sharing project in 1994. The experiment was conducted in two small plots in the wetland area. In one of the plots, resource users harvested grass by digging. In the other plot they harvested the grass by cutting using sickles. These plots were then monitored to see how the grass would grow in each of the plots. In the seasons that followed, no new broom grass germinated in the plot where grass was harvested by digging. Instead a new grass variety, which could not be used for making brooms, emerged. Stakeholders concluded that the best method to harvest their grass without depleting it was that of cutting. For two years after the experiment, no one dug broom grass in Batanai RMC. However, after the two years, some people resumed digging, despite the fact that they knew its adverse impact on the resource.

After the experiment, we realized that there were no opportunities for stakeholders to come together and discuss this undesirable change in behavior by resource users. We therefore organized platforms for resource users to discuss this problem and come up with solutions. During discussions held with broom grass resource users (who were diversified in terms of status, gender, ethnicity, and age), we learned that several factors had contributed to the sudden change in harvesting methods for broom grass. One of these factors was the continued market demand for dug brooms. In most places where people were selling their brooms, the customers wanted dug brooms, and these were said to sell faster than cut brooms. It was alleged that uprooted brooms lasted longer than cut brooms, because the grass did not loosen so easily. This market preference made many of the Batanai residents choose digging, even though they knew the adverse effects on the broom grass resource. One woman (who according to the wealth-ranking exercise was considered very poor) actually explained her experiences in trying to sell cut brooms from her scotch cart (an animal-drawn cart for ferrying goods) in Gokwe, a town about 20 km from Batanai. She narrated the following story:

> One day I went to Gokwe to sell my brooms, which were a scotch cart load. When I arrived in Gokwe, all the customers rushed to see the brooms and all they were saying was, "*une magaro here?*[4] *une magaro here?*" which means "Are they dug brooms? Are they dug brooms?" Not even a single broom was bought as the people discovered that I had cut brooms. I had to come back home all the way from Gokwe with all my brooms untouched. I was really pained at all the time and effort I had wasted.

The woman[5] just ended by shaking her head and saying, "*Ah, zvinorwadza veduwe*" meaning "Ah, it is very painful, I tell you."

Another woman had a similar experience, and she also told people her story during the same discussion. Her story went like this:

> Last year, I also went to Gokwe with a scotch cart full of cut brooms, and when I arrived a group of customers asked me to bring my brooms since they wanted to buy them. I pushed my scotch cart to where the customers were standing and as they were looking at the brooms and putting aside the ones they wanted to purchase, another seller came by and started shouting that she had dug brooms. All the customers who were about to buy my brooms threw them back into the scotch cart and rushed to the newly arrived seller. We actually had such a big fight, me and the newly arrived seller, that we ended up at Gokwe police station. I presented my case to the police and told them that the other woman was selling brooms that were illegal, since they were dug and not cut, and this was illegal in our RMC. The police however dismissed the case and said that there was no such law written down. I finally left the police station angry and disappointed.

After hearing these experiences, broom grass collectors decided to come up with strategies to ensure that their resource was used in a more sustainable manner.

These included the following:

- Everyone should help by telling the police whenever they see someone digging broom grass, rather than leaving the duty solely to the RMC police, who are sometimes too busy to monitor activities in the forest. According to the community members, the RMC police are not paid, unlike the Forestry Commission police who can devote most of their time to arresting people who transgress. The RMC police have to do the RMC work in addition to working in their fields in order to survive.
- Instead of getting a permit before harvesting grass, it was suggested that it would be better if people would pay after harvesting to enable the RMC members to inspect and check if the grass was harvested by cutting or digging. Resource users, however, thought that there would be a risk of people harvesting and disappearing before paying for their permits.
- Resource users suggested four methods of dealing with the problem of market demand. First, broom grass harvesters (within and outside the Batanai RMC) could decide to cooperate and only provide cut brooms. This strategy would force consumers to buy cut brooms, since they would be the only ones available on the market. Second, RMCs could negotiate with the Gokwe Rural District Council, so that a law could be passed to prohibit the sale of dug brooms. This strategy would then force all broom grass collectors to cut instead of digging the grass. Third, broom grass harvesters could come up with new bundling methods that could make the cut grass brooms more beautiful and longer-lasting. This strategy would make the cut brooms more appealing for customers, since these attractive brooms would be much better than dug brooms. Finally, resource users could advertise their brooms, so that customers would actually come and purchase the brooms in Batanai instead of the sellers taking the brooms elsewhere. This strategy would give a better opportunity for the RMC to inspect the brooms and make sure that all those being sold are cut and not dug.
- To deal with the problem of other RMCs harvesting by digging instead of cutting, resource users suggested that a "look and learn" workshop could be organized, so that other RMCs would come and learn from the Machije experiment. They could visit the plots and see for themselves, while one of the Batanai community members could explain how the experiment was conducted and what the findings were.

Sharing Information, Knowledge, and Experiences among Resource Users

In the various meetings organized by resource user groups, stakeholders discussed and jointly came up with problems they faced in each resource area, and together devised plans for solving them. During these meetings, there was remarkable sharing of information among stakeholders. For instance, a meeting was organized with beekeepers in Batanai. A small number of men had traditionally dominated beekeeping. Now with the formation of the resource user groups, some women and more men also joined and were very keen to learn how to keep bees. One of the issues raised at this meeting as a hindrance to beekeeping was lack of knowledge among users. When asked to specify the type of knowledge they were talking about, one young man said that he did not know how to harvest honey and this lack of knowledge caused him not to keep bees. He went on to say that most people in Batanai used some bad methods for harvesting honey that ended up by destroying bees. As soon as the man sat down, the RMC chair stood up and said that he had received training in effective honey-harvesting methods at a workshop organized and facilitated by the Forestry Commission. He immediately recounted 10 important things to do or not do while harvesting honey.

Beekeepers in Gababe shared their knowledge, too, as the following case illustrates.

At a meeting in Gababe with beekeepers, some people mentioned lack of a market as a very big problem for beekeeping, and this was identified as the main reason why people were not keeping bees. The chair of the beekeeping committee (a local businessman) stood up and told the other beekeepers that truly speaking, marketing was not a problem at all. He said that there was a huge market for Mafungautsi honey in Bulawayo where he normally sold his honey. He went on to say that most of the buyers there actually preferred honey from Mafungautsi Forest because it was thick, unlike the honey that came from gum trees which was watery. He then went on to advise the rest of the people that for them to attract good prices, they needed to package their honey neatly to make it attractive.

In some cases, it was difficult to bring knowledgeable people to group discussion meetings, because they were not willing to participate and share their knowledge with the rest of the people. For instance, there was an old Ndebele man who had about 110 beehives at his homestead and was well known for keeping bees. When we invited him to a beekeeping meeting, he sent a message that if the researchers did not pick him up he was not coming to the meeting. A driver had to go and pick him up, so that he would come to the meeting. At the meeting, this man did not say a word but just remained silent. At the end of the meeting, the driver had to drop him at his homestead.

There were instances also where, through the organized meetings, resource users had an opportunity to reflect on their decisionmaking process. This reflection helped them to come up with better management strategies that were more sustainable. The case below illustrates this kind of learning.

At a beekeeping meeting in Gababe, when beekeepers were discussing action plans for beekeeping, everyone suggested that they needed to have a project where they all jointly keep a certain number of beehives, either in the forest (with permission from the Forestry Commission) or in the village (with permission from the village head). When asked if they had similar projects in the village elsewhere, the villagers said they did not have any similar projects. With more discussion and critical reflection on why they thought this was going to work, one man at the meeting ended up admitting that such projects did not work because of the problem of free riding. He said that only a few people work on such projects while the rest just harvest the benefits without putting in much effort. When the man said this, everyone automatically agreed that such a project would not work. They later on referred to their cotton group where they had a committee and attended meetings occasionally, but each person had individual plots where they worked individually. In the end, it was agreed that for beekeeping, it would also be better for individuals to have individual beehives, and members could also meet occasionally to discuss issues concerning beekeeping. They also agreed to select a committee that would be responsible for organizing meetings and other things concerning beekeeping.

Feedback Meetings on the Status of Forest Resources

There was also sharing of knowledge and information at feedback meetings, where people questioned research results in trying to understand their implications for resource management. At one such meeting, to which all community members were invited, we presented our findings from the ecological surveys carried out in the forest. Most people were quite shocked by the rate at which trees were being lost due to honey harvesting. In the discussions that followed, people talked about how many years it took a tree to grow big and what loss it was for that tree to be cut just for the sake of harvesting honey. Participants said that people who cut trees to harvest honey did so in a hurry, since this cutting was illegal, and that most of them used smoke from burning rubber tires to chase away bees. It was alleged that this smoke actually kills bees instead of just chasing them away. Furthermore, participants remarked that the people who burned the tires were likely to be responsible for the many forest fires, as they were said to leave in a hurry without extinguishing their fires. After the feedback session with the researchers, stakeholders seriously discussed ways for dealing with the problem of cutting down trees to harvest honey. Stakeholders agreed to embark on beekeeping projects in their area, so that honey could be harvested without destroying the valuable trees.

In some cases during discussions and learning sessions, resource users disregarded contributions from some members, maybe because they did not trust them or they did not like their personality, and this hindered learning among the group. For instance, at one meeting in Batanai, a volunteer who had attended the Training for Transformation workshop was asked to give feedback on the workshop to the rest of the villagers. One man, Mr. Maraire,[6] volunteered to give the feedback. He stood up and started showing people a picture that was presented at the workshop of a well-known image that could be seen in one way as the portrait of a young woman and in a different way as one of an old woman. He told people that their faces were like that, since they were not honest. He used the meeting as a platform to complain about the past sins of other community members. All who were present started showing their displeasure and someone in the crowd actually shouted, "Is that what you learned at the workshop? Can someone else who went to the workshop retell what you really learned?" At that point, Ms. Ndoro stood up and articulated in a clear and passionate way what the workshop was all about. There was loud applause when she finished talking. Ms. Ndoro comes from a poor household and previously did not participate actively in resource management activities. In the TT workshop, she gained confidence in airing her views and also got the platform to do so. We were also amazed by the way she fluently presented her opinions; she presented herself as a very intelligent woman.

Changes in Perceptions and Norms after the Workshop

When we began our research, not a single woman came to the organized meetings. When the TT workshop was organized, we encouraged women to come, and this involved negotiating with the invited women's husbands. Some of the invited women still did not make it to the workshop. In meetings that followed, at first only women who attended the TT workshop came. Gradually, the turnout among women increased, and women currently dominate the organized meetings. There is no longer a need for the researchers to negotiate with their husbands.

It was extremely difficult to conduct meetings at first, especially in Batanai. The councilor and the ruling party political leaders had to be notified each time the researchers wanted to conduct a meeting; otherwise, they felt threatened. The following incident illustrates this:

> One day when Tendayi Mutimukuru was organizing a meeting in Batanai with resource users, a ZANU PF[7] chair stopped the research vehicle and peeped in the car saying angrily, "*Ungapi uNyirenda?*" meaning "Where is Nyirenda?" The researcher told him that Richard Nyirenda had not joined this trip to the village, as he was busy organizing other things at Gokwe center. Tendayi later asked the man if she could help. He explained that he was angry because he had heard a rumor that we were organizing a meeting in the village and had not informed him about it.[8]

At one of the meetings we organized in Batanai, it took an hour to do the introductions. The invited people introduced themselves with their political titles, such as ruling party chair, vice-chair, secretary, vice-secretary, and so on. During the meeting, anyone who wanted to speak had to chant the party slogan before and after saying something. This went on for several more meetings. Later, it became less pronounced. The same people became more serious when they came to meetings and concentrated more on how to manage their forest resources in a better way. Currently, except for new people who come to meetings, no introductions are conducted, as all the people now know each other.

Tracing Social Learning after the Workshop

This section highlights some of the findings of a survey conducted for tracing social learning after the TT workshop. Both participants and nonparticipants in the workshop were asked about what they had learned and how they had used the lessons in their day-to-day life. It was interesting to note that in Batanai, some people heard about the workshop while others did not know about it at all. Out of the 17 people who were interviewed and were nonparticipants of the workshop, 11 of them heard about the workshop, and 3 of these had learned something (see Table 9-1).

Table 9-1. *Responses from People Who Did Not Attend the TT Workshop, Mafungautsi, Zimbabwe*

Group	Sex	Learning	Number
Those who heard about the workshop	Males	Those who learned something	3
		Those who did not learn something	2
	Females	Those who learned something	0
		Those who did not learn something	4
Those who did not hear about the workshop	Males		3
	Females		5

Those who heard and learned something also changed something in their lives. For instance, someone said that he had learned that for one to develop, one has to work very hard. This lesson made him change his perception that it is only the government that brings development to an area; now he believed that, by working very hard, communities can develop their area on their own. This change in attitude will hopefully lead to a change in behavior and reduce the syndrome of dependency on donors in communities in Mafungautsi. In another example of social learning after the training, someone said that he had learned to be critical and to learn by asking questions. Being critical is important for stakeholders in natural resource management forums—it makes them question their norms and values, leading to better management strategies.

The survey also revealed that male workshop participants targeted their messages to men and not women. Two out of the five men who heard about the workshop from the workshop participants said that they had learned something, while the women who heard about the workshop learned nothing (see Table 9-1). Two of these women heard about the workshop when the male participants told them that they had learned a lot from it, but they did not explain the exact lessons, while the other woman only overheard one of the male workshop participants relating what he had learned to other men (see Table 9-2).

Of those who attended, all said that they had learned something during the workshop. All of them had used the lessons in their everyday life, and all of them had told someone about what they learned. It is important to realize that social learning processes and sharing of knowledge by stakeholders after the TT workshop took place without any outside facilitation. It was also noted that after the TT workshop, attendance at meetings by women improved greatly. Before the TT workshop, only 62 women on average used to attend meetings in Gababe, 28 in Ndarire, and 21 in Batanai. After the TT workshop, the average number of women attending meetings increased to 248 in Gababe, 174 in Batanai, and 102 in Ndarire (see Table 9-3).

Discussion

This section revisits the research questions and tries to identify the lessons with regard to social learning from the Mafungautsi case.

Table 9-2. *Responses of Women Who Heard about the TT Workshop from Male Participants, Mafungautsi, Zimbabwe*

Who told you about the workshop?	What did they tell you?
My husband	He just told me that they learnt a lot but he did not tell me exactly what he had learnt. I have seen him reading a book from the workshop, but I never asked him what it was talking about since I thought it was not important for me. I have access to the book since he did not hide it, and I can find out what it contains and I can ask for help from my husband.
Tongai Marufu (Male)	I cannot remember since he was reading the book to his friends and I also listened to his stories.
Nickson Sithole (Male)	He just told me that he had to leave the workshop because he had a toothache. He did not say even a thing about what exactly he had learnt, but I took a book that he had and looked at the pictures. No one explained these pictures to me.

Table 9-3. *Average Attendance of Women at Organized Meetings, Mafungautsi, Zimbabwe*

Site for RMC meetings	Average attendance before TT	Average attendance after TT
Batanai	21	174
Gababe	62	248
Ndarire	28	102

Note: Attendance includes meetings organized by both the CIFOR researchers and the community partners.

How Can Social Learning Be Enhanced in Communal Forest Management?

The Mafungautsi case has provided a number of lessons that may be useful to others interested in facilitating social learning in forest management contexts.

Training. Training is an important means for communities to critically examine their values and norms and learn together. After the TT workshop, more women started attending meetings and other training workshops. In this case, it can be said that the people of Mafungautsi Forest have reflected on their societal norms whereby only men were meant to attend certain meetings and not women. To some extent, people have also discovered the importance of training for women, as more and more women started attending training workshops. Women, as well as men, got a chance to reflect on the various issues concerning their resources, and they have ever since begun to think about ways of using those resources more sustainably. For instance, it was essential that women in Batanai, who dominated broom grass harvesting, be involved in coming up with action plans to sustain the valuable resource. Some of the measures they proposed show how they were trying to adapt to the changing environment.

The TT workshop demonstrated that people were open to new ways of think-ing. All the workshop participants learned new things, which have made them change their perceptions. Furthermore, all of the participants went on to share what they had learned with other villagers, and in turn some of the nonpartici-pants also learned important lessons that made them change their perceptions as well. The TT workshop was therefore important in generating new insights and challenging people to think critically, reflect on their values, and take action to develop themselves.

Experiments. Experiments are important social learning tools. Reflecting on the Machije experiment, it is clear that social learning has indeed led to adaptiveness in forest management. After the experiment, there was no digging of broom grass, because stakeholders had learned together the impact of digging. With so much evidence at hand, it was easier for all the stakeholders to collaborate and change their broom grass harvesting to methods that were sustainable. However, the Machije experiment also demonstrated that taking shared learning as a once-only event is insufficient. With an environment that was constantly changing and a market that continued to demand dug brooms, stakeholders had a continual need to meet, discuss, and reflect on what actions could be taken. In the absence of this conscious effort for people to reflect on the experiment and develop a way forward, some of the resource users resorted to uprooting the grass in order to meet the demand. When the effort was made to bring stakeholders together to discuss and reflect, they came up with very practical ideas on how to adapt their broom grass harvest-ing to suit the new demands. One young man suggested coming up with new bundling methods for cut brooms, so that they could compete equally with the dug brooms but also ensure that the broom grass resource was sustained.

Platforms. Creating platforms where communities share experiences and ideas is essential for enhancing social learning. The Mafungautsi case showed that, when given the platform, stakeholders could share their experiences and learn together in an effort to enhance sustainable resource management. During the resource users' regular group meetings, for example, people defined their problems and tried to find solutions. Stakeholders were willing to listen to those who had expe-rience in areas that the group had recognized as a problem. For instance, those at the Batanai beekeeping meeting who had a problem with harvesting honey paid a lot of attention when the RMC chair described what he had learned about har-vesting honey. In Gababe, everyone paid attention when the chair of the beekeep-ing committee explained about the Bulawayo honey market, since marketing had been identified as a problem.

Feedback. Feedback meetings enhance learning by encouraging discussions. Pre-sentation of the context studies by the researchers at a feedback meeting was an important way of initiating serious discussions on the status of the resource and the actions needed to sustain it. Resource users were alarmed by the rate at which they were losing trees in the forest because of honey harvesting, and this sense of urgency made them give serious consideration to the alternative of promoting individual beekeeping projects nearer to their homesteads. Beekeeping was considered as one of the measures that could discourage people from cutting down forest trees, since they could harvest honey from their own beehives.

Stakeholders. Involving all stakeholders, regardless of their differences in gender, status, age, and length of residence is crucial for learning processes that lead to collective action and adaptiveness in management processes. The Mafungautsi case showed that each individual possesses knowledge and valuable experiences of use to others. The creation of platforms for people to share their knowledge and experience with others has resulted in great learning opportunities, which in turn contribute to more adaptive resource management by stakeholders.

Can Social Learning Lead to Adaptiveness in Forest Management?

So far, the Mafungautsi case has shown that social learning does lead to adaptiveness in forest management. After the Machije experiment, there was adaptiveness in the way people in Batanai RMC harvested broom grass. For two years after the experiment, no one uprooted the broom grass as everyone had clearly learned that digging was a threat to the grass.

When discussing why people began resorting to digging again rather than cutting, broom grass resource users came up with several management mechanisms, which they intended to introduce to discourage digging. These serious discussions and implementation plans again show that through learning together and redefining their problems stakeholders can adapt their management strategies to sustain their resources. The Machije experiment has clearly demonstrated that there is need for continuous creation of platforms for stakeholders to discuss, share ideas, and come up with more adaptive management strategies.

The example of social learning about beekeeping also shows adaptiveness by stakeholders learning together. After getting statistics on the rate at which trees were being cut in the forest for the sake of harvesting honey and after discussing the consequent losses, stakeholders began to take beekeeping projects seriously. They recognized the problem and came up with a more sustainable solution. They agreed to initiate and promote beekeeping projects so as to reduce the cutting down of trees as people try to reach natural hives.

What Challenges Are Faced in Facilitating Social Learning?

- Institutionalizing the social learning process. Although a huge challenge for researchers, institutionalization is important for resource users as a means of ensuring continued shared learning and adaptation. Only an on-going process for discussion, feedback, and reflection will ensure that resources are sustained over the long term.
- Helping resource users to appreciate each other's contributions. Especially during discussion and reflection sessions, good facilitation skills are needed to help resource users learn to listen to ideas from those they look down upon. In the

case of Mr. Maraire, people disregarded his contribution to a discussion, maybe because of his personality.

- Not wasting resource users' time. The reflection and learning sessions are especially critical times for facilitators to keep the discussions moving, although it is always difficult to know how much "critical reflection" is enough.
- Involving all stakeholders. It is important to encourage all stakeholders to participate, without forcing them to do so, and to allow them to share their knowledge, regardless of their gender, age, ethnicity, or length of residence. For instance, social learning was lost in the case of the knowledgeable Ndebele beekeeper who had 110 beehives on his homestead, but did not want to attend the discussion meetings and share his experiences with the rest of the beekeepers.
- Changing the educational process. Transforming the educational process from one of domestication to one of liberation requires a lot of time and training, for this change involves changing people's attitudes and behavior. In Mafungautsi, in the initial stages of the research, only men attended meetings, never women: this was the tradition (see Chapter 8 for further discussion of this tradition). Even though they did not attend the organized meetings, women were also involved in harvesting and utilizing the resources, especially broom and thatch grasses from the forest. Thus, it was crucial, that these women also became involved in resource management processes, if the forest was to be sustained. The TT workshop was quite helpful in encouraging community members to reflect on their existing norms and values, and as a result the number of women coming to meetings increased tremendously.
- Cultivating the sense of interdependence among stakeholders. Without this sense of interdependency, stakeholders would continue to manage and utilize the resource independently, which would result in degradation of the resource. Attempts to make stakeholders realize that they needed to work together to manage their resource sustainably were made through training activities, such as the TT workshop. The workshop helped stakeholders to appreciate each other, despite their gender, ethnic, and status differences. They also recognized that persons could not solve their problems alone, but needed everyone to participate in collective action to manage the forest sustainably.

Conclusions and Recommendations

The Mafungautsi research has shown that the following tools are useful for enhancing social learning among stakeholders in forest management:

- Training workshops that break passivity among stakeholders, encourage them to be open with each other, equip them with analytical skills, and emphasize the importance of learning processes for sustainable resource management (for instance, TT workshops).
- Joint experimentation by stakeholders in participatory action research processes.
- Platforms for stakeholders to discuss, share experiences, and learn together.
- Feedback workshops, where stakeholders get feedback on research results and discuss their implications.

The research also showed that social learning, to a very large extent, leads to adaptiveness in forest management. Stakeholders are always confronted with changing economic, social, and political environments and must continuously adapt their management practices, if the resource is to be sustained. The research also highlighted some of the challenges faced in facilitating social learning, including institutionalizing the learning process, making stakeholders appreciate each other's contribution during discussions, involving everyone in the social learning process, and making the reflection and learning sessions brief enough not to waste resource users' time.

It is important, however, to acknowledge that learning is a continuous process, which happens with or without facilitation. People will always learn together, as shown by the willingness of participants at the TT workshop to share their lessons with other villagers without any outside facilitation. According to one researcher, "Learning is not something we do when we do nothing else or stop doing when we do something else" (Wenger 1998, 8), but it is a process that always takes place, whether we like it or not or whether we notice it or not. Wenger also stresses that, "Even failing to learn what is expected in a given situation involves learning something else" (Wenger 1998, 8). It is more rewarding if stakeholders can speed up this learning process by deliberately organizing situations in which they take note of exactly what they have learned in order to make decisions for sustainable resource use. Failure to tap lessons learned through sharing and reflecting on their experiences is a tragedy for both the resource users and the resource itself.

Our other major conclusion is that diversity among stakeholders can be a strength in resource management, because different stakeholders have different knowledge and experiences. The Mafungautsi case showed that if diverse stakeholders share their knowledge, learning processes are greatly enhanced. In this case, stakeholders of different gender, ethnicity, status, and length of residence had different experiences and possessed valuable knowledge that, when shared among each other in the resource users groups, greatly enhanced learning among themselves. This, in turn, helped the stakeholders to jointly reflect upon their experiences and come up with more adaptive management strategies for the forest resource.

Notes

1. A saying by one resource user at a Training for Transformation workshop, where emphasis was put on the importance of sharing knowledge and learning together in management of Mafungautsi Forest.

2. Mopane worms (*imbresia belina*) are edible and used as a relish when dried.

3. The research was conducted under the Center for International Forestry Research's (CIFOR) Local People, Devolution and Adaptive Collaborative Management of Forests (ACM) Program. Site selection was done by putting all the RMCs that fell under one chief, into a hat and selecting one. Three sites were ultimately selected in this fashion; these were Batanai, Gababe, and Ndarire. Gababe falls under Chief Njelele, Batanai under Chief Ndhlalambi, and Ndarire under Headman Chirima.

4. The literal translation for the phrase, *une magaro?* is "Does it have buttocks?"
5. The woman was from Batanai RMC, and she made this comment at an organized broom grass meeting.
6. People's names have been changed to protect their privacy.
7. ZANU PF is the ruling party.
8. Richard Nyirenda and Tendayi Mutimukuru are CIFOR researchers.

PART III

South America

THE SOUTH AMERICAN ANALYSES focus on four sites practicing adaptive collaborative management, two in Brazil (Porto Dias in Acre and six communities in Pará) and two in Bolivia (Salvatierra and Santa María, both in the Guarayo area). This collection also includes some comparative material from sites not specifically included in the ACM group (the Tambopata-Candamo region of eastern Peru and Maranhão in eastern Amazonia).

Chapters in this section address the importance of a household's stage in its domestic cycle for its members' involvement in the forest, as well as the limitations of outsiders' efforts to represent the interests of community members. Chapters also evaluate some mechanisms for sharing the divergent perspectives of community groups and consider the dangers of facilitators dominating in community meetings and priority setting. Gender issues are also examined, particularly the relationship between community stereotypes and efforts to involve women in formal forest management.

Brazil

Porto Dias, Acre (Chapters 10 and 12)

Porto Dias is an agroextractive reserve, meaning that the local rubber tappers have comparatively strong use rights to large areas of forest (about 300 ha per family). The reserve lies between two settlement projects near the Bolivian border in the far west of the Amazon basin. The Acre state government has taken a strong conservation stance and supports local attempts to maintain the rubber-tapping lifestyle. An NGO has been working with a segment of the local community to establish a small-scale, community logging project, as an environmentally friendly alternative income source, in the face of decreasing markets for rubber. The main social differentiation is between the forest-oriented rubber tappers and the farm-oriented settlers.

Pará (Chapter 11)

Six communities were involved in ACM research: São João Batista and Nova Jericó in Tailândia (Moju); Jaratuba, Recreio, and Nova Jerusalem in Muaná; and Canta-Galo in Gurupá. They include complex, rapidly changing communities of recent immigrants; traditional, fairly stable, riverine communities; and a community of ex-slaves. The communities also varied considerably in their capacity for and interest in collective action. Situated in Pará, in the eastern Amazon basin, all have been pressured by the timber industry, ranching interests, and government-sponsored resettlement and road-building.

Bolivia

Salvatierra, Guarayo (Chapter 13)

Salvatierra is located in a TCO, a legally recognized, indigenous community territory in central, lowland Bolivia. A large, USAID-funded project, BOLFOR, has been working with the community for some time to promote community timber management, and a number of NGOs have also been helping them deal with recent legal changes and other development issues. The community is composed primarily of Guarayo Indians with little or no experience in managing forests for timber.

Santa María, Guarayo (Chapter 14)

Located in the same TCO as Salvatierra, Santa María is a community of fairly recent immigrants. As in Salvatierra, BOLFOR is trying to help community members make a timber management plan that will take advantage of the opportunities provided by the TCO legislation, which established use rights for indigenous peoples in certain areas. There is some uncertainty, since most Santa María residents are not Indians, about their actual rights; they occupy a minority status within the TCO.

CHAPTER TEN

Intrahousehold Differences in Natural Resource Management in Peru and Brazil

Constance Campbell, Avecita Chicchón, Marianne Schmink, and Richard Piland

T HE CHANGES THAT TAKE PLACE within rural households over time—as new families are formed through marriage and children are born, grow up, and gradually leave to form their own families—are referred to as the family developmental cycle, the reproductive cycle, or simply the life cycle. These changes largely determine the livelihood requirements of any given household and the diversity that can be found across households within most rural communities. Intrahousehold composition is also influenced by demographic changes, such as migration, and by economic changes, such as depression or expansion of economic growth.

Nearly 100 years ago, Russian economist A.V. Chayanov was the first to analyze family farms as units of both production and consumption, whose first priority was to ensure the food security of their members (Chayanov 1986). Rather than maximizing profits, the farm family seeks primarily to provide for its own subsistence, a task that varies significantly depending on the balance between producers and consumers at any given moment in the family developmental cycle. Analyzing intrahousehold changes over time provides a useful framework for understanding natural resource management among different groups of people. So far, relatively little attention has been given to this important source of diversity in livelihood practices, although research has shown its significance in land use and farming systems (Litow 2000; Perz 2001; Sullivan 2000). This chapter presents tools and models for studying intrahousehold diversity in forest communities in Latin America.

The research reported here explored gender and intrahousehold differences in resource management strategies in two neighboring regions in Peru and Brazil: the Tambopata-Candamo region of eastern Peru and the western portion of the Amazonian region of Brazil (cf. Chapter 12 on Acre and Chapters 13 and 14 on the Bolivian Amazon). These sites were selected because project interventions were under way to help rural communities introduce reforestation and agroforestry as alternative forms of resource use that would be more compatible with conservation goals for the region than agriculture, ranching, and logging. The research experimented with methods for analyzing intracommunity and intrahousehold differentiation in these changing resource use strategies.

The selected communities exemplify many of the challenges facing forest management initiatives. The Peruvian and Brazilian research sites in this chapter both focus in part on colonists who have arrived in the Amazon region within the last three decades. The strategies and practices of these migrants are often quite different from those of "native" populations. The migrants' lack of experience with forest management accounts for some of the variance between the two groups, but intrahousehold differences between migrants and natives are also critical and often overlooked as a factor affecting variance in resource management practices.

The two case studies reported here explored, in different ways, how the dynamics of intracommunity and intrahousehold differences affect the adoption of alternative resource use strategies. One (Brazil) used longitudinal analysis of sample households to study the dynamics of decisions about adoption of agroforestry practices as part of a grass-roots project. The other (Peru) used linear programming to model decisions about alternative resource use strategies by households with different labor and capital constraints in two communities heavily dependent on timber extraction, where nongovernmental organizations wished to explore viable alternative strategies. Using different methodological approaches, the case studies explored ways to analyze internal differentiation among households and communities in conservation and development projects.

The research built on baseline descriptive information previously collected in both sites on the projects, the communities, and the key resources and user groups.[1] In Brazil, this data was part of an eight-year working relationship between the community and Constance Campbell, who conducted complementary dissertation research (Campbell 1996a) and participated in technical assistance through a USAID-funded University of Florida agroforestry project. In Peru, Avecita Chicchón's previous work in the region as Peru Country Program Director for Conservation International (CI) provided five years of background research. The research was linked to broader regional projects working with communities to design and implement resource management strategies in collaboration with local nongovernmental organizations.

The research reported here experimented with specific methods to analyze how resource management strategies differed among social groups and within households of different types in the sample communities. The analysis focused primarily on decisions about intrahousehold allocation of labor and income, comparing households of different size, life-cycle characteristics, and migratory status.

Timber Extraction, Reforestation, and Conservation in Peru

Madre de Dios is the third largest state in Peru and the least populated. Although it is slightly larger in size than Costa Rica (over 85,000 km²), its population is about 67,000 (INEI 1999). Madre de Dios is the Peruvian state that grew the most (5.7%) from 1982 to 1993 because of a large immigration wave from the highlands. Migrants arrived in Madre de Dios to extract gold and timber and establish agricultural landholdings. One-third of them settled in Puerto Maldonado, the capital city of Madre de Dios.

Madre de Dios is a frontier area for Peru, and men from other regions often arrive first to establish their livelihoods before they bring their families. People keep coming from the highlands and settling along the road that was opened in the 1960s, linking the city of Cusco in the highlands with Puerto Maldonado. From Puerto Maldonado this road turns north into the state of Acre in Brazil, the site of the other case study. The road is built on firm ground away from the main rivers. The rainy season from December through March has a large impact on road conditions, limiting travel and marketing of local products. When the road is closed, market goods have to be flown into Puerto Maldonado, and prices soar.

In the past two decades, some colonists have abandoned their settlements along the road because they began to see diminishing returns in their agricultural output. Soils became exhausted from agricultural practices that were suited to the highlands (monoculture and annual crops). Availability of forest resources, such as timber, also diminished. These changes were enough reasons to move to unoccupied areas along the road or away from the road further into the forest; some also moved to Puerto Maldonado, giving up agriculture altogether.

In this context, FADEMAD (the Agrarian Federation representing the colonists in Madre de Dios) and conservation organizations began to look into ways that would help the colonists to stay on their lands along the road. Cutting trees for timber provides a steady source of income for local farmers, who are the target beneficiaries of conservation projects. The research results presented here aim to improve our understanding of the importance of timber extraction in two communities in Madre de Dios—Santa Rita Baja and Monterrey—and to examine labor and capital constraints to alternative resource use systems.[2]

The Setting in Santa Rita Baja and Monterrey

Colonists from the Andean highlands established Santa Rita Baja in 1963 and Monterrey in 1987. The colonists from Santa Rita Baja arrived shortly after the completion of the road that connects Cusco with Puerto Maldonado.

The average annual income per household was similar in Santa Rita Baja and Monterrey (Table 10-1). The sources of income varied because of differences in ecological habitats and proximity to favorable markets. The farmers from Santa Rita Baja marketed their agricultural products in Mazuko, where they obtained cash mainly for rice and lowland tubers. Maize did not grow well in Santa Rita Baja, but it did in Monterrey, where soils were more productive. Patches of Brazil nut forest were unavailable in Santa Rita Baja, but they were abundant in Monterrey. Lacking maize and Brazil nuts, farmers in Santa Rita Baja depended more than those in Monterrey on cattle for income.

Households tended to have a diversified and flexible set of economic activities available to respond to changing market conditions. When asked, farmers identified agriculture as their main economic activity, and women identified housework as their principal occupation (although other data clearly showed their participation in farm activities as well). People declared that they needed to be involved in

Table 10-1. *Average Household Income from Various Economic Activities in Monterrey and Santa Rita Baja, 1999*

Activity	Monterrey (in soles)	Santa Rita Baja (in soles)
Rice cultivation	545	1,673
Maize cultivation	885	245
Tuber cultivation	6	237
Other crop cultivation	220	970
Poultry-raising	392	384
Cattle-raising	0	3,353
Off-farm labor	1,204	873
Off-farm labor (Brazil nuts)	328	0
Brazil nut collection	1,121	0
Timber extraction	6,709	3,433
Total	11,370	11,168

Note: US$1 = 3.1 soles.

timber extraction because timber was a product that was in steady demand. Repeatedly, farmers declared that the cash obtained from timber was used to send their children to school. During 1999, colonists in Monterrey shifted toward a more marked dependence on woodcutting as a source of cash income, in part because of the relative failure of the Brazil nut harvest that year. Usually a nut gatherer earns six times more from that activity (around 6,000 soles, according to Dourojeanni 1997) than from woodcutting.

In addition to cash income generated by agricultural effort, most families covered their own needs for rice, maize, and tubers. Still, most families bought most of their food. Estimated yearly expenditures for food, schooling, and medical costs were 3,718 soles, while production input costs, which included tractor rental, fuel, and labor, were about 4,253 soles. Those major categories of expenditure accounted for about 70% of spending; the remaining 30% (some 3,000 soles) was not accounted for, but was likely spent on clothing and other goods. Almost 70% of households reported that they did not save, while those that did reported an average of only 443 soles per year.

As expected, timber extraction was one of the main sources of income for both communities, although not a favored occupation.[3] To be involved in this work, a man had to have enough strength to manipulate the chainsaw to cut the trees, open trails in thick forest, carry large pieces of wood, and get the tractor out of the mud if it got stuck. A woman's participation in timber extraction was limited to parallel activities: finding the personnel to help her spouse in the forest, preparing food for the team who went to the forest, and occasionally marketing the wood to the middlemen who came to the community requesting the product. "Women are more responsible handling money," male colonists said. Thus, although logging was seen as a male activity, women also participated in support activities and in decisions to undertake this undesirable activity.

According to the census carried out in Santa Rita Baja in early 1998, 30 of the 32 households in the community of 150 people reported participating in timber extraction in some form to supplement their income. All but three had

Ministry of Agriculture permits or worked for other farmers in timber extraction. Most of the timber extracted was so-called common wood (*madera corriente*): *lupuna* (*Ceiba* spp.), *copaiba* (*Copaifera* spp.), *pashaco* (*Schizolobium* spp.), *catahua* (*Hura crepitans*), and *oje* (*Ficus insipida*). Because of Santa Rita Baja's longer history of settlement and logging, compared with Monterrey, the timber extracted in Santa Rita Baja was not as valuable. During the same research period (1998), loggers in Monterrey extracted more high-value species than common species. Colonists there stated that, because it was so much work to extract timber, they would rather focus their efforts on extracting the high-value timber still available. From 1988 to 1996, 65% of the timber production in Madre de Dios included only three species: tropical cedar (*Cedrela odorata*); mahogany (*Swietenia macrophylla*); and *tornillo* (*Cedrelinga catenaeformis*). By 1999, people mainly harvested *moena* (several species from the Lauraceae family). This shift in species, necessitated by the depletion of the higher valued trees, underscores the ephemeral nature of timber extraction as an income source.

In Monterrey there were a total of 55 households (221 people); 41 households were interviewed in 1999. The main economic activity in this community was Brazil nut extraction, although timber extraction and agricultural activities also were important. According to the 1999 survey, 19 of 41 households interviewed participated in Brazil nut gathering, while 14 of the 41 participated in some aspect of the timber extraction process. Most extracted timber from within their assigned agricultural land with a ministry permit, or worked as wage laborers in different phases of the extraction process. Young people chose to work as wage laborers in timber extraction because it was a quick way to obtain cash.

People in both communities who were involved in timber extraction were, on average, younger (at 33 years old) than those who were not involved (who averaged 42 years old). Those who were more heavily involved in rice production in both communities were less involved in logging. This inverse relationship was more evident in 1998 before the price of rice went up from 0.60 soles per kilo to 0.90 soles: loggers had planted an average of 1.67 ha of rice, while non-loggers had planted an average of 2.45 ha. In 1999, loggers invested more effort planting rice as a result of a more favorable market price for rice.

Linear Programming Models of Different Household Production Systems

The study experimented with linear programming as a technique to model household differences in labor constraints as they might affect optimal economic strategies.[4] The exercise was intended to produce useful information about the role of timber in small-scale systems and to test the potential for substitution by alternative natural resource management strategies that were being promoted by FADEMAD.

Household composition by age and gender, a set of variables closely linked to family reproductive cycles, has important consequences for resource use strategies. Linear programming allows researchers to model these relationships in many different ways. As a first step in application of this technique, the study evaluated the potential of the current economic system among households with different labor and credit constraints.

The production environment and mix of economic activities were different for the two study communities. In general, Santa Rita Baja had a higher yielding agricultural environment and higher market prices for agricultural products than did Monterrey. Families in Monterrey had access to Brazil nut forest not available in Santa Rita Baja. Few cattle were raised in Monterrey; thus this activity was not considered in its linear programming model. Because the economic system in Monterrey emphasized both Brazil nut and timber extraction (activities that require capitalization for the purchase of inputs, such as gasoline, labor, food, tools, and transport), start-up capital was higher there than in Santa Rita Baja.

Tables 10-2, 10-3, and 10-4 summarize production factors and constraints placed on production inputs that were used to construct the linear programming models, comparing estimates by women and men. Eight scenarios were modeled for each community, optimizing the economic activity mix to maximize discretionary income after meeting family consumption needs. These scenarios were just examples of the many further tests that would be necessary to determine the precise factors involved in changing production systems.

The scenarios first captured different labor constraints based on the availability of children's labor differentiated by gender, with and without credit. It was

Table 10-2. *Production Factors in the Linear Programming Model for Monterrey*

Variable	Rice	Maize	Tubers	Brazil nuts	Wood	Off-farm work
Unit of analysis	ha	ha	ha	shelled kg	board-ft. sold	daily wage
Yield (kg/ha)	721	970	3,250	—	—	—
Male labor (days/unit)	40	20	20	0.025	0.00483	1
Female labor (days/unit)	30	20	50	0.050	0.00161	0
Price of inputs ($/unit)	160	160	0	0.769	0.05011	0
Selling price ($/unit)	0.277	0.338	0.071	1.612	0.13850	4.31

Note: Dashes indicate not applicable.

Table 10-3. *Production Factors in the Linear Programming Model for Santa Rita Baja*

Variable	Rice	Maize	Tubers	Cattle	Wood	Off-farm work
Unit of analysis	ha	ha	ha	1 animal	board-ft. sold	daily wage
Yield (kg/ha)	1,025	820	2,954	—	—	—
Male labor (days/unit)	40	20	20	2.35	0.00354	1
Female labor (days/unit)	30	20	50	2.35	0.00119	0
Price of inputs ($/unit)	49.20	49.20	0	65.39	0.05561	0
Selling price ($/unit)	0.469	0.338	0.123	299.07	0.13850	4.31

Note: Dashes indicate not applicable.

Table 10-4. *Assumptions of the Linear Programming Model for Both Communities*

Factor	Assumed maximum units available to households
Land for cultivation	3.5 ha
Pasture	15 ha
Cattle	1 animal per ha
Rice production for subsistence	750 kg per year
Maize production for subsistence	250 kg per year
Tuber production for subsistence	500 kg per year
Chicken production for subsistence	12 animals
Cattle production for subsistence	1 animal
Wood extracted	25,000 board ft. per year
Male labor, father only	200 days per year
Male labor, father plus other males age 12 or older	320 days per year
Female labor, mother only	150 days per year
Female labor, mother plus other females age 12 or older	270 days per year
Off-farm labor	90 days per year
Capital in Santa Rita Baja (no credit)	US$769.23
Capital in Monterrey (no credit)	US$984.62

assumed that children age 12 or older begin to make an economic contribution to family economies. For these models, such children add 120 days a year of labor; adult males add 200 days a year of labor; and adult females give 150 days a year. The rest of the female labor available during the year would be invested in childcare and other activities not incorporated in the linear programming model. Scenarios included the following household situations: a couple with no children age 12 or older; a couple with one female child age 12 or older; a couple with a male child age 12 or older; and a couple with one female and one male 12 or older. Family situations with more than one male or female child 12 or older did not change the results of the model. In addition, each family situation was modeled for two credit scenarios: one in which families had only minimal capital available to reach the average pattern for economic activities in the two different communities; and another in which families had access to additional credit of $925.

Monterrey. Figure 10-1 shows the actual average economic activity mix taken from survey data for Monterrey, as well as four optimization models for different family situations. The most striking aspect of the figure is the overwhelming importance of timber over all other activities, even more so in the optimized models, in which Brazil nut collection virtually disappears and maize cultivation doubles. This factor is so significant that there are few differences among the different family situations.

The linear programming optimization models show that under the current economic situation, the income generation potential in Monterrey is only minimally higher than actual levels. The mix and relative importance of economic activities does change in the optimized model. Rice and tuber (manioc and *uncucha*) agricul-

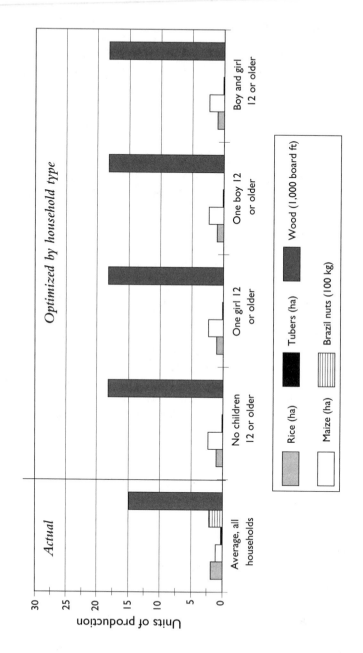

Figure 10-1. *Actual and Optimized Economic Production, by Household Type, in Monterrey*

ture is reduced to levels of purely subsistence production. Maize production is maximized to the constraints placed on the availability of cultivable land area. The model phases out Brazil nut extraction and emphasizes timber extraction.

Assuming the availability of credit for income maximization and optimization of economic activities in Monterrey leads to a fuller and more differentiated use of available female and male labor (Figure 10-2). Agricultural activities remain unchanged from the situations with no credit modeled for Monterrey. The linear programming model tends to increase the utilization of female labor in the extraction and processing of Brazil nuts. Available male labor is utilized in timber extraction. Increased available male labor results in maximizing timber extraction to the limits of the constraints placed on the model (this constraint complies with the legal maximum amount of timber cut with a permit—5000 board feet). The model fails to maximize female labor in Brazil nut processing over male labor in off-farm labor and timber-cutting. This optimization indicates that timber-cutting is a more efficient use of capital to generate income. The data used in the model for production factors in Brazil nut extraction, however, reflect the abnormally poor Brazil nut season of 1999.

Increasing the availability of children's labor in these models does not lead to large increases in income generation. The models do show, however, that increasing the capitalization of the economic system in this community is much more likely to produce higher incomes. Without credit, optimizing the pattern of economic activities yields family incomes between $3,300 and $3,600. Increasing the capitalization of these economic systems by $925 yields family incomes between $5,000 and $5,750, more than enough to cover the cost of this credit. Moreover, credit availability increases investment in the more sustainable forest extraction activity, namely, Brazil nut collection that involves women's labor.

Santa Rita Baja. According to survey data applied to the linear programming matrix, the average yearly family income in this community should be about $3,100 (reported yearly income is about $3,400). The linear programming model for the optimization of family economic activities for maximizing income in Santa Rita Baja, in contrast to the model for Monterrey, shows a significant potential for increasing family income, even without the application of credit. Most of the increase from the actual average income to an optimized income of around $6,300 is derived from changes in the production and marketing of cattle (Figure 10-3). Although the optimized model calls for maximizing the amount of cattle to the constraints of available pasture (15 ha), only four head of cattle more than the reported average, the model calls for all but one head to be sold. Families in this community generally sell only three or four head of cattle per year (usually cattle production represents an emergency savings account for families).

The optimization of agricultural activities, as in the noncredit model, calls for producing only enough rice to meet subsistence needs of the family. However, agricultural surpluses destined for the market are generated for both maize and tubers. The availability of male or female labor determines which of these two crops is emphasized. If there is only one economically active female in the household, the model emphasizes maize production. With increased female labor, tuber production is maximized. Timber extraction remains roughly equal among all family

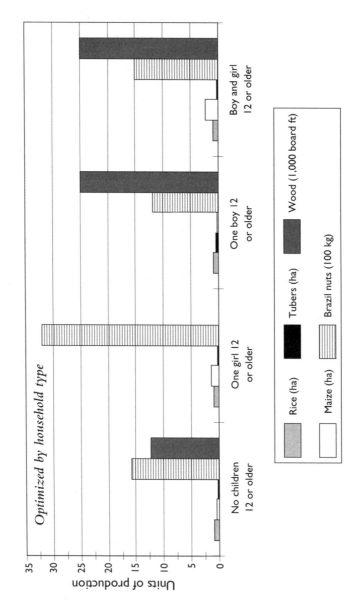

Figure 10-2. *Optimized Economic Production, Assuming Availability of Credit, by Household Type, in Monterrey*

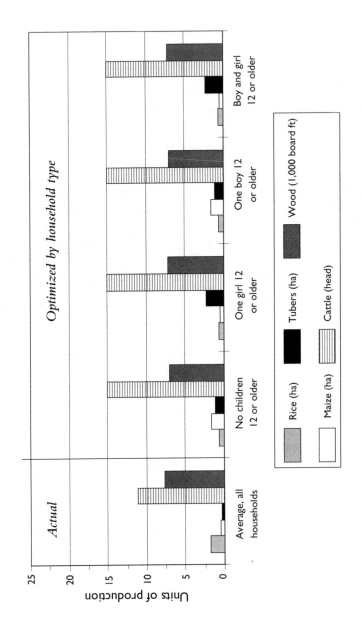

Figure 10-3. *Actual and Optimized Economic Production, by Household Type, in Santa Rita Baja*

situations, with optimized scenarios resembling the actual scenario closely. The amount of labor used and the amount of off-farm employment is also similar in the reported and optimized scenarios.

Optimization of the mix of economic activities in Santa Rita Baja for income maximization is also quite dramatic when these activities are capitalized through credit (Figure 10-4). Since cattle production is constrained to the utilization of only 15 ha of pasture by the model, most of the credit is invested in supporting timber extraction. Agriculture follows the same pattern as the noncredit situation: if only one economically active woman is present in the household, agricultural endeavors will favor maize; if there is extra female labor then it will be invested in tuber agriculture.

As in the noncapitalized case, different optimized family situations in Santa Rita Baja are not very different in their potential to generate income. The range for the optimized credit scenarios in this community is between almost $8,300 and $8,700. Some $2,000 more income is generated by scenarios with the $925 credit.

In both communities, modeling of the survey data by linear programming indicates that, given present economic options and market conditions available to families, timber production is the most efficient use of male labor and capital. The models show that optimization of the existing system would entail expansion of timber extraction and (in Santa Rita Baja) cattle, both activities unfriendly to the environment. Brazil nut collection could be expanded with access to credit, but under the conditions present during the survey, Brazil nut collecting still was an inferior economic choice when compared with timber extraction. Many of the activities available to women for generating or maximizing family income are in some way constrained. Agriculture, for example, is constrained by the practical limit of a family's inability to manage more than 3.5 ha without access to mechanization or large lines of credit. Brazil nut production also depends on male labor to collect the nuts that are then processed by women. Because of these constraints, the models do not show significant differences in optimization for different family situations.

That actual households do not maximize income generation in ways indicated by the linear programming models examined here can be explained along several lines. As mentioned earlier, timber extraction is a risky undertaking that many men do not prefer. The lack of credit also significantly limits activities such as Brazil nut and timber extraction, which require capitalization. In the communities studied here, options for maximization of income depended more on access to credit for expansion of timber extraction and cattle-raising—two undesirable activities from the standpoint of conservation—than to differences in the availability of labor in the household.

The findings suggest, first, that households in the study communities are pursuing the best possible strategies given their current constraints and, second, that viable alternatives to timber and cattle that would have more favorable conservation outcomes have yet to be presented by organizations working with these communities. These kinds of cautions from the results of the linear programming analysis hint at its potential usefulness in projecting household differences and orienting research and extension programs for diverse groups' needs and constraints. Al-

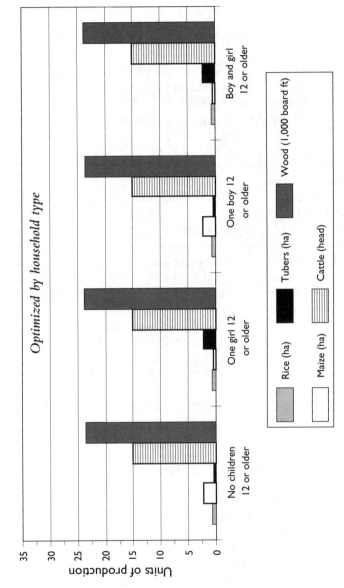

Figure 10-4. *Optimized Economic Production, Assuming Availability of Credit, by Household Type, in Santa Rita Baja*

though in this study, life cycle differences (as measured by availability of adult and child labor) were overshadowed by other factors, such as land and labor constraints, ecological conditions, and availability of credit, linear programming models could be adapted to address a wide range of issues related to intrahousehold diversity.

Agroforestry and Smallholder Livelihoods in Brazil

Just across the border from Bolivia and from the Peruvian department of Madre de Dios lie the extreme western Brazilian states of Acre and Rondônia. As in Madre de Dios, this region of the Brazilian Amazon has a long history of rubber tapping (*Hevea brasiliensis*) and Brazil nut gathering (*Bertholettia excelsa*). Many of the workers from Brazil's northeastern region who were brought into the Amazon to tap rubber during the boom of the late nineteenth century settled permanently in Acre and Rondônia. Their descendants constitute part of the population surveyed in this research project.

As the rubber boom ended in the beginning of the twentieth century with the introduction of cultivated latex from Malaysian plantations, rubber tappers and their families shifted from purely extractive use of natural resources to a combination of slash-and-burn agriculture, hunting, fishing, and extraction of a variety of non-timber forest products. This combination of production strategies, which continues today, resulted in modified gender roles and gendered natural resource use within these households (Campbell 1996b). Since the mid-1980s, the market value of natural rubber has fallen to such a degree that many rubber-tapper households increasingly rely on slash-and-burn agriculture, state subsidy payments, and wage labor for their livelihood (Campbell 1996a). As in the Peruvian communities, the challenge for the households and communities in this region is to find viable alternative production systems that are compatible with environmental goals for the Amazon region.

Some of the households in the research presented here are former rubber tappers who have participated in agroforestry production and marketing since 1989. Participation has required them to modify their resource use from a combination of forest-based extraction and small-scale agriculture on their traditional forest plots of 300–500 ha to a more intensive cropping combination of annual and perennial species on 100-ha lots within a government-sponsored agrarian reform project. Natural resource use and agricultural practices in these households continue to shift significantly in response to credit fluctuations, changes in the market value of their products, and differential access to public services such as education.

Another important change in the political ecology of the western Amazon took place in the mid-1980s with the arrival of colonist farming households from southern Brazil. These families, most of whom were seeking secure land tenure, migrated into the Amazon region as part of the Brazilian government's federal programs to populate the basin and to promote capital-intensive agroindustrial enterprises, such as cattle ranching (see Schmink and Wood 1992). Upon their arrival in Acre and Rondônia, most of these producers initially planted cacao and coffee and attempted to raise cattle.

Some migrant households adapted their production strategy to the forest resources on their 100-ha lot, learning from their Amazonian neighbors how to tap rubber and gather Brazil nuts. Thus, some men and women who had practiced different cropping and livestock techniques in southern Brazil became engaged in full-time forest extraction. These families later shifted into cacao and coffee production once they attained sufficient cash income and cleared enough land for planting. As with the former rubber tappers, colonist farmers found that the production levels and market values of their crops were unable to provide sufficient returns for a sustainable livelihood. This hardship situation gave rise to the RECA project, which is the focus of this research.[5]

The RECA Project

Part of RECA's membership is composed of former rubber tappers who opted to receive a 100-ha lot when their traditional forest areas were divided into smallholder parcels under Brazil's agrarian reform program. The majority of RECA's members are migrant farming families from southern Brazil. When the traditional practices of both groups of producers failed because of poor crop performance, market incompatibility, and lack of capital, they formed an autonomous producers' association named RECA (*Reflorestamento Econômico Consorciado e Adensado,* or Partnership for Economic Reforestation).

After seeking out technical assistance from governmental research and extension agencies for production alternatives, RECA members decided to plant commercially valuable, native fruit trees in an agroforestry consortium with their traditional annual crops of rice, beans, manioc, and corn. The three fruit species they selected were Brazil nut, *cupuaçu (Theobroma grandiflorum)*, and *pupunha* or peach palm *(Bactris gasipaes)*. With assistance from the Catholic Church, RECA was successful in securing funds from various international and national donors for assistance with producer organization, agroforestry production, and post-harvest processing. In 1989–1990, RECA planted 154 ha of the three-species consortium with an initial 81 families participating; each family planted from 0.5 to 3.0 ha of the agroforestry system on their lots. As of 1998, the project had 560 ha planted on the lots of 274 families.

In negotiating outside financing, in-kind matching support in the form of producers' land and labor was an important component of RECA's philosophy and practice. The original concept of RECA relied heavily on communal labor days, organized by the producer groups on each feeder road in the project. At the inception of the RECA project, neighbors and relatives informally exchanged labor days among themselves and gathered for more formal communal workdays when needed for larger efforts, such as repairing bridges or schools. This practice of sharing labor among households and in larger groupings on the feeder roads changed dramatically over the course of the project, however. As the political ecology of the region shifted with the economic crisis in 1994, shifts in household income led some RECA and non-RECA households alike to migrate to cities and to use more hired labor.

A variety of programs provided in-kind support or new credit lines to producers through the RECA association or direct bank loans. RECA households selected which credit programs to participate in based on their analyses of risk, the availability of labor, and the financing conditions of each program; for example, one program promoted a policy geared to address gender and intrahousehold resource allocation in agroforestry, increasing the participation of women in these activities by financing plantings in their names.

The analysis in this chapter compares the household labor allocation of 14 families between 1992 and 1998. Changes over this period illustrate how use of credit and other income sources, labor deployment, and access to public services changed as the project developed, more credit lines became available, and the households themselves matured. The analysis will focus on the linkages between changes in project conditions and in household composition over time, comparing groups from different cultural backgrounds.

Changes in Household Labor Allocation

Examination of household shifts in labor allocation from 1992 to 1998 among the 14 households revealed a diversification of income sources, an increasing reliance on hired labor, and a decreased participation in labor-sharing practices. These shifts were due to a combination of national changes in economic policy, community shifts on communal labor, and on-the-ground changes in labor requirements for the agroforestry plots. Labor requirements rose as the project matured: in 1992, households were involved in three to four weedings per year; by 1998, labor expanded to include weeding, pest control, harvest, and transportation. Several families indicated that they lost part of their 1996–1998 harvests because they were unable to hire labor during fruitfall. Labor shortages arose from four factors—a lack of available labor for hire, limited cash resources with which to contract such laborers, a lack of household labor (due largely to schooling), and a decrease in the number and frequency of *mutirões* or communal workdays.

Communal workdays within the RECA producer groups were a common practice at the outset of the project. Initial surveys in 1992 indicated that most RECA households benefited from roughly one day per month of communal labor on their fields (the equivalent of 10–15 day laborers) on a rotating schedule in their feeder group. In addition, the elected leader of most groups received an additional one to two days of communal labor on his or her fields to compensate for time devoted to project duties. Each feeder road group made decisions on the organization and allocation of communal labor days. By 1994, this communal work was restricted to two days per year, only for the elected leader of the group. The rotating workdays on each producer's lot had been eliminated in most groups. This reduction and shift in communal labor was a common decision among most of the RECA producer groups in the same period. Several informants attributed this shift to the economic crisis generated by a government economic plan (introduced in July 1994), which forced producers to concentrate on their own plots or to seek off-farm wage labor, thus reducing their interest and availability for communal labor work days. Several households indicated that they preferred to share a day or two of labor with neighbors or relatives on an ad hoc basis as needed for specific tasks.

Within the household, the findings showed that women's participation in agroforestry activities declined from 1992 to 1998, as they got older, retired, or migrated to urban areas so their children could go to school. The access to new social payments from government retirement policies allowed older women to substitute their labor with monetary income.[6] In some households, these payments became the major source of cash income, often used to pay hired labor in the agroforestry plots. As the household composition changed over the six years of data collection, and younger children reached school age, several female heads of household moved to urban areas for their children's schooling, thus becoming urban housewives and/or part-time domestic servants. This shift in household residence patterns was a major factor in changing resource management strategies among the RECA producers.

As shown in Figure 10-5, 50% of women interviewed in the sample reported housework and livestock to be their main labor activities in both 1992 and 1998. What changed dramatically over the course of the study was the diversification of labor activities among the remaining 50% of the respondents. Women's participation in agroforestry fieldwork declined dramatically as households relied increasingly on hired labor. Furthermore, women dropped forest-based work (e.g., gathering Brazil nuts) completely from their labor activities over the course of the study.

A new role emerged for several women over the course of the study—that of urban housewives. There also was an increase in women's participation in wage labor, as these newly urbanized housewives sought out the limited employment opportunities available in town, mostly as domestic servants. Despite this rural–to–urban migration, many urban housewives continued to work on the household's agroforestry fields, spending the weekend out on the lot or, in the case of better-off households, working in the fields at key planting or harvest times to supervise hired labor.

Figure 10-6 shows labor allocation trends for the male heads of household over the same period. As with women, men reported a decrease in forest extraction and an increase in wage labor from 1992 to 1998. Nonetheless, there was an important

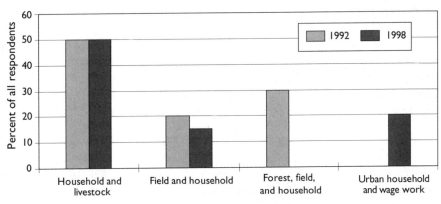

Figure 10-5. *Women's Labor Allocation, by Reported Main Activity, in the RECA Project, 1992 and 1998*

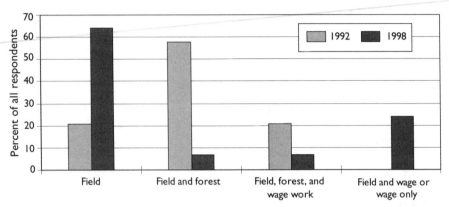

Figure 10-6. *Men's Labor Allocation, by Reported Main Activity, in the RECA Project, 1992 and 1998*

increase in the percentage of male household heads (from 21% to 64%) who reported that they were able to devote all of their time to their agroforestry fields and to livestock production, without having to work as day laborers to bring in cash income. Most of these men reported that they hired day laborers throughout the year to complement the household's own labor inputs.

With the exception of one respondent who held a salaried position with the RECA project in 1998, all of the wage labor reported for other male heads of household in both 1992 and 1998 was from day labor. Most of this work was carried out on the agroforestry plots of RECA neighbors.

Intrahousehold labor shifts were opposite for men and women over the course of the study: women tended to decrease their agroforestry work and increase their wage labor activities, while men tended to increase their field time and decrease their off-farm labor commitments. All households increased their reliance on hired labor. The attraction of urban schooling for children was a major drain on women's agroforestry labor.

Over the course of the study, increased reliance on hired labor and on the male head of household as the primary agroforestry laborer was tied to educational investments. Of those households with school-age children in 1992, none had children living in town to attend school. By 1998, all of the households originally from the Amazon region had at least one child in an urban residence for access to school, and in half of these households the women had moved to town with their children. This investment in better education for the next generation left the household with severely restricted labor resources.

These educational investments and shifts in labor allocation altered the households' resource management strategies over the course of this study. Several households (particularly Amazonian households that had moved to town and had insufficient on-farm labor) indicated that they decided against taking out additional financing because they lacked the labor resources to handle more perennial crops, and they were uncomfortable with the prospect of relying on hired labor.

In general, women who took on debt tended to be younger and also to have more formal education than those who did not. Perhaps more important, these women also had some source of cash income. Almost all women who took out a loan did so knowing that they would have to rely on hired labor for the planting, weeding, and harvesting of their plot.

Discussion and Future Directions

These case study examples have experimented with two approaches—linear programming models, and longitudinal ethnographic and survey data—that show promise for understanding the complex dynamics of production decisions among households of different kinds. The combination of the two types of information provides a useful basis for efforts to improve conservation and development projects with rural communities (Kaya et al. 2000). The findings of our study, their implications, and recommendations for forest management are summarized in Table 10-5.

In the Peruvian case, access to credit was shown to have a potentially significant impact on households, but not necessarily in the desired direction away from timber harvesting. These findings suggest that efforts to substitute other activities for logging (e.g., Brazil nut collecting and processing) are unlikely to be successful under current market conditions, unless labor constraints are specifically targeted in credit and technical assistance programs.

Results from the Brazil study also show that the degree of household investment in agroforestry plantings depends in part on variables—such as the age and composition of the household, and cultural origins—that affect labor availability and the propensity for households to participate in credit schemes. Colonists of Amazonian origins were less likely to take advantage of credit for agroforestry. Similarly, the women-focused credit program in the RECA project provided benefits to a particular group of women within the sample, namely, those who were non-Amazonian, younger, and better educated and who had a complementary source of cash income. These findings suggest the need to involve people directly in the design of credit programs appropriate to distinct client groups within rural communities. A better understanding of these intracommunity and intrahousehold differences—including further linear programming analysis and longitudinal studies—could help to spread benefits more widely. In particular, programs could be designed to address the needs of younger women, those with less market experience, and families with school-age children.

The linear programming approach used in the Peruvian study provides a useful tool for projecting the realistic possibilities for improvements in production systems and the impact they will have on households of different types. Rather than working with "average" households, this method specifically permits analysis of different client groups by introducing different goals and constraints into a real production model. In the Peruvian case, the models focused on two variables: availability of children's labor and access to credit. The results showed little variation among household production systems with different degrees of access to children's labor. Among these households, other life cycle variables, such as presence of

Table 10-5. *Household Diversity and Recommendations for Forest Management Projects*

Research findings	Implications for forest management	Recommendations
Environmentally damaging activities (timber and cattle) may be the most economically profitable.	Economic incentives promote depletion of high-value species.	Alternatives require further research and creation of credit and technical assistance programs tailored to diversity.
Credit is required for women's economic activities.	Greater access to credit can stimulate non-timber forest management.	Special credit programs for forest management are needed, especially for women.
Linear programming shows significant variance in real systems.	Linear programming allows projection of returns by gender for alternative practices.	Diagnostic tools need to be sensitive to intrahousehold variance.
Significant shifts in labor and natural resource management take place over time among families.	Forest management plays different roles for different households at different times in the life cycle.	Natural resource management programs need to adapt to the diversity of household needs and changes over time.

young children who require much of their mother's labor time, may prove to be more significant constraints to production shifts.

The longitudinal data from Brazil showed that the importance placed on secondary schooling for children age 12 and older places specific constraints on the ability of households to mobilize the labor of these household members, especially at certain times of the year. This constraint suggests the need to link species selection for agroforestry projects with different life-cycle stages of the households. Fruit crops that ripen during school vacation time, for example, could allow households to keep their children in school while still taking advantage of their labor during harvest seasons. More broadly, future agroforestry projects for communities and households of similar diversity could allow participants to select from alternative packages of financing and combinations of species, tailored to their particular needs. Likewise, credit programs could be more effective in stimulating sustainable and equitable shifts in production by providing options designed for different kinds of household systems.

The longitudinal and qualitative information from the Brazil case study reveals the rich nuances of intrahousehold dynamics over time, many of which could be used to inform further analysis using linear programming and to provide specific recommendations for project design and implementation. The study revealed that, as agroforestry systems matured, households were able to move away from traditional production systems and women shifted their labor, and even their residence, away from rural production. Despite efforts to involve women in agroforestry by providing them special credit lines, men became more, not less predominant in that activity over time. With new forms of cash income available, women and children chose to leave rural settlements for educational and wage opportunities in cities. Linear programming models that incorporate these variables, including more complex analysis of labor by age, gender, and season, as well as changes in income streams and credit sources, could help to inform the design and implementation of agroforestry projects, so that they address the diverse and changing needs of households and social groups.

The viability of small-farmer households in the Amazon is intricately tied to the region's complex policy and economic environment. Experiments with alternative production systems more compatible with conservation goals for the region will require greater attention to the impact of policy and market shifts. Those shifts affect the viability of forestry, agroforestry, agriculture, and other production options for distinct groups within rural communities, and they change the distribution of benefits and incentives for adopting alternative production systems through time, across households, and among social categories, including gender. The approaches presented in this analysis suggest some relatively low-cost methodological tools that, when combined with other kinds of technical information, can help to analyze the impact of changes at the household level and assist in the design and implementation of more effective forest management projects.

Notes

The authors wish to acknowledge the generous support for this research from the North-South Center, the International Food Policy Research Center, the John D. and Catherine T. MacArthur Foundation, and the University of Florida.

1. The research methodology in the two regions combined qualitative and participatory methods with quantitative techniques. Systematic methods included random household surveys, time allocation studies, analysis of income streams, and linear programming. Qualitative techniques included interviews, participatory mapping, oral histories, and focus groups (Slocum et al. 1995). In this chapter, we report results from only a portion of the information gathered in the two sites. These sites are not ACM main sites.

2. In early 1999, four anthropology students from the Catholic University (Lima) participated in applying a follow-up survey to the test survey carried out in 1998. The sample within the communities was chosen using a table of random numbers applied to the community census in order to ensure that different groups were represented. We aimed to interview a high proportion of households within each community (29 out of 32 in Santa Rita Baja; 41 out of 55 in Monterrey) to avoid bias due to small overall numbers. In the discussion that follows, we are mainly using the results of the follow-up survey, and we make a few references to the 1998 survey when there are differences worth highlighting.

3. To extract timber in Madre de Dios in areas less than 1,000 ha, an individual must present a timber management proposal to obtain a forestry contract or permit or must request a permit from the Ministry of Agriculture office to extract timber within his/her agricultural land. Most small farmers obtained annual permits to cut valuable trees within their own landholdings, because this timber was a convenient source of cash, and they did not have to present a management plan. Only eight of the more capitalized farmers in Santa Rita Baja and Monterrey had multiyear contracts to extract timber. Similarly, to gather Brazil nuts or grow crops an individual must obtain a permit from the local Ministry of Agriculture office. Thus, for the same piece of land, different individuals might hold the forest concession, the Brazil nut concession, and the agricultural permit. To complicate matters, yet another person might have a mining concession over the land.

4. Linear programming is a mathematical method for determining an optimal combination of inputs and economic activities to maximize or minimize an objective in which the inputs are

subject to constraint. It is often used in agricultural economics to identify cropping patterns that will maximize income (Gittinger 1982). Linear programming can offer recommendations for specific farms but must take into account the particular conditions of those farms. The method is less useful for generalizing recommendations based on average values for production factors spread over wider areas (Hildebrand and others 1998). Thus, the following exercise is not meant to be a basis for recommendations, but a tool for examining the role of household labor patterns, the optimality of timber for household income generation and the effect of credit on household economies. The optimization of family economic activities through this model depends on market conditions. These conditions fluctuate, and optimization patterns offered by the models would change accordingly.

5. The results presented here focus on data gathered from 14 households, which were among the first to join RECA in 1989, its initial year of operation, and the evolution of their participation in the RECA project over the subsequent nine years. The 14 households vary in household age and size composition, degree of participation in the project, cultural origin (colonists versus former rubber tappers) and transportation conditions (distance from trafficable roads and urban areas). Project maps, membership rolls, and RECA staff were consulted in determining which feeder roads to select and, along those, which households would comprise the sample. Data presented in this section focus on comparisons between the results of two formal questionnaires carried out in 1992 and 1998. In 1992, the 14 households had three years of experience with the RECA project. By 1998, they already had mature systems producing fruit for market.

6. Over the course of this study, legislation was enacted that granted agricultural pension payments equivalent to a monthly minimum salary to women age 55 or older and to men age 60 or older. In July 1998, one minimum salary was Reals $130 per month. Collecting one's pension required a trip to Acre's capital city, Rio Branco, which cost Reals $20 round trip (US$1=$1.13 Real).

CHAPTER ELEVEN

Improving Collaboration between Outsiders and Communities in the Amazon

Benno Pokorny, Guilhermina Cayres,
and Westphalen Nuñes

P ROFESSIONAL RESEARCHERS AND FACILITATORS are powerful
actors in initiatives for local development. Even in participatory approaches,
they tend to dominate communication and decisionmaking processes. This chapter, reporting our experiences with communities in the eastern Amazon of Brazil,
focuses on how professional researchers and facilitators can better recognize and
manage the power differentials that shape their own interactions.

The success of development activities aimed at the alleviation of poverty and
increased social equity in the rural areas of developing countries, both principal
goals of national and international development agencies and research organization (CIFOR 2001; Skarwan 2002), depends on the active participation of local
actors and the consideration of local knowledge and structures (Arnold 1991; Cernea
1991; Hobley 1996). Various studies have stressed the importance of participatory
learning processes and effective strategies to give more control to local communities (Chambers 1997; AGRITEX 1998), and various tools have been developed for
extension agents and researchers to adapt participatory processes to local conditions (Carter 1996; Cerqueira 1997; Rees 1998; Ingles et al. 1999; Brito 1999).

But in practice most community work still shows strong deficits regarding the
aspect of local participation, because of the shortage of time, the strong pressure for
results, the inadequate training of community workers and researchers, as well as
the lack of adequate methods. We suspect that in a lot of cases local values and
views are not sufficiently respected. Usually project objectives—whether equitable
development, gender balance, or sustainable forest management, for example—are
defined by governmental and non-governmental institutions and donors and not
by the communities. Believing that everybody is sharing these "good" goals, project
leaders often fail to explain and discuss their approaches adequately with the local
actors. Thus, from the beginning, there is no local ownership of these ideas. Even

Table 11-1. *Characteristics of the Communities in Three Districts of the Eastern Amazon*

Characteristic	District where the community is located		
	Muaná	Tailândia	Gurupá
Settlement type	Traditional, riverside communities in the Brazilian Amazon (ribeirinhos)	Illegal settlers in the frontier area, often with only 1–5 years in residence (colonos)	In more remote riverside areas, descendants of African slave refugees (remanescentes de quilombos)
Biophysical environment	Less degraded floodplain forests, with some farms nearby	Strongly exploited forests	Less degraded floodplain forests
Main economic activities	Subsistence, mainly fishing and extraction of forest products, in particular Açaí palm (Euterpe oleracea Mart.)	Mainly farming of rice and manioc production (farinha)	Some farming of manioc (farinha) and subsistence, mainly fishing and extraction of forest products
Living conditions	Low level of education, low family income, no schools or medical facilities, rampant malaria, difficult access to local markets		

when profound discussions take place, leaders may be at risk of manipulating local communities by using participatory methods or promises of benefits to ensure collaboration. Only a narrow line separates sensitization and facilitation, on the one hand, and manipulation and exploitation of communities, on the other. This risk is especially relevant for researchers, if applying participatory methods mainly to extract information for short-term academic purposes rather than long-term community empowerment.

The lack of respect for local values and interests can undermine development and research objectives and damage longstanding cultural structures by establishing dependence on outsiders. In this chapter, we present findings from our work on how to promote "real" participation and collaboration between outsiders and communities. The reported experiences have been worked out with families and communities that are representative of some of the wide variety of cultural, biophysical, and economic characteristics of the rural population in the eastern Amazon (Table 11-1).

This chapter has five main sections. First we discuss the usefulness of criteria and indicators (C&I) as a means for perceiving and evaluating others' views and values. We then discuss the empowering effects of some of the participatory methods we tested. The third section focuses on an evaluation of the collaborative potential of those methods, particularly on the issue of how to keep researchers from dominating the process. In the fourth section, we analyze participatory processes in relation to women. Finally, we discuss the local view on intracommunity heterogeneity and its implications for effective adaptive collaborative management.

Defining Criteria and Indicators

Community forestry projects can potentially alter the local communities' entire system of livelihood strategies and social structures (see Chapter 4). Rarely is sufficient attention paid to anticipating the possible impact of projects on complex, local systems. As a result, many indirect effects are not even considered when methodological approaches are determined, monitoring systems are put in place, and projects are evaluated.

Because so many different actors are involved, project planning and implementation reflect a conglomerate of underlying values and interests. Also community workers and researchers have individual interests, values, and expectations, which are seldom properly articulated. The result is a lack of self-consciousness and transparency, which can increase the already existing communication barriers with the project partners and may provoke misunderstandings and frustrations on all sides. Additionally, communities, because of their precarious economic situations, tend to accept external support without fully understanding the implications. If the articulation among the partners is incomplete in the initial project phase, the viability of the project as a whole can be drastically diminished.

A first important step to diminish potential problems caused by lack of transparency is for the project team itself to articulate its expectations and define the possible indirect effects of the planned activities. We have found the definition of project-specific C&I useful. C&I are tools to deliver information required to conceptualize, evaluate, and implement sustainability (Prabhu et al. 1998). They denote a hierarchy of linked items (principles, criteria, indicators, and verifiers), where the information accumulated at lower, more concrete levels is used to assess the related items of the upper, more abstract levels (Prabhu et al. 1999). By accepting sustainability as a leading principle, participants can explore long-term impacts in their discussions.

We used this C&I strategy while searching for a way to replace the externally driven background studies (a planned component of CIFOR's adaptive collaborative management program) with, instead, a diagnostic survey carried out by the communities themselves supported by researchers (Pokorny et al. 2003a). As part of this approach, the team—five researchers and students of forest science, sociology, geography, linguistics, and agricultural science—defined a C&I set built on the institutional and individual expectations of team members in relation to sustainability (Pokorny et al. 2003b). Following the logic of the ACM approach, the team defined three categories of sustainability (Figure 11-1). The first category described all issues related to social processes, like quality of communication and collaboration. The second category considered the operational aspects of natural resource management, including planning, monitoring, and adjustment. The third category incorporated the expected results of adaptive collaborative management of natural resources, which at the same time reflects the conditions under which management takes place.

Our experiences with this exercise were positive. It was possible to discuss individual expectations as well as the probable implications of project activities. Through the discussion process, the whole team developed a better understanding of the

Figure 11-1. *Schema of Sustainability, as Defined within the Approach of ACM*

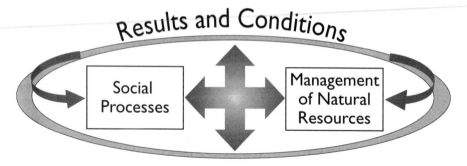

Note: In this schema, sustainability depends on (1) social processes, such as quality of communication and collaboration; (2) operational aspects of natural resource management, including planning, monitoring, and adjustment; and (3) expected results of and conditions for collaborative and adaptive management of natural resources.

project's theoretical background and a more transparent view of its objectives. Additionally, the C&I set facilitated our communication and negotiation with the project partners and served as a base for defining work plans and methods. Finally, we expected the C&I set to be useful over time to monitor the project and evaluate the outcomes.

Our initial idea to support communities in assessing the C&I set, however, was unachievable. For the communities the evaluation of an entire set of C&I was too complex, abstract, and time consuming. Thus, the definition of C&I was more important as a preparatory step for our project and the working team. The exercise helped us to develop a more integral understanding of our planned activities. Most important, the discussion of our own viewpoints and biases started with a critical evaluation of our observations and information, as well as the identification and acceptance of (differing) opinions and values from the communities. Thus, it was a first step to avoid manipulation in the communities.

Making Participatory Methods More Participatory

A wide variety of participatory methods have been developed as tools to support researchers and extension agents in their attempt to empower local actors to find their own ways toward sustainable development. But in practice, the application of these methods is often disrupted by the effort to achieve externally defined project outputs. Against this background, we wanted a clearer picture of what kind of empowering effects might be expected from the application of participatory methods (cf. the empowering results of participatory methods reported in Chapter 4).

We first categorized a wide variety of possible empowerment effects and evaluated how different participatory methods could potentially contribute to them.

Table 11-2. *Potential Empowering Effects from Participatory Methods, by Empowerment Categories*

Empowering category	Empowering effect
Organization	Planning and distributing tasks, learning how to make decisions
Confidence building	Gaining self-confidence, respecting the group, respecting local knowledge, gaining technical training, guaranteeing ownership
Learning processes	Enhancing creativity, getting access to new ideas, learning to learn, finding stimulation for own initiatives
Sensitization	Critical perception of one's own environment, self-positioning, reflection
Communication	Information exchange, articulation of one's own opinions, constructive discussion

Participatory methods obviously affect the ability of local people to articulate and organize themselves in various ways. According to their specific way of contributing to empowerment (e.g., by influencing social structures, management processes, physical results, etc.), these effects were organized into five empowerment categories (Table 11-2).

In the discussion of empowerment effects, we realized the tendency to overemphasize the importance of direct, visible, and measurable results of participatory methods, such as the drafting of maps, the undertaking of certain activities, or the creation of associations. Instead, more attention has to be given to stimulating the indirect effects of participatory methods.

To evaluate the potential contribution of different participatory methods to the empowerment effects, we analyzed and tested 42 participatory methods commonly used in the field (Table 11-3) (Colfer et al. 1999a; Colfer et al. 1999b; Carter 1996; Ingles et al. 1999). After the field testing, each team member independently judged the empowering potential of each method by ranking the potential effect from 0 (no contribution) to 10 (very strong contribution). When strong discrepancies arose among the individual evaluations, the team discussed and reached a consensus score. Figure 11-2 shows the results of this process, in percentages of methods strongly contributing to each of the five defined categories of empowerment effects. Methods with a mean ranking above 8 were considered to be strongly contributing and therefore were included here. This exercise revealed that most of the tested methods affected more than one category of empowerment effects. The tested methods contributed especially to confidence-building and communication, but relatively little to organizational issues.

Through the evaluation process, the team increased their awareness about the possibilities and aims of participatory methods. This distinct, shared understanding about participation helped to reduce the team's attention to direct visible products of the participatory processes. Consequently, we worked more intensively on processes to promote empowerment and local initiatives. The exercise also increased understanding of the effects to be expected from different participatory methods, so that we were able to more adequately select participatory methods for the field work.

Table 11-3. *Participatory Methods, by the Principal Purpose of Application*

Principal purpose	Participatory methods
Acquaintance and confidence building	Methods and activities that aim to get actors to know each other and to create trust among them: • team presentation; • presentation of participants; • games; • interviews for presentation; • accompanying daily activities; • working with children.
Presentation of contents or results	Methods with the principal aim of presenting the knowledge of participants: • presentation of group results to plenary; • theater.
Packages of methods	Methods consisting of a complex sequence of methodological elements: • future workshop; • future drawing; • ideal community; • Objective Oriented Project Planning (ZOPP); • problems versus solutions/What to do?
Methodological elements	Elements that could be applied in different methods: • focus groups; • smaller working groups; • meeting group interview; • election of representatives; • drawing; • simulation; • preparation of material.
Extraction of information	Methods that primarily aim to extract information: • formal interview (structured); • informal interview (semi-structured); • informal conversation; • accompanied field trips and visits; • interview guidelines; • observation.
Elaboration of information	Methods that primarily aim to elaborate information together with the community: • brainstorming; • spider net; • individual participatory mapping; • collective participatory mapping; • mapping of boundaries; • mapping of resources; • historical-ecological matrix; • paper tree; • timeline; • Venn diagram.
Discussion	Methods that primarily aim to stimulate discussion among the actors: • discussion in plenary; • sociodrama.
Evaluation	Methods that primarily aim to evaluate specific questions: • informal daily evaluation; • distribution of points; • ranking; • systematization.

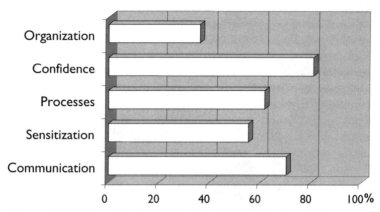

Figure 11-2. *Contribution of Participatory Methods to Five Categories of Empowerment*

Note: Percentages are based on individual rankings by the research team for 42 participatory methods tested in the field.

The Collaborative Potential of Participatory Methods

One of the most problematic aspects of applying participatory methods arises from the tendency of outsiders, because of their formal qualifications and perceived knowledge, to dominate the project's communication, content, working rhythm, and results. This dominance diminishes the local communities' ownership of the results and increases the danger of creating a relationship of dependency (Freire 1980; Cernea 1991; Saywell and Cotton 1999). To overcome this tendency we investigated the possibility of letting the communities themselves apply participatory methods. In this case, local actors by taking on the role of facilitators could become owners of the process and increase their influence on the contents and working rhythm. To evaluate this possibility, we reanalyzed the participatory methods tested in the field in relation to: (a) the complexity of the method, (b) the intellectual capacity needed by the facilitator to systematically understand and process the information, and (c) the level of guidance needed to keep the group dynamic going.

Figure 11-3 shows for all three aspects that only a relatively small percentage of the tested methods presented characteristics that enable an easy transfer of leadership to community actors. Most of the methods were complex, needed a high degree of educational experience and facility with abstraction, and a high level of guidance. This made specific training of facilitators necessary.

In a next step, we tested the possibility of training local actors as facilitators. The training was provided in small groups of interested community representatives. The lessons began with a short presentation and discussion of the principles of participatory methods. Then, we explained and demonstrated some selected methods. Finally, the group applied the methods themselves. Afterwards, the experiences were intensively discussed and evaluated. Based on the critiques and recommenda-

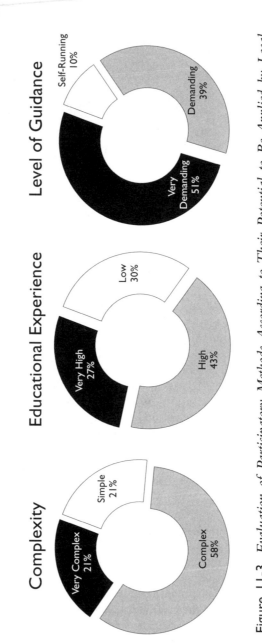

Complexity

Very Complex 21%
Simple 21%
Complex 58%

Educational Experience

Very High 27%
Low 30%
High 43%

Level of Guidance

Self-Running 10%
Demanding 39%
Very Demanding 51%

Figure 11-3. *Evaluation of Participatory Methods According to Their Potential to Be Applied by Local Facilitators*

Note: Percentages indicate the share of 42 participatory methods falling into each of three levels of difficulty for the facilitator, according to (1) the complexity of the method, (2) the intellectual capacity needed by the facilitator to systematically understand and process the gathered information, and (3) the level of guidance needed to keep the group dynamic going.

tions, the group repeated the application and evaluated the progress. Finally, the trained community representatives applied the methods successfully in a plenary with other community members.

Based on the experiences gained in this training, the team became convinced that local actors, if adequately trained, could take over the facilitator role for nearly all methods. To make use of this enormous opportunity to diminish external dominance and enhance local ownership of participatory processes, effective training strategies need to be developed.

Participatory Processes and Women

Studies have confirmed the fundamental importance of including all local groups, particularly the marginal and depressed ones, in participatory processes to ensure that decisions, resources, and benefits are shared broadly, rather than accruing only to local elites (Freire 1981; Ostrom 1990; Chambers 1997; Colfer et al. 1999a). Nonetheless, many community projects have failed to achieve this fundamental precondition for successful development, because of insensitivity, poor logistics, inadequate communication, or simple idleness (Chambers 1997).

Women, in particular, are often neglected in participatory approaches. In the Amazon region, as in most rural communities, women contribute significantly to the subsistence of the family. In addition to their domestic tasks, women participate in the fieldwork, extract non-timber forest products, care for livestock, and tend home gardens (Meggers 1977; Santos 1982; Simonian 1995; Castro and Abromovay 1997). In contrast to men, who stop working when they arrive home from the fields, hunting, or fishing, women continue working on domestic duties, such as cooking and cleaning. As a consequence women work longer and have significantly fewer rest intervals than men (Maués 1993; Wolff 1998).

Men usually represent the family in meetings and collective processes; women, during men's absences, have to maintain the domestic and extra-domestic activities (Campbell 1996b). As Cayres (1999) observed in an NGO-driven development project located in the traditional community of Nazaré in Paragominas near Belém, this arrangement had drastic effects on the acceptance of new initiatives. There the men participated intensively in a long-term project to identify and realize new income opportunities, while the women became overloaded with maintaining traditional subsistence activities (cf. Chapters 13 and 14). Concerned by their increased workloads, the women started to agitate against the project, although they had initially agreed in general terms to its contents and objectives.

For the same project, Cayres and Costa (2000) documented another typical impact of insufficient participation by women. The authors described the example of a training course about using hoes instead of the traditional but less efficient *ferro*-equipment to eliminate weeds for field preparation. As "official" representatives of the family, the men participated in the training courses, although weeding is traditionally one of women's responsibilities. The women, because they had not participated in the training, continued to use the traditional technique, without considering the potential for improvement.

These examples are not exceptions. Women's presence in community forestry groups and executive committees is typically low. Women are often disadvantaged in the distribution of costs and benefits of collective actions, which diminishes the viability, effectiveness, and sustainability of community forestry projects. There is a wide consensus that more active participation of women is important, not only because women's involvement promotes equity, but also because women's specific knowledge, facilities, and competences contribute to the efficiency and sustainability of collective initiatives (Kainer and Duryea 1992; Souza 1997; Agarwal 2000).

However, attempts to increase women's participation have to deal carefully with the existing local culture, so as not to destroy traditional systems that ensure social and economic stability. If outsiders remain ignorant of existing culture, they are in greater danger of manipulating local actors to accept social norms and cultural arrangements to fit outsiders' imaginations (see Chapters 8 and 12). It is fundamental to adequately understand the existing social structures and their advantages and disadvantages before any intervention.

Local Views on Heterogeneity

Outsiders conducting community forestry research often project their own understanding onto communities, taking for granted that most locals have the same values for social equity and economic development. These researchers, unlike classical social anthropologists who aim to understand local value systems, tend to neglect the existence of local values.

Below we provide an example of the differences between locals' and outsiders' perceptions of heterogeneity. The example deals with a participatory approach to identify and discuss the heterogeneity of groups developed in collaboration with three communities located in the districts of Muaná and Tailândia mentioned in Table 11-1 (Pokorny et al. 2003c). The method used a relatively abstract level of communication by focusing on the identification of types, in which the different participants placed themselves. This avoidance of individual characterizations prevented personal debates and supported constructive criticism. To diminish difficulties caused by deficiencies in reading and writing, the group contributions were documented by a symbolic language.

As a first step, the groups were asked to define criteria relevant to differences in thinking, doing, or evaluating. To facilitate this exercise a simplifying poster of a community was presented showing rivers, church, soccer field, grassland, agricultural land, forest, and people of different sex, age, and social status. The groups were asked to compare the model with their own community and imagine differences and similarities between the people shown in the poster and themselves.

All groups crystallized their understanding of heterogeneity on personal characteristics. A discussion about more abstract units, such as woman and man, young and old, rich and poor, was unsuccessful, because the group's vision of reality was strongly related to the characteristics of the specific persons. In their views, the basic unit for typology was the family, as a summarized result of the attributes of all family members. Special importance was given to the person who represented the family in the specific group, independent from his or her gender.

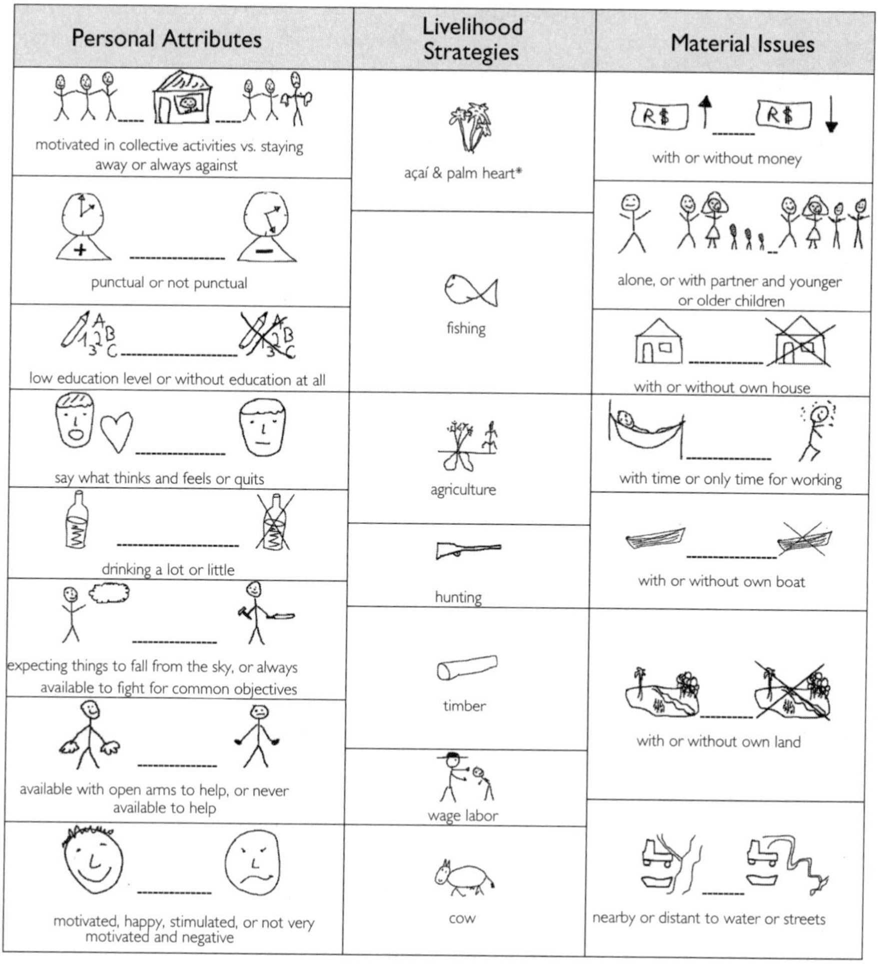

Personal Attributes	Livelihood Strategies	Material Issues
motivated in collective activities vs. staying away or always against	açaí & palm heart*	with or without money
punctual or not punctual	fishing	alone, or with partner and younger or older children
low education level or without education at all		with or without own house
say what thinks and feels or quits	agriculture	with time or only time for working
drinking a lot or little	hunting	with or without own boat
expecting things to fall from the sky, or always available to fight for common objectives	timber	with or without own land
available with open arms to help, or never available to help	wage labor	
motivated, happy, stimulated, or not very motivated and negative	cow	nearby or distant to water or streets

Figure 11-4. *Symbolic Representations Used to Discuss Values in Local Communities*

Note: These symbolic representations were developed by three local community groups in the Eastern Amazon (districts of Muaná and Tailândia) to aid in discussions that identified criteria relevant to differences in thinking, doing, or evaluating, structured into three categories: (1) personal attributes, describing individual characteristics of the group member, representing his/her family; (2) livelihood strategies, expressing the conditions and economic activities of a family; and (3) material issues, describing the issues related to wealth and property.

The development and discussion of the heterogeneity criteria was lively. The groups used the opportunity to articulate, apparently for the first time, difficult topics such as alcohol or negative social behavior. Figure 11-4 shows the most important criteria relevant to differences in thinking, doing, or evaluating, as articulated by the three community groups. For each criterion a symbol was designed and drawn on a flip chart. Because all three community groups defined similar criteria, the results are not presented separately. The heterogeneity criteria were structured into three categories: (a) personal attributes, describing individual characteristic of the group member representing his or her family; (b) livelihood strategies, expressing the conditions and economic activities of a family; and (c) material issues, describing the issues related to wealth and property.

In all three community groups, the category of personal attributes played a key role. Two dimensions of personal attributes were identified: a descriptive dimension, including attributes like happy, punctual, and problems with alcohol, and another one expressing the way a family and its representatives are participating and contributing to community life, focusing on negative social behavior.

In relation to the criteria related to personal attributes, the categories of livelihood strategies and material issues were given relatively low importance. Many participants had difficulties accepting money and property differences as relevant. In two groups, intensive discussion took place about this issue.

The criteria defined by the groups differ from outsiders' perceptions of diversity. Gender, age, and power issues, generally emphasized by outsiders, do not have the same priority for locals. This discrepancy can provoke miscommunication and conflicts between outsiders and locals, as well as inside the community. This example indicates the need to pay more attention to local values in community forestry.

Conclusions

Like many other rural development projects that use participatory methodologies, community forest projects face the dilemma of choosing between the wish to achieve outsider-defined results and the need to respect local values, demands, and working rhythms and to actively involve local actors in decisionmaking processes. Outsiders, who have formal qualifications and own the project resources, risk dominating locals in the definition of objectives, the selection of methods, the definition of contents, the rhythm of work, and consequently the results. But, it is possible to diminish this risk, if outsiders become aware of it.

Knowing one's own point of view is a precondition for perceiving and evaluating others' views and values. Thus, researchers and extension agents, before starting their work in communities, need to become aware of their own expectations, interests, and perceptions. We found it useful to define a set of C&I that revealed our own values in the initial phase of the project. We also considered it helpful to develop a shared set of objectives related to the application of participatory methods. This exercise increased understanding about the possibilities and aims of participatory methods and the importance of processes to promote empowerment, which may not always generate direct, visible outcomes.

One key strategy identified to diminish outsiders' dominance was the training of local actors in applying participatory methods themselves. This strategy increases local influence on the content and rhythm of work and supports local ownership. We think that most of the participatory methods are transferable to interested locals by adequate training.

Special attention has to be given to incorporate women in participatory processes, because women's competences and knowledge are often indispensable for success. Nonetheless, outsiders need to be aware that such intervention affects existing social structures and may contrast with local views and values. To avoid conflicts and social instability, projects that implement externally defined objectives have to be based on a profound analysis of the existing social arrangements (see also Chapter 12) and the local willingness to change them.

To make communities do what outsiders want must not be the objective of community projects. Rather, projects should limit their contribution to supporting locals in developing and realizing their own initiatives. It seems worth concentrating more on the question of how outsiders can participate constructively in local processes rather than continuing to ask how the local actors can participate in our projects.

CHAPTER TWELVE

Diversity in Living Gender
Two Cases from the Brazilian Amazon

Noemi Miyasaka Porro and Samantha Stone

T HIS CHAPTER CONTRASTS TWO SITES in the Legal Brazilian Amazon[1] to discuss how gender issues have been addressed in different types of
community-based forest management. Given the spatial distribution of deforestation in the Amazon, an "arc of deforestation" that advances northward and west-
ward (see map in Wood 2002, 5), we selected an Extractive Settlement Project
(PAE)[2] in the westernmost state of Acre as our forest-rich site and a conventional
Settlement Project (PA)[3] in the easternmost state of Maranhão as a forest-poor site.[4]

In forest-rich sites of the Brazilian Amazon, increasing rates of deforestation,
beginning in the 1970s as a result of extensive cattle ranching and national and
foreign consumption of Amazonian wood, finally prompted the government in
the 1990s to look for alternative policy instruments for improving forest management (Cunha dos Santos and Salgado, undated). Community-based forest management for the sustained production of commercial timber was one of the principal
means identified to reduce deforestation (Amaral and Amaral Neto 2000). Since
1993, community-based timber management projects have been initiated in the
Brazilian Amazon in national forests, extractive reserves, and agricultural colonization areas, involving a diverse array of organizations including federal and state
government agencies, nongovernmental organizations (NGOs), and community
associations.

Access to and control over commercial timber, one of the economically most
valuable resources in the country, historically has been restricted to those with
political and economic power (Bunker 1985; Hecht and Cockburn 1990). Community-based timber projects represent an effort to reverse this situation by transferring some government forest management decisionmaking and administrative
power to local people and by providing them with a means to improve their
standard of living through the production of an economically valuable product
(Amaral and Amaral Neto 2000). The failure to address gender issues in these

projects, however, has resulted in men having almost exclusive access to and control over the management and selling of the timber, because harvesting is considered a male activity. Women, as a result, have been systematically marginalized in these efforts, and gender-based inequities have been sustained, if not exacerbated.

By contrast, in forest-poor sites in the eastern Amazon, especially in the state of Maranhão, deforestation began much earlier, with the introduction of cattle, cotton, and sugar plantations beginning in the colonial seventeenth century (Droulers and Maury 1980). In continuously slashed areas, as a "subsidy of nature" (Anderson et al. 1991), forests of babaçu palms (*Attalea speciosa*) emerged as a second chance offered by nature: a highly homogeneous, but very stable, secondary growth. In the current local peasant economy, men direct agricultural activities in which men, women, and children cultivate mostly rice, beans, cassava, and corn. Women, named *quebradeiras de coco babaçu* (women who gather and break open fruits of the babaçu palm fruits), direct extractive activities carried out mostly by women and children (Porro 2002).

Kernels extracted from the fruits of these palms provide local forest dwellers with oil for domestic consumption and cash, as well as providing raw material for industries of edible oils and margarines, soaps, and cosmetics. This nontimber resource, exported since the 1910s, was an eagerly sought product in the international market, especially from the 1950s to the 1970s (May 1990; Almeida 1995). Maranhão was also the first state to host the so-called expansion frontier (Velho 1972) and related development policies provoking agrarian conflicts, deforestation, and political clashes between indigenous and peasant and capitalist economies (Schmink and Wood 1992). As a result, the babaçu economy began to decline in the 1980s, and in the 1990s it was further penalized by the entrance of competitive oils in the Brazilian market because of the so-called economic liberalization (Almeida 2001). In this context, deforestation of babaçu palms increased.

In contrast to an increasing interest in promoting low-impact timber harvesting in western areas of the Amazon, the national and international awareness that came with denouncing the destruction of palm forests did not result in a search for alternative policies to protect this nontimber resource. Rather, current investments in the development of the state of Maranhão, now focused on the goal of alleviating poverty, continued to minimize the importance of babaçu forests as an environmental issue. The investments in environmental conservation are mostly directed toward ecotourism and parks. Community-based initiatives targeting conservation of the remaining forests and babaçu palm forests are not the object of any major policy or substantial governmental and nongovernmental investments.[5]

In keeping with local conceptualizations of gender defining the division of labor, women are the major protagonists of both this extractive activity and the political mobilizations for the protection of the babaçu forests. In this context, community-based forest management initiatives have been directed, or have had great participation, by women. In spite of some success, however, the extractive activities involving these nontimber resources continue to be regarded as a female domain and relegated to a "second sex" type of sustainable development. Gender fairness is still far from reach.

Despite the many significant differences between the forest-poor and forest-rich sites, we observe in both sites equally that initiatives in non-timber forest resource management have failed to truly improve men and women's lives, because of a lack of substantial funding and policy support. Although initiatives in timber management have had greater support, most have been slow to improve gender equality, because they have not included women in decisionmaking and activities (cf. Chapters 13 and 14).

These contrasting aspects of our sites demonstrate that both the relative absence of gender considerations and the proliferation of them in project discourses have not resulted in equitable benefits in gender relations within community-based forest management in the Amazon. We suggest that a more critical conceptualization of gender, based on local situations and definitions, is necessary for the sustainability of community-based forest management initiatives (see also Chapter 8). We demonstrate our evidence for this perspective by using materials from two cases: Porto Dias in the western Amazon (Acre) and Ludovico in the eastern Amazon (Maranhão).

Porto Dias

The Porto Dias extractive settlement project (PAE) is located in the southwestern corner of Acre in the municipality of Acrelândia. It encompasses an area of 22,145 ha of forest throughout which are dispersed approximately 90 registered families. The majority of the residents are descendants of migrants from the northeast of Brazil who came in the 1940s as peon workers to tap rubber and, later, also to collect Brazil nuts. They share a long history of exploitative patron–client relationships, political marginalization, and poverty. Also living in the area are more recent immigrants, agriculturally oriented colonists who originally came to the region during the extensive colonization resettlement projects of the 1970s and 1980s.

Previously an abandoned rubber estate, PAE Porto Dias was created in 1989 after much pressure from local residents. Instead of transforming the area into a conventional colonization project, the National Institute of Colonization and Agrarian Reform (INCRA), influenced by the ideas promoted by the National Council of Rubber Tappers, took into consideration the area's history of extractivism in designating it a PAE. As a PAE—federally owned land designated for sustainable extraction of forest products and conservation of renewable natural resources by resident populations—Porto Dias was designed to integrate the conservation of forests with improved livelihoods. To meet this objective, the Inter-American Development Bank provided considerable funding from 1991 to 1996 to develop infrastructure (roads, health posts, school buildings) and to assist in the implementation of a multiple-use forest management project.[6]

With additional funding from the International Tropical Timber Organization (ITTO) and the Pilot Program to Conserve the Brazilian Rain Forests,[7] a local nongovernmental organization—the Amazonian Workers' Center (CTA)—began working in Porto Dias in 1995 to plan and implement the multiple-use forest management project. Although originally intended to include the management of nontimber forest products, this project has focused almost exclusively on timber.[8]

Originally, 10 families were selected to carry out the timber project. In 1996, plots of 10 ha of forest per family were inventoried, but timber was not extracted until 2000 because of infrastructural, bureaucratic, and technical difficulties.[9] As of 2004, 10 families are involved in the project (6 of the initial families and 4 new families).

Men are the primary participants in the project. When the timber project was first planned and implemented, gender was not explicitly taken into consideration. Both CTA and locally established cultural norms took it as a given that the management of this highly valued product is a male-only activity. There is one exception, a woman with usufruct title to the land and the only woman who is a member of the directorate of the local rubber tappers' association. With the exception of this woman, the majority of the individuals who actively participate in decisionmaking processes and carry out project activities are male heads of household. While other women are not forbidden from becoming involved in the overall project, they remain virtually absent. Women seem not to disagree with men's carrying out these activities, but women do express concerns regarding the impact of these activities on themselves and their families (cf. Chapters 13 and 14) (Stone 2003).

The relative absence of women's participation in the project reflects both the novelty of addressing gender in timber management initiatives and local, culturally grounded definitions of gender roles and relations. Although there is no formal prohibition to women's participation, women are not expected to participate in the monthly association meetings and the project participants' meetings where all decisions regarding the timber project are made. When asked why women did not participate, many women said either that they were not interested in getting involved in the heated debates typical of these meetings or that their husbands told them that it was sufficient to have one household member (the husband) in the association and too costly to add another.[10] In addition, to have access to the association's resources and to credits, the titleholders to the usufruct rights—almost all male—are required to become members of the association, which further keeps women from joining.

Women are similarly not involved in carrying out the timber management activities, which range from assisting in inventories of forest plots to accompanying the processing of wood in the sawmills in the cities. In general, these activities are concentrated either in Porto Dias, primarily within the forest plots where timber is extracted, or outside Porto Dias, in the city. In Porto Dias, gendered power relations are clearly expressed in men and women's differentiated access to and control over physical spaces and resources (Stone 2003).[11] Both the forest and the city are, by traditional local norms, relatively unfrequented by women.

The forest is implicitly defined as men's space and place. Traditionally, men have carried out the majority of activities in the forest, predominantly rubber tapping, Brazil nut collection, and hunting.[12] Men have also been the ones to take care of any business or necessities in the cities. By contrast, women have been more involved in productive and reproductive activities (cooking, raising the children, agricultural work) concentrated in or near the homestead. Several women expressed fear of being in the forest, often making allusions to the dangers of mythical creatures. In this context, these myths are cultural systems, which bring coherence to and mediate community values regarding gender relations.

At first glance, it seems that women consent in their own exclusion from both public decisionmaking forums and project activities. It remains unclear, however, whether women in Porto Dias feel excluded from the project or are satisfied with the status quo. On the one hand, when asked if they would want to be more involved in the project, many women said no, citing that timber is a dangerous activity and a man's activity. On the other hand, several women also expressed a desire to participate in the management training courses and were curious about the activities taking place in the forest. However, few had ever gone to visit the forest plots. What is clear is that one consequence of women's nonparticipation is that men, who have almost exclusive access and control of the management of trees for timber, end up having exclusive control over both the product that they manage—timber—and the money they receive from it. At the same time, women are negatively affected by the project, primarily through the withdrawal of male labor from nontimber production activities, such as agriculture and rubber tapping.

Since the onset of the project, the majority of efforts to address gender issues have been limited to attempts to "bring in women" into activities thought to be acceptable to them, without tackling underlying gendered inequities. In 1997, as part of the project's original concept of management based on multiple uses, a workshop for crafting nontimber products was held, to which both men and women were invited. Men brought leaves, roots, fruits, and barks to the workshop to transform into medicinal products. Women were trained to work with handicrafts, such as exotic jewelry and curtains made of seeds. However, the project made no provision for a business plan or follow-up funding and relied largely on the marginal time and effort of the female agents CTA hired for the overall project. Thus, the initiative quickly faded away. Other similar efforts, including women's collection of seeds of native species and production of small wooden crafts, have also been short lived.

In an effort to train community members to carry out activities previously done by CTA and other professionals (for example, doing inventories of the forest plots), CTA has held many workshops, which have been open to both men and women alike. They have been met by apathy, and few women have shown an interest despite the CTA female project coordinator's particular interest in getting more women to participate. This raises the question, Why are women not participating when given an opportunity? We suggest that in community-based timber management, the institutions involved are encouraging women's participation without changing the foundations of the male-dominated model of forest management. In other words, if women are to participate, they will have to wear "men's hats," which may conflict with women's and men's own views on locally accepted norms of gender relations (cf. Chapter 8).

Women's apparent lack of interest may be a form of resistance to externally defined and threatening conceptualizations of appropriate gender roles. Or it may be a form of submission to local, gendered power relations, with which women may or may not agree. Another reason may be the possible impact on the gender division of labor. Men traditionally spend time in the forest tapping rubber, collecting Brazil nuts, and hunting. The timber project, although it does interfere with some of these activities, does not prevent them from being carried out. For women, however, involvement in the project takes them away from the homestead and from production and reproductive activities for which they have traditionally been responsible. Leaving the house to get involved in the project means that food will not be cooked, children will not be taken care of, and agricultural activities will not be completed. For some women, these activities (and the spaces where they take place) may represent the only activities they have control over and the only bargaining chips that give them power to negotiate with husbands. Some women may not want to give these up to participate in the timber project; other women may not have the choice. The scarcity of older children and difficulties in hiring outside labor means that women have little time to do anything but continue carrying out their traditional activities.

The recent efforts to "bring in women" have been complicated by the different perspectives of gender held by CTA and the community. The most notable case was the initial disapproval, among some families, of CTA's effort to get one of the community forest managers' daughters involved in carrying out inventories in the forest. This disapproval, however, has given way to greater acceptance and, as of 2004, another young woman was being trained in a variety of forest management activities related to the project. This reflects the ways in which CTA's and the community's views of women's role in the timber project are in the process of simultaneously acting upon and altering each other. It is a dynamic process of negotiations (although, almost entirely implicit), which, we think, will eventually inform and redefine how women (as well as men) will be involved in the project in the near future. Current gender relations are slowly changing, a result not just of the project's impact but of larger changes occurring in Porto Dias, including the construction of a road and the education of young girls, as well as changing attitudes in the city regarding women.

Ludovico

Also a former private estate turned into a Settlement Project (PA) through an agrarian reform act in 1988, Ludovico presents very contrasting conditions. Located in the municipality of Lago do Junco, in the easternmost Amazon, its resource-poor environment is mostly composed of pastures, species-poor secondary growth, and *babaçu* palm forests. Although Ludovico, like Porto Dias, has about 90 families, only 32 families, those who were actively involved in the violent agrarian conflict that resulted in the agrarian reform, benefited from their efforts with a meager 536 ha of land. They became the few with legal access to land among a majority of landless. Throughout Brazil approximately half a million squatter and landless families are living on agriculture and extraction in babaçu palm forests covering approximately 20 million ha (Anderson et al. 1991).

By the first decades of the twentieth century, immigrants, who were suffering from land concentration and droughts in the northeast, began to search for lands that were more suitable and, more important, free of landlords. In the humid valleys and forests of Maranhão, they joined former slaves, who were freed in 1888, and formed peasant villages, practicing shifting cultivation and extractive activities.

Beginning in the 1960s, land concentration due to agrarian policies favoring powerful sectors of the society eliminated many of these villages, for the expansion of cattle ranching, the major goal of the regional development plan. As one of the resisting villages, Ludovico finally managed to recover its rights to the land, but by that time pastures already surrounded it, and the remaining forests had to support a higher population density.

With progressively reduced fallow periods, labor became more intensive due to the increase of weeds, and the size of the agricultural plots decreased. Dependency on income from babaçu extraction increased, and *quebradeira* women were not only gathering babaçu from their own lands, but needed to search for fruits underneath the scattered palms throughout the cattle ranchers' pastures (Porro 2002). Therefore, the women's claims in defense of the babaçu palms now are not restricted to the reformed lands, but include palms on private properties.

In spite of their economic and ecological importance, the *quebradeiras'* activities have not been the object of adequate investments. Ludovico has received scattered resources from PROCERA, a national program for credit to agrarian reform areas, which excludes extractive activities. Without the appealing natural resources of a rainforest, Ludovico does not have any support for environmental conservation. In general, the scattered resources destined for this region are mostly related to "poverty alleviation" projects, which regard women as tools to tackle poverty, rather than addressing gender as a means of empowerment. A good illustration is the US$80 million World Bank rural poverty alleviation project for the entire State of Maranhão.

For areas such as Maranhão, which are perceived as marginal to the main Amazonian environmental concerns, poverty relief and not environmental conservation is definitely the core theme of sustainable development. In the World Bank's project, environmental aspects were treated in only two paragraphs out of 144 pages. "Because of their relatively small size, … subprojects would not have a significant effect on the environment" (World Bank 1997, 33). Within this development initiative, women are treated as a separate topic. In the only paragraph addressing them, it is suggested that the project must "target groups and activities in which female participation has proven constructive" (1997, 34).[13]

> Women interviewed in Ludovico did not think there was a single group or activity in which female participation has not been constructive, and their practices can hardly be separated from overall production. Reading about this project and comparing it to materials released by gender programs and units at the World Bank, it seems that, "the Bank has indeed made some strides towards gender-sensitive social policies, but so far these appear more cosmetic than real. Bureaucratic inertia, an ideological commitment to economic rationality, and an internal masculine culture resistant to feminist reforms have resulted in marginalizing both social and gender concerns in development policies" (Kurian 2000, 131).

Certainly, people of Ludovico are well aware of the impoverishment of their social and natural environments after more than four decades of policies favoring development through land concentration and cattle ranching. Even so, they keep calling babaçu palms the "mother of the people" and their species-poor forests the "wealth of the poor" and the "security of the poor." Because in the gender division of labor women are in charge of extractive activities, they have been motivated to participate in political mobilizations for sustainable management of forest resources. This has led women to participate more in regional and municipal movements as well as in their own local associations. Today, membership in the Association of Ludovico is by family, and both men and women have their names under a single membership. Women of Ludovico are active members of the Women's Association (AMTR) and the Cooperative (COOPALJ) of their municipality, Lago do Junco.

In 1991, a Catholic agency funded COOPALJ to build a small processing plant to extract oil from babaçu kernels. This project could be seen as part of the anti-poverty approach[14] to incorporating women into development. Women from Ludovico, along with women in 10 other villages, began to commercialize kernels and oil through the cooperative. In 1994, UNICEF provided funding to the AMTR to build a small processing plant for handcrafting soaps made of babaçu oil. Because of its central position among the villages of the municipality, Ludovico was chosen to host the processing plant. The women's objective was to integrate their economic and political struggles against the destruction of their forests and livelihoods by producing and selling soaps as a means of promoting their cause. This project can be seen as part of the efficiency approach.[15] With tremendous difficulties and efforts, the oil from COOPALJ and the soaps from AMTR have been sold in national and international markets.

Engagement in these economic challenges reinforced women's political activities. In 1997, by their massive presence in several public hearings at the city hall of Lago do Junco, women managed to win approval for a municipal law protecting the babaçu palm and favoring free access to its fruits, even on private properties (Shiraishi 2001). They call this practice the Free Babaçu law, which has extended benefits to people who have not managed to gain access to land. In 2000, women managed to elect a *quebradeira de coco* from Ludovico as their representative in city hall. The situation of Ludovico is also influenced by the *Movimento das Quebradeiras*.[16]

Through the case study of Ludovico, we learn that women are struggling to transform the standard discourses on gender, which often view women as means to alleviate poverty. Women are nonetheless overcoming these dominant gender conceptions and redesigning antipoverty projects through their initiative and political mobilizations. Even when women manage to participate, according to Rocheleau (1990), too often they are visible only as mere "resources," as poorly remunerated (if at all) "fixers" of badly planned and implemented projects.

Although excited about the initiatives in which they are involved, Ludovico's women leaders volunteer their time to this work. Mentioning several examples, Kabeer (1994, *269*) criticizes the visibility brought about by women's participation in education, health, and overall welfare, which assumes "their natural willingness to undertake more work in the interests of family and community 'with more knowledge, but little more time or money.'" Citing Bruce and Dwyer (1988, *18*), Kabeer further notes: "Invisible women of the economic theorist become the all-powerful mothers of the health and welfare advocates" (1994, *269*). Unfortunately, these "all-powerful" women, who manage to carry out triple or quadruple roles in domestic, production, community, and advocacy spheres, become exhausted. Taking on so many roles gives these women more visibility, but at the same time the labor they invest in these roles is often interpreted as being flexible, infinitely expandable, and in the end, disposable.

Therefore, in Ludovico, while women celebrate their great accomplishments, they also know that gender equality has not been achieved to their satisfaction. Women are still struggling with gender inequalities that emerged from their own locally constructed concepts of gender and from development policies that impact men and women differently. Their accomplishments have been overshadowed by the marginal attention given to the environment and the nontimber resource they are willing to manage.

Discussion

Gender issues, at least in discourses and project statements, have increasingly gained space in development initiatives. This is similar to what Foucault (1990) observed about sex in seventeenth-century Western societies. According to Foucault, until the eighteenth century the main approach to sex was repression, to control the alliances and transmission of wealth, through marriage and kinship. From then on, though, repression was not enough to control economic and political processes. In the nineteenth century, there was an exhaustive scrutiny of sex in all social spaces. In the twentieth century, along with a supposed tolerance toward sex, a proliferation of discourses emerged, and a compulsion to talk and think about it. The authorities enunciating these discourses deployed and controlled sexuality itself, suggesting powerful ways to take control over it (Foucault 1990, 106–115).

The reverberating discourse now is that sex is nature's differentiation of males and females, and gender is society's construction to differentiate men and women. After fieldwork in Acre and Maranhão, we learned that this discursive notion of making gender a social construction is a very important first step, but definitely not the final answer to the contradictory practices of inequality between men and women. Without further qualification and fine-tuning, without a diligent and passionate hunt for what exactly gender differences mean in each interconnected and specific situation, these discourses are not only of little help, but may also cover up development policies and practices that further gender inequalities. Gender is indeed a social construction.

While it is satisfying to know that gender issues are no longer ignored, standardized discourses on gender, based on concepts spelled out by international agencies of development, are attempting to control the diversity of ways gender differences are practiced. The repetitive discourses that link gender and predefined forms of development, propagated in manuals, booklets, conferences, and workshops, may end up controlling how gender relations must be lived. In our fieldwork, we learned that local conceptualizations of gender are strongly connected with the way local forest dwellers conceptualize their environment. In the case of the rubber tappers, spaces and activities in the forest are distinguished as male and female domains, and these are part of the very definition of being a woman or a man. In the case of the *quebradeiras,* despite the fact that major international agencies and the government ignore babaçu, a nontimber resource of low market value, the women keep defining their social identity as *quebradeiras de babaçu* and identify babaçu as a "mother" and the "wealth of the poor."

In addition, powerful extraneous discourses and practices driven by market development have permeated local ones, daunting local gender relations and development efforts. Through the direct translation and extrapolation of the terminology of European forestry, which focuses on engineering the use of forests for selected marketable timber, development practitioners have reduced the status of a multitude of species and complex relations, labeling them as nontimber forest products (Niekisch 1992).[17] In this way, entire forests of babaçu palms and numerous undervalued species in rainforests have been dislocated to the margins of the attention of "green" investors and donors. Thus, women, who mostly participate in "nontimber" extraction, are turned into "nonparticipants," as they are "nonmen," being defined by what they were not. Even in the community-based forest management projects, flora and fauna are designated as "timber and nontimber," the diverse social groups in the projects are reduced to "managers or nonmanagers," "participants or nonparticipants," "organized or disorganized," establishing a common language and a single history homogenizing tropical forests and peoples around the globe.

During our fieldwork, we realized that this same common language and homogenization process was often molding our views and thoughts, permeating our interactions with the community-based forest management initiatives we were researching. We realized that our views and knowledge, as well as those of the project agents we met in the field, were greatly affected by the major international development agencies, such as the World Bank and the UN. These institutions spell out their gender discourses within an overall development discourse through conferences, policies, programs, and decades of projects, affecting governments and NGOs, and consequently, community-based forest management. According to Escobar (1995a, 9),"this apparatus came into existence roughly in the period of 1945 to 1955 and has not since ceased to produce new arrangements of knowledge and power, new practices, theories, strategies, and so on. In sum, it has successfully deployed a regime of government over the Third World, a space for subject peoples that ensures certain control over it."

Throughout the decades of development, this focus on women, initially advocated by international institutions through their programs on Women in Development (WID), gradually transferred to Gender and Development (GAD).[18] In initial stages, WID addressed women as homogeneous and isolated targets, seeking to integrate them more efficiently in the development process. Taking development as a given and women as homogeneous, the WID approach intended to understand the specificity of women's roles, especially their responsibilities in production and reproduction. WID aimed to increase productivity by improving their access to and control of resources and benefits. The main idea was to make the process of development more efficient.

After about a decade, GAD emerged, approaching women in their socially constructed relations with men and taking into consideration other social relations (such as ethnicity, class, age, and race). This perspective potentially results in a more conflicted approach, in that it addresses inequality and challenges power relations between men and women not only within the household, but also in the development process itself. GAD, since its conceptualization, has aimed to introduce social change (Moser 1993). Currently, the groups that are still labeled as WID use mostly the same conceptual frameworks and practices as GAD, leading us to think of the two approaches more as phases than contrasts. Through these approaches, we had several focuses: welfare, equity, antipoverty, efficiency, and lately empowerment (Buvinic 1983; Moser 1993).

We understand that many relevant concepts arose from the contexts of UN and World Bank efforts to implement gender issues and that they inform aspects of our own research. However, in community-based forest management, we must investigate why these gender discourses do not always "fit" in local conceptualizations of gender and how the automatic reproduction and imposition of these discourses have not provided adequate space for negotiations in gender relations. We must be suspicious of definitions of gender, neatly confined in one of the many text boxes of manuals, determined a priori in desk studies in the international agencies of development. In this sense, women who do not perceive themselves as "underdeveloped" or in need of being "developed" have to find their own ways to conceptualize gender and their own forms of development. We believe that both WID and GAD are overall approaches to resolve the UN's and the World Bank's projects, and not necessarily people's projects.

Conclusions

As the case studies of Porto Dias and Ludovico illustrate, the inadequate attention to gender and the application of simplistic, standardized conceptions of gender in forest management projects *both* fail to offer constructive lessons to deal with gender inequities and to empower men and women alike. We argue that changes in the way gender issues are tackled on-the-ground (in projects) and in theoretical discourse (in policy shifts) are necessary if forest management projects are to address gender effectively.

The application of homogeneous gender frameworks, now a standard requirement in international and national development policy, is as problematic as not paying attention to gender. "Bringing in women," as WID did in its earlier approaches, or "engendering" development, as GAD now does, is not sufficient. Imposing homogeneous (primarily Western) understandings of gender risks exacerbating existing gender inequities in the communities or creating new ones. Gender relations are far from constituting a static or uncontested set of relationships. Men and women's "appropriate" roles in communities are more often than not heterogeneous, dynamic, and under constant negotiation, as they are interconnected with broader economic-political contexts. Such complexity implies that approaches to incorporating gender in forest management projects using a standardized, homogeneous concept of gender may be too simplistic, if not harmful.

Therefore, practitioners dealing with community-based forest management (foresters, agronomists, anthropologists, etc., including community leaders) need to change the way they have traditionally incorporated gender into projects. For example, development practitioners and social movements alike have paid little attention to gender relations at family and village levels, and the linkages between them. These social relations may be viewed as forms of resistance, reinforcing the community against external domination, and are foundations for genuine political and social advancements. Rather than focusing on how to transform local gender relations so that they *fit into* preestablished development frameworks, development practitioners should build on locally defined gender relations that reflect the realities and complexities of women and men in those communities without (and this is the challenge) reproducing local gender inequities.

We are not arguing for discarding prevailing gender frameworks in development, such as GAD. Rather, we suggest using the accumulated knowledge gained from both gender frameworks and local conceptions of gender as platforms to construct new and context-specific gender frameworks for forest management projects. In other words, what is needed are reflexive, adaptive, and reiterative gender frameworks that are constructed jointly by development practitioners, community leaders, and community members. The role of development practitioners (such as ACM facilitators) must be to *facilitate* discussions and negotiations, not just among themselves and community constituents but also between men and women in the specific community. This will require identifying mechanisms and tools (e.g., workshops, meetings) to achieve two objectives: first, to strengthen and build on both women's and men's knowledge, skills, and resources to enable them to effectively define their own participation in the projects; second, to reverse control from external authorities (including development practitioners) to community organizations and individuals to define and implement men's and women's participation (cf. Chapter 11). The goal is to facilitate and promote self-mobilization among men and women alike to define the conditions under which they want to participate (or not) in forest management projects, without permitting either men or women to impinge on the rights and freedom of the other.

Notes

1. The Legal Brazilian Amazon covers 5 million ha that include the states of Acre, Amapá, Amazonas, Mato Grosso, Pará, Rondônia, Roraima, Tocantins and the state of Maranhão to the longitude 44°W.

2. These projects, *Projetos de Assentamento Extrativista* (PAEs), are similar to extractive reserves, which are supervised by IBAMA (*Instituto Brasileiro do Meio Ambiente e dos Recursos Naturais Renováveis*, the Brazilian Institute of the Environment and Natural Resources). Both PAEs and extractive reserves are federally owned, forested lands where resident populations have secure usufruct rights. However, PAEs fall under the supervision of INCRA (*Instituto Nacional de Colonização e Reforma Agraria*, the National Institute for Colonization and Agrarian Reform) and, while legally they are not designated as conservation units, they adhere to many of the same natural resource management regulations that apply to extractive reserves.

3. Although the majority of the areas that benefited from the agrarian reform in the state of Maranhão have been occupied for three generations or even since colonial times, they have been

named as settlement projects (*Projetos de Assentamento*). These projects are created and adminis-tered by INCRA until their emancipation. In terms of environmental conservation, they fall under supervision of IBAMA, and as any other area, 50% of their area should be destined for legal reserves. However, as the areas were already degraded before the Agrarian Reform Act, supervision is practically nonexistent and only activated when an interested party denounces the destruction of babaçu palms.

4. In this chapter, we are using forest-poor and forest-rich categories as proposed by Ravi Prabhu, "forest rich refers to a landscape resembling a 'sea of forest with islands of people' and forest poor refers to 'a sea of people with islands of forest' (Colfer and Byron 2001, 28–29). In 2000, the state of Acre had a population density of 3.6 inhabitants per square kilometer; 92.6% of its area is covered by rainforest (20.3% is dense tropical forest while the rest is a mix of open forest with bamboo and palm) (Government of Acre 2000). The state of Maranhão had 17 inhabitants per square kilometer, with more than 40% of its population in rural areas, which were covered by increasingly fragmented tropical forests in the northwest, babaçu palm forests in central areas, and shrub lands (*cerrados*, savannah type of vegetation) in the south.

5. Although the *Movimento das Quebradeiras* has existed since the end of the 1980s, it was only in 1999 that the first substantial project, Sustainable Economic Alternatives for Poverty Eradica-tion in the Babaçu Region, began to be elaborated with DFID support. However, so many condi-tions and adaptations were demanded by DFID and the Brazilian Agency of Cooperation that, in 2002, the project presentation was finally cancelled by DFID, without any official explanation to the Movement (correspondence and verbal communication from the *Movimento Interestadual das Quebradeiras de Côco Babaçu*).

6. This funding was distributed among all the PAEs in Acre.

7. The Pilot Program to Conserve the Brazilian Rain Forest, otherwise known as the PPG-7, was initiated in 1992 with funding from the Group of Seven (G7), the European Union, and Holland (MMA and SCA 1996). The Porto Dias multiple-use forest management project is one of PPG-7's Demonstration Projects (PD/A).

8. See original project proposal (CTA 1995).

9. The proposed volume of extraction originally was 1000 m³/year (nine families, collective-ly). In 2003, the scale of production was reduced to 150 m³/year.

10. Each member pays a monthly payment of 2 Brazilian reals (less than US$1.00 in June 2002) to the association.

11. See Rocheleau et al. (1995) and Rocheleau and Edmunds (1997) for a rich discussion on men's and women's differential access to and control over different niches in landscapes and natu-ral resources.

12. The extent to which women carry out activities in the forest varies greatly depending on the extractive social group. In some other reserves, women spend a significant amount of time collecting Brazil nuts, collecting medicinal plants, tapping rubber, and in some cases even hunting.

13. This project has problems stemming from its approaches to poverty alleviation and equity. These approaches, which viewed women as "useful tools to fix poverty" and equity as a matter of simply "add women and stir," were already discussed and addressed by the current WID approach. Nonetheless, there is still confusion regarding WID and GAD approaches within the World Bank, as some of its agents ask why they are now calling it "gender," if they keep working only on women (Moser et al. 1999). This project is an example of how not all sectors within the World Bank are updated or adopting the lessons learned throughout the process of incorporating gender issues in development.

14. The antipoverty approach was the second approach to the issues of women in develop-ment, adopted after the 1970s, which aimed at increasing the productivity of poor women, on the premise that poverty, not subordination, was the key issue in tackling underdevelopment. Through income-generating projects, it seeks to resolve women's practical needs.

15. The efficiency approach argues that development will be more efficient and effective with women's contributions. Although it addresses some strategic needs, the approach strongly relies on women's triple roles and deems women's time to be elastic.

16. This movement involves women working with babaçu in the states of Maranhão, Pará, Piauí, and Tocantins. Its goals are "to interconnect *quebradeiras* as women, agroextractive workers, and citizens, in their struggles for "Free Babaçu" and for the agrarian reform; to share, systematize, and document the experiences in their diverse forms of organization; to search for alternatives in agroextractive activities, in economic, social, political and environmental terms; to create adequate space and momentum for their political vindications" (*Movimento Interestadual das Quebradeiras de Côco Babaçu* 1993).

17. Niekisch (1992) talks about how Europeans' views of nature have been imposed on the management of tropical forest ecosystems that originated from diverse peoples and historical relations with the environment.

18. Though both WID and GAD originated in the context of United Nations (UN) conferences and were adopted by the World Bank, GAD emerged in 1995, informed and departing from the experiences of WID, which originated in 1975.

CHAPTER THIRTEEN

Gender, Participation, and the Strengthening of Indigenous Forest Management in Bolivia

Peter Cronkleton

COMMUNITY FORESTRY INITIATIVES are becoming increasingly prevalent in Bolivia's indigenous territories as ways to improve the well-being of local families and maintain standing forests. For many indigenous people, the introduction of community forestry requires a shift in production strategies from household-based swidden agriculture to cooperative efforts to commercially manage communally owned forests. Such changes are likely to produce new patterns of income and wealth and could disrupt existing socioeconomic processes, which could cause conflict—especially if the outcomes are not seen as equitable. Project planners need to be aware of the processes they set in motion, not only to avoid the unintended marginalization of segments of the local population, but ultimately, to prevent debilitating conflicts that could undercut the long-term success of forest management efforts. Because it is difficult to determine the course of future change or the resulting impact, it is important that the widest possible range of community stakeholders participate in defining the forest management plan and actively monitoring and evaluating its implementation.

To explore strategies for encouraging broader participation, I will examine an on-going community forestry management (CFM) project in the Guarayo village of Salvatierra in Eastern Bolivia. Since 2001, Salvatierra has been developing a forest management project with the assistance of the Bolivian Sustainable Forestry Project (BOLFOR).[1] In this case an early focus on gender issues led to a reorientation of technical assistance and greater participation. This chapter will examine the nature of change facing the community and describe the efforts made to broaden participation and improve the community's ability to control and adapt the project.[2]

I will begin this chapter by providing some background on indigenous community forestry in Bolivia and then provide more specific detail to place the Salvatierra CFM project in context. The second part of the chapter will describe how adjusting the approach used in Salvatierra in response to the gender plan allowed us to increase active participation in the project by a broader range of residents. I will present two examples of activities used to expand participation in monitoring and planning for the distribution of benefits produced by the forest management plan.

Communal Lands and Commercial Forestry

The Salvatierra community forestry project is part of a wider trend that is transforming forest management in Bolivia. Less than a decade ago, Bolivia's indigenous people gained rights allowing them to embark on projects to manage forest resources commercially within a system of communal land rights. To understand what is happening in Salvatierra today, it will be helpful to quickly review the policy changes that created these opportunities for indigenous people and place the village in its local context.

In the mid-1990s dramatic policy change by the Bolivian government devolved rights to territory and forest resources to indigenous people. Two laws ratified in 1996 accounted for most of the new opportunity offered to indigenous groups: an agrarian reform law (*Ley INRA*) and a new forestry law (*Nueva Ley Forestal* 1700). Ley INRA, named for the *Instituto Nacional de Reforma Agraria*,[3] attempted to resolve rural Bolivia's chaotic land tenure system by confirming undocumented land rights and providing land to the landless. In response to increasingly activist native populations, whose territorial demands and rights had long been ignored, the law created a class of indigenous homelands called *Tierras Comunitarias de Origen* (Indigenous Community Territories), or TCOs. TCOs provide indigenous people with nontransferable communal title to their traditional territories.

While the agrarian reform law was an attempt to give indigenous people territorial rights, the new forestry law offered greater control over forest resources on their land. The forestry law was a comprehensive attempt to reform Bolivia's timber sector and realign standard practices of forest user groups under the rubric of sustainable management by regulating "the sustainable use and protection of forests and forested lands for current and future generations in harmony with the social, economic, and ecological interests of the country" (MDSMA 1997). The law created enforcement mechanisms and promoted sustainable management based on orderly, long-term plans with improved silvicultural treatments intended to lower impact, encourage natural regeneration, and protect environmentally sensitive areas. The law also attempted to expand resource rights to more stakeholders in possession of forestlands to give them more incentive to manage these resources sustainably (see Pavez and Bojanic-Helbingen 1998).

The forestry law provided two important rights to indigenous communities: the right to use forests for subsistence and the right to commercialize forest products, although only under a management plan approved by the government's Forest Superintendence (Mancilla and Andaluz 1996; Pavez and Bojanic-Helbingen 1998). These provisions radically changed the relationship indigenous people had with their forests, at least in the eyes of the state. The recognition of subsistence rights did not introduce new opportunities but validated practices that were already in place. The legal right to commercialize forest products, however, was a revolutionary break with past policy. Before these policy changes the state only allowed the sale of forest products by industries that held forest concessions (although illegal timber sales were widespread). Indigenous people's efforts to harvest and sell forest products, even those extracted from lands where the people had traditionally lived for generations, were not legal. Under its new derivation, Bolivian policy saw indigenous people as stewards and potentially active players in forest industries.

In Bolivia, CFM projects usually focus on efforts to manage forest products, especially timber, commercially and are distinguished from subsistence use by indigenous people. Because subsistence uses do not require the development of government-approved management plans, elaborate projects are not required (although clearly these subsistence systems are a type of management done by communities). To gain approval of a forest management plan, a community must identify a discrete forest management unit free of claims by others and conduct a comprehensive inventory. Then, before any timber can be cut, the community must submit annual operating plans based on commercial censuses of the selected harvest units. The management plan for all types of forest user groups must conform to technical norms, such as cutting cycles of at least 20 years, protection of environmentally sensitive areas, retention of seed trees, and minimum cutting diameters. Indigenous communities also are required to demonstrate that they have reached a consensus on implementing, organizing, and administering the plan and on distributing the benefits that result from forest management. These additional rules are intended to protect indigenous groups from manipulation by outsiders or from attempts by minorities to seize control of collective resources.

CFM is attractive to different groups for a variety of reasons. Indigenous political leaders see CFM as a means of demonstrating land use and thus strengthening claims to sparsely populated lands within TCOs. Local people with few economic alternatives are attracted by the possibility of generating income from the sale of forest products. Development agencies believe that showing forest management to be a viable option for communities would not only validate another tool for poverty alleviation but also create a strong incentive to maintain natural forests rather than transform them to agricultural uses. Often, efforts to assist community forestry initiatives focus on building technical capacity to carry out forest inventories and censuses. However, as important for the long-term viability of these projects, if not more important, are efforts to help communities develop organizational strategies, rules, and processes that will allow them to implement and administer the management plan in the future.

Once the 1990s policy changes were in place, increasing numbers of indigenous communities in Bolivia began experimenting with commercial forest management. The number of CFM projects has jumped from 9 in 1999 to 13 approved by the Forest Superintendence by 2001 (Stocks 1999; Superintendencia Forestal 2002a). By July 2003, there were 17 approved CFM projects and another 15 in progress. The approved projects represent 575,500 ha of forest managed by indigenous communities that six years earlier had no legal right even to use forest resources (Superintendencia Forestal 2002a). One center of indigenous CFM activity is the Guarayo TCO, and a promising new project in that region is located in the village of Salvatierra.

Salvatierra's Community Forestry Project

Salvatierra is a small Guarayo village located in Bolivia's Santa Cruz department in the Guarayo TCO (see Figure 13-1). The Guarayo TCO could potentially cover 2,205,370 ha, or 81% of the Guarayo Province, and if the requested area is ever fully titled, will place hundreds of thousands of hectares of forest in the care of Guarayo people. Before 1996 nine timber companies controlled these forests, but they reduced their holdings by 70% in response to the forestry law's area-based tax on timber concessions (Vallejos 1998). As they decreased the size of their holdings, the companies usually relinquished control over forests that had low commercial value or that had been logged several times previously. Nonetheless, there were still areas with commercial timber available to Guarayo communities because the companies had been interested primarily in high-value timber species such as mahogany (*Swietenia macrophylla*) and cedar (*Cedrela fissilis*). The titling process has been slow and controversial. The TCO was originally slated for titling within 10 months after the Guarayo political organization, COPNAG,[4] submitted the demand in 1996 (Tamburini 2000). Today only two of the TCO's five sections defined by INRA have been titled. While the titling process runs its course, new land claims are not allowed, although diverse interest groups jockey for position to ensure their land claims will be acknowledged. Fortunately for the residents of Salvatierra, their lands are within one of the titled sections.

The Salvatierra village consists of 59 households with a total population of 333 residents. These are some of the poorest families in one of the poorest municipalities in the Santa Cruz department (INE 2002). Salvatierra has a very young population. Only 7% (24 individuals) are older than 49, while 51% are under 14. Households rely on swidden agriculture to meet subsistence needs, but the low-value, bulky crops (rice, corn, manioc, bananas, and peanuts) do not allow families to accumulate much wealth. Market conditions are complicated by the lack of an all-season road, which limits access to other market towns such as Urubichá, the municipal capital. Often, rather than dealing in cash, residents simply barter with local merchants for salt, cooking oil, ammunition, and other essential items. Families depend on the surrounding forests to supplement their near-subsistence economy by extracting building materials, tools, food (through hunting and gathering), and medicinal plants. Having seen others profit from logging operations on their land during the past decade, community members were very interested in the possibility of timber sales as an alternative income source.

In 2001, community representatives from Salvatierra sought out the assistance of BOLFOR's regional office in Guarayo, and together they began to plan a forest management project. BOLFOR organized a series of meetings to introduce management rules and explain the approval process, although some Salvatierra residents were already familiar with it. The year before a group of men from Salvatierra had participated in a CFM project organized by residents of Urubichá, but had split off after the first timber harvest because of internal conflicts over management rules and control of project funds.[5]

The Salvatierra forestry management project is the newest of three CFM initiatives in the Urubichá municipality. The Urubichá project had been organized in

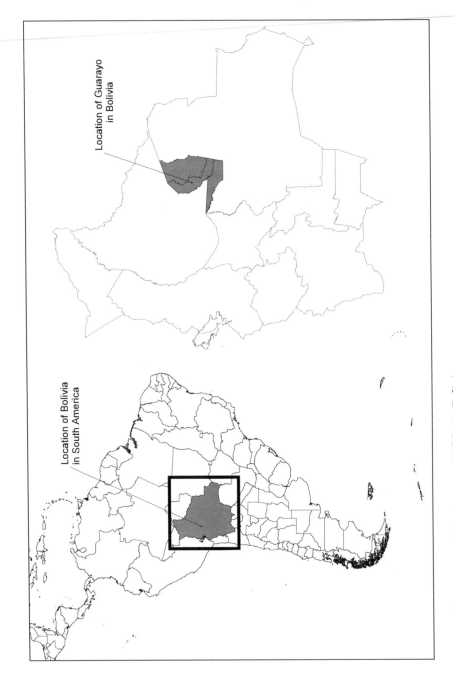

Figure 13-1. *Map of the Guarayo TCO in Bolivia*

1999, and another nearby village, Cururú, started a project in 2000. The search for a management unit for Salvatierra was complicated because the Cururú and Uru-bichá management areas are north of Salvatierra, just outside of a zone villagers had set aside for agriculture. To find unclaimed production forests that did not conflict with agricultural lands, the Salvatierra inhabitants would have had to travel 40 to 50 km north of the other community management units. The villagers decided it was preferable to work close to home and sacrifice a portion of their agricultural lands to designate 40,000 ha of forest about 10 km north of the village as a management unit (see Figure 13-2).

CFM projects on communal land are ambitious initiatives that will require large-scale collective action over extended periods, and the Salvatierra project is no exception. The entire community will have to adjust, and the end result will depend on their actions, responses, and attitudes. Adding timber management to a village swidden agriculture system based on household production means that residents will need to develop new arrangements to ensure their subsistence needs are met, while freeing labor for the new activities. This will not be easy since there is normally little surplus labor remaining after families complete agricultural and domestic chores in the traditional system.

In Salvatierra households, although there is a distinct division of labor, both men and women work in the fields. Men clear forest for swidden fields, hunt, and travel to market. Women gather firewood, fetch water, and take care of most domestic chores (child rearing, cooking, and laundry). Women tend to stay closer to home, especially if they have infants; however, if the family's agricultural field is far, the entire household may move to the field during key points in the season. Some forest management activities require men to spend extended periods away from the village, which could produce additional burdens for the rest of the family. Normally when men are away from home for more than a day, they are expected to leave the family with enough food to get through their absence. It becomes harder for men to leave sufficient food if they have to be gone for a week or more.

Such extended absences are problematic for households because women need to take on chores normally performed by men, in addition to their normal productive and domestic work, and some tasks, such as the maintenance of agricultural fields, may have to be postponed. Families are not likely to jeopardize subsistence to invest in an unproven management project. Therefore, the organization of a management project must ensure that extended absences do not become problems and that the perspectives of all those affected by changes are taken into account (see Chapter 14, for a broader discussion of variation in male and female views toward forest management in nearby Santa María). Those with intimate knowledge of how the village system functions and who will have to shift their routines to accommodate the new demands—namely, the villagers themselves—have the best chance of avoiding conflicts caused by inadequate plans and organization.

Indigenous CFM projects are required to define benefit distribution plans because all residents have communal rights to benefits from community resources. However, planning benefit distribution and actually implementing these plans are two very different beasts. Salvatierra's plan defines a system of direct and indirect benefits: direct benefits are distributed as wages for labor invested in the project;

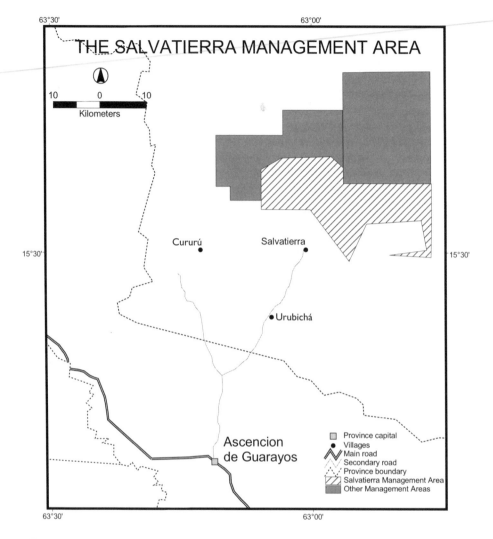

Figure 13-2. *Map of the Forest Management Areas within the Guarayo TCO*

indirect benefits are investments made at the end of the season to meet some communal need. The design of the distribution plan is intended to avoid problems with "free riders," individuals who receive benefits without contributing to the generation of those benefits.

Participation in the CFM project is largely motivated by the payment of decent wages for labor investment. Those residents not receiving a wage can benefit from communal investments made with funds remaining at the end of the year (after paying operating costs and debts and setting aside money for start-up capital for the following season). When the community's management plan was submitted, the distribution system existed only on paper, and few knew how it would work (or even how it was supposed to work). The challenge to the Salvatierra residents now is implementing the system so that it functions and is seen by all to be fair.

By involving the local stakeholders from the beginning, the derived management plan is likely to reflect their needs and desires and to incorporate their knowledge of local conditions. If some residents do not participate in the project's development and, as a result, know less about how it works, they are unlikely to respect or conform to the management system. Such an outcome could undercut the project's assumptions of sustainability, or if conflicts erupt between villagers over the plan, the government could withdraw approval of the management project.

Neither outcome would be desirable. The development of workable institutions and practices will require much negotiation, experimentation, and monitoring by villagers and other interested actors. For a management plan to work, all residents must understand and accept the rules, or at least agree not to undermine them. Although there are no guarantees of success, there are practices that can facilitate learning and help stakeholders to develop adaptive management mechanisms.

"Mainstreaming" Gender into Community Forest Management

While Salvatierra's CFM project was being initiated, BOLFOR was embarking on an effort to promote greater gender awareness throughout its programs. Gender mainstreaming—as the process was called by USAID—entailed a comprehensive effort including the development of an internal gender policy, the evaluation of project impacts by gender, and the reorientation of regional strategies to improve assistance to local communities based on greater consciousness of gender issues. I was able to view this process firsthand as a member of an evaluation team charged with helping BOLFOR develop a gender policy and action plan, and was later invited to participate in BOLFOR's gender committee responsible for implementing both the policy and action plans.[6] In this process I assisted BOLFOR's Guarayo regional team in reorienting their approach to communities such as Salvatierra. As I will explain later, the Guarayo team (myself included) was able to incorporate insights about gender to build local capacity for self-monitoring and collective learning.

One of the first challenges in initiating BOLFOR's gender plan (Paulson 2001) was convincing the organization's foresters that the new focus was worthwhile. Initially some BOLFOR's staff opposed the gender initiative because they believed the new policy would create additional chores and distract them from their main goal of promoting forest management. The effort was further hampered by confusion and misunderstandings over the term "gender," which had been widely used by other institutions and was often vaguely or variably defined depending on the user's agenda or the specific contexts. For many of the foresters, at best, gender was seen as a women's topic primarily dealing with reproductive health; at worst, it was seen as a code word for a radical feminist political ideology. Because of these misconceptions, some foresters feared their mission would be diluted and they would be required to lecture families about domestic violence or birth control instead of silviculture.

Rather than debating the validity of these misconceptions, we decided to reframe the argument by showing how awareness of gender concepts and issues was fundamental to understanding local social dynamics and how such understandings would play an integral part in developing stable, functional management institutions with potential for long-term continuity. We avoided using the term "gender," because of its overuse and negative connotations. Instead, we talked of how communal projects needed broad participation by all residents, both male and female. We explained that rather than working with solitary individuals, the technicians were dealing with people embedded in networks of social relations that define what is acceptable social behavior. We pointed out that management arrangements that were being introduced would be more acceptable if the process of defining them considered the opinions and needs of all members of the family, since they all have rights and will feel the impact of those management arrangements. We wanted to show that, rather than creating busy work, incorporating gender concepts would help technicians and villagers reach their goals and improve the project.

While there were some vocal opponents, we realized that for most foresters resistance was not based on male chauvinism; many simply did not think forestry (i.e., timber) was a relevant topic to discuss with women. The assumptions among project planners had been that forest management was not a traditional activity for women and that management tasks entailed heavy labor inappropriate for them. These assumptions were based on poor understanding of community dynamics and on an overly narrow view of management activities. In one sense the planners were correct: forest management, as defined by the law, was not a traditional practice of women. In fact, it was not a traditional activity for anyone. The conventional wisdom seemed to view forest managers as synonymous with loggers operating chainsaws. At the same time, the technicians knew that logging was not the only task involved in forest management.

Management activities that take place in the forest involve such things as measuring and tagging trees, recording data, and monitoring harvested timber volumes—tasks that require skill but not brute force. The men who participate in the inventory and census had to learn the skills required to measure trees and record the information. Furthermore, many tasks fundamental to management, such as planning and decisionmaking occur outside the forest. Women could easily do all these tasks. This is not to argue that women need to work in the forest and earn a wage for the project to have an equitable distribution of benefits. If a male earns a wage and channels that income back into the household, that outcome can also be positive. Many women, especially those with young children, may not want to work in the forest. However, women should make their own choice, and to make that decision, they need information.

Some foresters questioned whether we had the right to change traditional village norms. In response we argued that BOLFOR programs were already provoking change, although often without conscious awareness of the implications of modifying household labor allocation, access to knowledge, or income patterns. Since change was already taking place, the technicians needed to be aware of how it was occurring and attempt to avoid introducing negative impacts or entrenching the marginalization of segments of local populations.

BOLFOR foresters, concerned that the CFM projects continue once BOL-FOR assistance ends,[7] quickly grasped that entire families would need to work together as communities to maintain the management system. Seen in this light, strategies focusing technical assistance on a small segment of the male population would not build the local institutions or processes necessary to sustain the management plan. The long-term viability of these projects will depend on the technical capacity and organizational skills of community members to carry out project activities. These skills will have little impact, however, if family members oppose the shifts required by the project or if other members of the community decide they have no interest in cooperating.

The best-laid technical plan can be debilitated by conflicts among local stakeholders struggling to control a contested resource. Foresters realized that the year and a half remaining for implementing the project would pass quickly, so they wanted to leave the community with the skills necessary to keep learning. To that end, we needed to involve more of the community, so they would better understand and support what was happening in the management plan. The processes used to distribute benefits, especially, required a high level of participation and transparency. If a project did not provide equitable benefits, or was perceived to treat some segments of the population unfairly, opposition to the project could develop. Thus, including women in the project plan was not only a question of equity; it was crucial for project stability and for promoting local control. Once the foresters realized that the gender policy would not increase their workload (only required changing some practices) and saw that a gender focus offered useful concepts that could improve their understanding of village dynamics and increase the project's chances of long-term success, they were convinced.

Reorienting Our Approach to Salvatierra

One of the first steps in developing a local gender strategy for Salvatierra was to quickly appraise the situation. After interviewing project staff and visiting village families, it was apparent that women were not participating actively in the project nor did they know as much about the project as men (Selaya 2002). Some men in Salvatierra resisted female participation, and in the beginning BOLFOR technicians had ignored this tendency. The technicians had not excluded women, but they had not gone out of their way to invite them either.

Women were unsure about whether they were allowed to participate and even whether they wanted to, because the usual morning meeting time frequently conflicted with domestic chores. When women did attend project meetings, it was not always easy for them to speak up and participate. Since Spanish fluency is lower among women, their opportunities to understand or comment on the project were essentially blocked. This was a vicious cycle: because women participated less, they knew less about the project, and this lack of knowledge limited their ability to participate in decisions and management activities.

Changing this situation required changing the way technicians were interacting with the community. From that point on we went out of our way to specifically invite women to attend meetings BOLFOR organized. We emphasized that technical assistance was available to all residents and that forest management was not just a project for men. Having women present in meetings was not enough; we encouraged their active participation and validated their attempts to join discussions. Often this only required simple gestures like waiting for their arrival before starting a meeting, looking at them while addressing the group, or asking their opinions. The BOLFOR technicians began working with translators or at least changing the pace of meetings to allow time for translation into Guarayo.

In some cases we changed meeting dynamics to ensure that women's opinions were heard. For example, when possible, the technicians started breaking meetings into subgroups for discussion and later uniting the groups to have representatives present reports for general discussions. Women were more confident to speak in smaller groups, and the presentations to the whole group created a space that allowed women to insert their points of views. The technicians also began changing meeting schedules to make meetings more accessible to women. Finally, because some members of the community had not participated during the initial months, we organized review meetings to allow these individuals (mostly women) to strengthen their understanding of the management project.

Although there is still room for improvement in participation, there has already been major progress. Rather than a small number of women shyly peeking in windows during meetings, management meetings now count on a vocal contingent of women. Although outnumbered by men, they clearly understand they have a right to participate and a role to play contributing to the project. For progress to continue and for the participation to have positive impact, we needed to engage people in monitoring what is going on and using their observations to develop future plans. Ideally, these activities will lead to mechanisms that will continue to work into the future. Initially, we focused on benefit distribution issues. The income generated by the project is a highly charged issue. Residents are motivated by the potential income, but are not sure how it will be distributed and are concerned that it be handled properly. Planning and monitoring the benefit distribution was sure to generate interest and participation by community members. Two good examples of our attempts to create broad-based participatory mechanisms can be seen in our use of collective monitoring and future scenarios methods developed by CIFOR.

Promoting Transparency through Collective Monitoring

The sale of timber could bring substantial financial benefits to Salvatierra, but it could also provoke conflicts. More cash would be in circulation than ever before, and its likely distribution would result in some families having more than others because of differences in time and effort invested. For the communal project to work, the benefit distribution process needed to be seen as fair and equitable by all residents. The apparently simple system of direct and indirect benefits still required basic accounting and administrative skills. If the community could not keep track of which individuals were working, how much they were owed, or how much has been paid, the project could break down. Attaining the necessary skills is a difficult challenge for people with low levels of literacy, scant experience managing money, and almost no experience in administration or formal accounting. These are critical problems as the sudden influx of cash into this impoverished community could provoke conflict, create debilitating inequalities, and aggravate internal divisions.

At the outset, the project faced a dilemma: how could Salvatierra's residents create a transparent system they could use to monitor the distribution process when there was nothing to distribute? Since the project had not started harvesting timber, there were no sales and no income to monitor. If the project postponed defining a system until cash was in hand, there was a greater chance of misunderstandings and conflict. Fortunately, Salvatierra's leaders had documented the distribution of funds BOLFOR provided as start-up capital, and analysis of this distribution could be used to evaluate the process.

BOLFOR, like most projects in Bolivia, provides funds as a subsidy to allow communities to invest labor during the initial stages of their management plan (usually during the inventory and first census). In Salvatierra, BOLFOR's regional team turned the funds directly over to the community with the understanding that future assistance would depend on good administration and accounting. The Salvatierra residents determined that 25 bolivianos (B$25)[8] would be a reasonable daily wage for those working on the project. BOLFOR funds were used to pay B$10, and the organization would pay the remaining B$15 once timber sale had taken place. The community's project leaders documented when and how the funds were spent but saw these records as obligations to BOLFOR and did not use them for anything else. Outside of the leadership, most residents were unaware the records existed. This was a unique opportunity: because the records showed who had worked, what they had been paid, and what debts were owed—all aspects the group should monitor. The records gave us a chance to evaluate how the monitoring system was working and discuss how it could be improved in the future. In addition, because the community members had gathered these records themselves, they could build confidence in their ability to monitor as they used the records.

To evaluate how these funds were distributed, we organized a community meeting where we placed the accounting information on a wall chart with columns listing all activities, the days invested by each individual, and totals for days and wages earned. We also prepared cards with each individual's personal data on labor and wages received. During a morning-long meeting we presented this information and facilitated a discussion of how it had worked. Many of the community members were surprised that they had recorded so much information. They were very interested in comparing themselves to others and noticed several trends that will be discussed shortly. Much of the community participated, including those who had not worked in the forest yet, because wages were topics that interested everyone. People wanted to know how the project was going in general, but more specifically they wanted to have a better understanding of how much household members had earned to date and what they could expect in the future.

At that point, the community had carried out three activities: a survey to define the boundaries of their management unit; an inventory; and a census of commercial timber species in the harvest area. Thirty-four different individuals had worked, and in total the community had invested 789 days in the project between August and October 2001. This represented an income of B$19,905 to the community members, although only B$7,890 had been paid and the remaining B$12,015 would only be paid after the timber was sold. The average period worked was 23 days; the most worked by any individual was 47 days, the minimum only 9. Three households had two members earning wages during these activities, which meant that these households received slightly more than average income (working 28, 9, and 49 days, respectively). The evaluation noted trends that could lead to future conflicts. Direct benefits were not reaching all households or individuals: The 34 people who had worked during the initial activities came from 31 different households, meaning that just over half of the 59 households had participated and earned direct benefits. Only one woman had earned a wage at the time of our evaluation.

We chose to stress several points. Although it is a communal project, some individuals will earn more than others because of the way it is designed. One point of discussion was the relationship between the amount paid to individuals and the number of days they had worked. This seems obvious, but the period between labor investment and eventual payment was likely to stretch over a year, and it would be easy to forget the amount different individuals had invested. Since some individuals will get more than others and some will get no direct benefits at all, the situation could provoke jealousy and discontent. Furthermore, the activity leaders earn the most because they work all of the days their crews are in the forest. This needed to be clear to avoid suspicion that leaders were unfairly benefiting.

Wage payments are the first tangible benefits that the village will receive, and the villagers need to begin evaluating how the system works as soon as possible. Over time, the difference in wage income could open a larger gap between households. This is a likely outcome of the strategy to exclude free riders, but the community needs to examine and prepare for such impacts. Some participants wanted to experiment with paying a higher wage in the future. A higher wage would better compensate those who work on the project, but it could also intensify differences between households even more and surely would decrease the funds remaining at the end of the year for communal investment. Such a decision should be based on an evaluation at the end of the season.

Women who attended were pleased to know how much their households would earn from the labor invested by their husbands. Some women complained that on previous market trips their husbands had spent income on alcohol and coca instead of other household needs. They suggested that wages be paid in the presence of couples rather than to individual men. That way they will know how much has been paid and will have more influence over how it is spent. Originally, it had been difficult to find women willing to spend extended periods in the forest. However, once they saw how much the project's cook had earned, a number of women wanted an opportunity of their own. The participants discussed a biweekly rotation that would allow more community members to work and also allow those who were working to leave the forest to take care of other chores.

All participants made it clear they wanted to have regular meetings of this type, although in the future they wanted to simplify the system so that it would be easier to manage. During the next cycle, we will work with community leaders to help them prepare and present the information, gradually diminishing our role. If all works well, this type of public discussion of financial records will continue and will strengthen the project.

Creating Scenarios to Plan Indirect Benefit Distribution

For many residents of Salvatierra who do not work in the project, the primary benefit will be indirect, as a result of the investment of project surpluses at the end of the season to meet some communal need. Although these indirect benefits provide an excellent opportunity to improve local conditions, they also pose the difficult task of making a communal decision to allocate relatively large sums of money. With such important decisions looming, most residents were unsure how the distribution would work or disagreed on how the money should be used. Our initial attempts to discuss how the community should use this income did not seem to engage residents: it was difficult to move beyond a few general themes volunteered by leaders, and these were usually too vague to be used for planning.

We knew people had diverse opinions, some strongly held, but these were not being enunciated. Because having income to distribute is an unprecedented opportunity, Salvatierra's residents needed to explore a wide variety of options to ensure they were able to do the most good for the most people. To help them start the planning process and to clarify expectations and goals, we created a vision scenario of how these funds should be invested. The vision scenario method is ideal for defining the distribution plan.

A vision scenario is a simple tool in which individuals are asked to imagine an ideal future and attempt to explore details and implications of that future situation (Wollenberg et al. 2000). As in other scenario methods, focusing on a hypothetical future point removes constraints imposed by concerns for current realities and provides an opportunity for people to examine less conscious expectations and desires. Normally, the process of developing a vision scenario takes place in stages. Participants begin by developing a personal vision of an ideal future, and then these individual visions are shared among participants. By comparing and discussing individual visions, the participants attempt to reach consensus on a single version that unites their diverse opinions and views. The process encourages people to define their expectations and beliefs explicitly, and by sharing and integrating these views the group is more likely to develop a plan they can all live with, avoiding conflicts resulting from misunderstandings. Although reaching consensus can be difficult, the vision scenario method has the critical advantage of identifying differences in opinions and expectations early, when choices are less constrained by previous decisions and investments.

The vision scenario was developed over two days in August 2002 during a general community meeting. The meeting was well attended because the village men and women knew we would discuss benefit distribution. We started the scenarios exercise by using a technique called guided imagery (Borrini-Feyerabend 1997), because it is sometimes difficult to engage participants and get them to make the effort to imagine a detailed future. To start, we had the participants relax and close their eyes. Working with a translator, we asked the residents to imagine a point five years in the future and assume that the management plan had worked perfectly, meeting all their expectations.[9]

While thinking about this future point, we asked them to imagine walking through their community, visiting their home and the houses of neighbors and even walking through their forests. We told them to note what they were seeing, what they felt, what they liked. We asked people to identify things that had changed and things that had stayed the same. After doing this for a few minutes we wanted them to focus on how the benefits from the management plan had been used to produce that situation. Finally, we asked people to remember what they had seen and open their eyes.

After the guided imagery exercise, we separated the community into subgroups consisting of men, women, and a mixed group of adolescents, and gave each group paper and markers. We asked the participants to discuss their individual visions of the ideal future and then try to agree on common features in a drawing of their households, the community, and the forest depicting the collective view of the future. Once the drawings were finished, the meeting resumed and representatives presented each group's drawing.

The three groups' drawings were each quite distinct, although all shared characteristics. The men's vision spanned all three domains of household, community, and forest, not surprising since most had been actively involved in the project from the beginning and had more time to think about the benefits. The women had very detailed household drawings, more so than the men, but with less detail in the community and almost none in the forest. The young people had an extremely detailed view of the community, but few details of the forest management plan.

Several topics were repeated in all three drawings: new health posts reflecting a common concern with health issues; other infrastructure such as wells and water tanks to provide potable water; an improved plaza and public meeting spaces; and an expanded school building indicating a common desire to improve education. In earlier meetings, women had complained that men usually did not use their cash income to make domestic investments in their homes. The women wanted such things as wooden doors, glass windows, and kitchen improvements for their houses. Now all three groups included household improvements that reflected the earlier demands of the women. For example, all three house drawings included outbuildings that served as kitchens, although the men depicted a building with few details while the women and adolescents drew in gas stoves, pots and pans, and even running water.

The scenarios activity generated debate over household improvements because it exposed a difference of opinions between those who thought investments should be made with communal funds (i.e., indirect benefits) and those who thought the investment should be paid from the wages earned by each household. Regardless of the community's ultimate decision, the fact that this proposal was publicly discussed decreased the potential for conflict later over the use of project funds. Furthermore, the fact that the debate has gotten to this stage is a good indication that an interest of female residents that had not been considered earlier is now likely to be met. By including the women, we learned of new demands. Women wanted funds set aside to purchase raw materials for the women's weaving group, which they depicted with stick figures weaving in the houses. This proposal had not been considered earlier but was well received by all, because this investment would generate additional income.

The scenario activities were only initial steps, and follow-up exercises will be necessary before the community reaches a final plan. They will need to prioritize investments and define them in greater detail. For example, before spending on the village's health situation, they need to identify the types of health problems they want to address. If the primary problems are diarrhea and parasites, then investing in potable water would be a good strategy. This may seem obvious, but if the distribution process is going to work, community members need to make these connections and agree that the improvement strategy is worthwhile.

Conclusions

Community forestry holds much promise for Bolivia's indigenous people, potentially allowing them to improve their well-being and maintain their forest resources. However, this potential will not be achieved if the new activities solidify patterns of inequitable distribution, exclude subgroups from their rights, or provoke internal conflicts in communities. Building the technical capacity of the community is important, but is unlikely to lead to sustainable outcomes unless efforts are also made to build local institutions. For example, the management groups need to establish rules, practices, and processes that are participatory, inclusive, transparent, and democratic. Local management institutions develop gradually, so project technicians must attempt to encourage participation and collective monitoring to build the capacity of communities to learn and adapt. The participating technicians must also be prepared to listen and observe, so they can also learn and adapt to make their efforts more appropriate in the local context. These efforts must be an integral part of project strategy beginning with the design of approaches and methods that will be used, not included as an afterthought once the technical approach is defined.

In the example discussed here we showed how we were able to incorporate gender sensitive methods in a forestry project, overcoming initial resistance of some of the technical staff. At the outset, some foresters in the project thought their work had little to do with women and expected that the gender plan would entail additional chores besides their forestry work. We quickly defused this tension by refocusing the discussion on the participating communities. Whether we liked it or not, the project being introduced would affect entire families, not just men working in the forest. If the work generated conflict or hardships at the household level, it had little chance of success. We needed to be in-tune with these local dynamics and involve a wider spectrum of the community to minimize misunderstandings and conflict. Once foresters saw that the gender strategy would not entail new activities, just modifications in their approach to adapt to local contexts, they incorporated the goals into their work plans.

Reorienting the assistance provided to Salvatierra with a gender focus had a dramatic impact on how the project functions. When it began work in the community, BOLFOR had assumed that all interested residents were participating. By

272 • Peter Cronkleton

opening opportunities for women, however, the project attained a higher profile, and more people began to participate in general. Community members responded positively to the opportunity to attend, ask questions, and hear discussions. We involved people who had played only a peripheral role in the project up to that point who had limited understanding of the project, and who had different opinions about the project's direction and goals. Encouraging their involvement increased the likelihood that the project will respond to broader needs in the community and lessened the potential for disputes and conflicts.

The collective monitoring and future scenarios exercises were opportunities to promote greater participation in discussions on a controversial topic (money) that affects the entire community. BOLFOR's efforts seem to have increased understanding and have set off continued cycles of discussion in the community. Broader participation brought new opinions to discussions and required others to examine their expectations of the project. For example, including women's opinions of communal investments seems to have changed the plans for use of the indirect benefits, possibly leading to future domestic improvements and for support for female weavers. Since then, we hear of these debates and receive questions almost every visit we make to the village.

It was easy to generate participation because we chose a topic that had general appeal and broad impact. Everyone was interested in understanding and developing the benefit distribution plan. Villagers could see how this plan would affect their lives. In the future, we will need to help the community to expand monitoring and learning approaches to other aspects of the management plan. While community members are enthusiastic about monitoring economic benefit, it is less clear how to generate interest in monitoring aspects of forest management that do not have clear direct benefits or quick resolution.

There is still much progress to be made in monitoring and learning from the management activities. The team needs to continue to evaluate the methods and approach they are using with the community, as well as the results they are producing. As successes and failures are observed, the villagers need to modify their management strategy accordingly. The community members need to treat their management plan as an on-going experiment and continue efforts to improve it, so that their plan better meets their needs and the future needs of their children.

While these efforts to develop learning and monitoring mechanisms are a good start, they were heavily influenced by outside facilitators. For community members to see the potential of monitoring and understand how cycles of collective learning work, they need a catalyst to set the process in motion. Our efforts as catalysts were necessary to get a project cycle going. However, if outsiders play too strong a role, the process will not stand alone once they pull out (also a concern in Chapter 11). To function autonomously in the future, the community will need to take over these mechanisms. We have started a process; in the future we need to step back and allow the community to take over.

Notes

1. BOLFOR is a joint project of the Bolivian Government and U.S. Agency for International Development (USAID) and has been instrumental in developing forest policy in Bolivia and assisting the forest sector to comply with the new forestry regime. It has, in its final stage, emphasized assistance to rural communities attempting to manage their forests commercially.

2. Observations presented here are based on personal experiences in Bolivia as part of the Adaptive Collaborative Management (ACM) program of the Center for International Forestry Research (CIFOR). In this role I have actively assisted BOLFOR to develop a gender action plan, and my research in Guarayo has allowed me to closely follow efforts to reorient social forestry activities in response to this plan.

3. National Agrarian Reform Institute.

4. *Central de Organizaciones de Pueblos Nativos Guarayos*, or the Center for Native Guarayo Peoples.

5. Shortly after the division of Urubichá's management organization in 2001, the Urubichá project was temporarily suspended by the Forest Superintendence pending an investigation into the misuse of community funds. In 2003, the Forest Superintendence's local office allowed the group to finish harvesting timber from the 2000 census, but the full suspension of their management plan has not been officially lifted.

6. The other members of the evaluation team were Dr. Susan Paulson (Miami University), Raúl Lobo (BOLFOR), and Bolivar Tello (Randi Randi, an Ecuadorian NGO).

7. The BOLFOR project closed in December 2003. USAID subsequently initiated the BOLFOR II project but under contract to a consortium led by The Nature Conservancy.

8. At the time, the exchange rate was approximately B$7 (bolivianos) to the dollar. Wage labor is not easy to find in the region and often pays less than B$25 per day.

9. Vision scenarios do not require taking such an optimistic tone and could be used to examine other types of situations; however, because we wanted to examine people's expectations of benefits, we wanted them to assume that there would be benefits in the future.

CHAPTER FOURTEEN

Women's Place Is Not in the Forest

Gender Issues in a Timber Management Project in Bolivia

Omaira Bolaños and Marianne Schmink

SINCE 1993, THE BOLFOR PROJECT in lowland Bolivia has sought to reduce the degradation of forest, soil, and water resources, to protect the biological diversity of Bolivia's forests, and to build Bolivian public and private-sector capacity to develop and implement programs for sustainable forest use (http:// bolfor.chemonics.net/). In 2001, the project formed a new Community Forestry Unit (CFU) to work with several community forestry groups in forest management efforts. Shortly before that time, in June 2000, CIFOR began to collaborate with the BOLFOR project through an Adaptive Collaborative Management (ACM) program carried out in parallel with the University of Florida (UF) research reported here.

The UF and CIFOR had been working together on issues pertaining to adaptive collaborative management in Brazil since 1997, and this research represented a continuation of that collaboration in a new site. The UF research shared the focus of CIFOR's ACM program to link the well-being of local forest peoples with protection and sustainable management of forests. UF researchers have used participatory approaches to carry out applied research in communities where people have a long-standing cultural and practical connection to forested areas. Working in collaboration with partner organizations involved in timber management projects, the research has focused on bringing local peoples' interests and voices into the learning and decisionmaking processes related to these projects.

Timber management projects in Latin America are frequently designed without considering the many other important uses of forests for local communities, and their monitoring systems often overlook the complexity and dynamism of socioeconomic systems, including the multiple views and interests of different stakeholder groups within local communities. In Brazil and Bolivia, UF researchers are collaborating with CIFOR researchers and local nongovernmental organizations to carry out applied research in local forest communities and to use the results to promote better communication and collaboration with other stakeholders in the local setting.

In this chapter, we discuss the complexity of community expectations about the implementation of the BOLFOR forest management project in Santa María village, a *campesino* community within an indigenous territory. We analyze the ways different groups perceive the forest, the forest project, their potential participation, and the internal conflicts over shifts in productive activities that may arise and affect the implementation of the project. We conclude with a discussion of how differential benefits and conflicts may motivate or constrain community participation in forest management.

Background

The Guarayo TCO (indigenous territory; cf. Chapter 13) is located in the northwestern part of the Department of Santa Cruz, Bolivia, in the Guarayo Province and in the western part of Nuflo de Chávez Province. The TCO is a demarcated 2,205,369-ha area that is divided into five areas called polygons. So far, the TCO has obtained title to the first and second polygon with the assistance of COPNAG (*Central de Organizaciones de los Pueblos Nativos Guarayos*), which is the organization that politically represents the Guarayo Indians. Most of the communities of the Guarayo TCO are campesinos and colonists who settled in the area years ago. The indigenous Guarayo population is concentrated primarily in 6 of the 24 TCO communities: Salvatierra, Cururú, Urubichá, San Pablo, Momené, and Yaguarú. In the other communities, the population is composed of a mixture of indigenous peoples (Guarayos, Aymarás, Mojeños, Quechuas, and others), campesinos, and migrants (VAIPO 1998).

The community of Santa María is located in the northwestern part of the municipality of Ascención. The community is on the border of the departments of Santa Cruz and Beni, along the interdepartmental highway that connects the cities of Santa Cruz and Trinidad. Santa María is one of the six campesino communities included in the fifth polygon of the Guarayo TCO (see Figure 13-2).

The settlement was founded in 1949, when missionaries of the World Gospel Church explored the area and established an agricultural program, an elementary school, and a church to preach their religion to the Sirionó Indians. In 1970, amid much political commotion in Bolivia, the church was relocated, and the farm and the school were closed. Only five families remained in the village; the rest abandoned the area. These five families are still living in Santa María and consider themselves natives of the area. Only one of the families has members who are descendants of the Sirionó Indians.

In 1971 the first private sawmill, named "Chajtur," was established. It brought many loggers who stayed in the village and worked in the forest only during harvest times. In 1983, the extractive area was sold to Milton Parra, a timber company that extracted timber until 1988. After 1988, the Laminadora Roda Timber Company began working in the forest, until timber became unprofitable. According to the original settlers, it was during this time that the population grew and rendered Santa María a fully established community. Timber operations and the opening of the road attracted many colonists to the village.

In Santa María there are three *sindicato* organizations that distribute the land among their associates. *San José Obrero sindicato,* with approximately 50 associates, was founded in 1987. (It was led by a woman until May 2001.) Santa María I was founded in 1998 and has 33 associates, while Santa María II was founded in 2001 with 85 associates. There is a conflict between Santa María II and the other two *sindicatos* concerning the overlap of the area designated as the "second phase of the land distribution project." Each side claims this area as its own. Underlying this dispute is a split between those who favor inclusion of the village in the Guarayo TCO as a strategy to secure land and titles, and those who believe that this inclusion will diminish and restrict control over land. This situation has created a division among the members of the community, which has spilled over to concerns with the forest management project. One group sees participation in the forest management project as an indigenous strategy to consolidate territory, and therefore a threat to campesino interests, while another group is actively participating in the project.

A census taken during mapping activities done with the community in July 2001 showed that there were 93 families in Santa María, for a total population of 365 (203 men and 162 women). Most families in Santa María are composed of parents and children, although some families include grandparents, grandchildren, and other relatives. Of these families, 73 lived permanently in the village and the remaining 20 lived elsewhere and visited the area during harvest time. Only 20.5% of the permanent families were from the Department of Santa Cruz, and 76% were from other departments—Chuquisaca (34%), El Beni (18%), Cochabamba (12%), Potosí (7%), Tarija (4%), and La Paz (1.3%)—the origin of 3% was unknown. Of the total population, only 6.8% were natives of Santa María.

Each family in the village of Santa María has, in addition to its dwelling, a crop field called a *chaco* in which the family can grow rice, corn, plantains, and pineapples. Although these are primarily subsistence crops, part of the surplus is sold in the cities of Trinidad and Ascención. These surplus grains and fruits are also traded for other food (usually oil, bread, coffee, sugar, and salt) in the *tiendas* (stores) in Santa María. The *chaco* is usually between 3 and 7 ha in size, although some families have *chacos* as large as 50 ha.

The management project began in Santa María in May 2001, when BOLFOR contacted the group of people who had participated in a training program, Forest Management in TCOs, sponsored by the Ascención de Guarayos municipality during May and July 1999 (Municipio de Ascención 2000). Of the 29 people who had participated in this program, 7 were women and 22 men. BOLFOR carried out a first meeting with some of the participants from this training program, but later decided to involve the whole community in the project. The 13-member forest group Directorate now includes 7 of the training participants; 2 women are part of the Directorate, one involved in marketing and the other in administration.

Our research sought to understand women's and men's views of the forest and of forest management, the different roles of community members, and their opportunities to participate in the forest management project. Examining these questions would help to understand the dynamics of specific gender relationships in the Santa María community that facilitate or prevent participation by different groups.

Methods

To collect information regarding gender differences in expectations regarding the forest management project, in 2001 Bolaños interviewed all 19 women and 16 men who had been involved in the development of the forest management plan. She reinterviewed these same people in 2002, except for three women who had left the village and moved to Santa Cruz. The results presented here represent the views of the people interviewed (project participants), not of the whole community.

Workshops carried out in 2001 and 2002 were devoted to conducting a gender analysis and an analysis of the impacts of the new forest management practices on the daily activities and labor division of the family and community. The first session analyzed the family division of labor, and the second focused on the likely impacts of the forest project on this division of labor. Members of the BOLFOR team also participated in the second session. Community participants were those directly or indirectly involved in the forest project (wives or husbands of the forest group members).

In this chapter, we present our analysis of the information gathered in the interviews and gender analysis workshops from 2001 and from follow-up interviews and workshops in 2002. From the interviews, we focus on questions related to the importance of the forest to women and men, the traditional use of forest resources, the interest of men and women in participating in the forest management project, and the ways in which they would like to participate. Data from the gender analysis workshops allowed us to examine the gender division of labor within families, as well as peoples' views of the likely impacts of the forest project on both family and community social dynamics.

Expectations of the Forest Management Project

Interviews were carried out in 2001 and 2002 with Santa María participants in the forest project, to explore community expectations of the project. In 2001, the project had not yet begun timber extraction, but had held meetings and training with BOLFOR on technical matters. To carry out the interviews, Bolaños visited 25 households, and in each one she interviewed at least one person; in some of the households the wife, husband, and some neighbors participated. In this paper, we examine 4 of the 15 questions asked of the community members. These questions relate to how people view the importance of the forest, how they use the forest, how interested they are in participating in the forest project, and how they want to participate.

Importance of the Forest

People were asked, Why is the forest important? Table 14-1 shows the differences in responses in 2001 and 2002, ranked according to those cited most often, comparing women and men. In 2001, men and women generally agreed about the

Table 14-1. *Percentages of Women and Men Who Cited Importance of the Forest in Providing Various Resources, 2001 and 2002*

Resource mentioned in responses	2001		2002	
	Women	Men	Women	Men
Food	84.2	56.0	37.5	25.0
Timber	42.1	62.5	37.5	75.0
Game	42.1	56.3	31.3	31.3
Firewood	42.1	12.5	18.8	18.8
Environment	15.8	18.8	31.3	25.0
Cropland	15.8	12.5	31.3	12.5
Employment	0	18.8	6.3	12.5
Water	26.3	6.3	18.8	12.5
Life[a]	10.5	0	25.0	6.3
Number of respondents	19	16	16	16

[a]Refers to the capacity of the forest to provide all the essentials to stay alive (food, water, oxygen, shelter, etc.).

Source: Interviews with members of the Santa María forest group.

most important aspects of the forest: they both indicated that the most important aspects of the forest are to provide food, timber, and animals.

Responses in 2002 did not show much difference between women and men in their views about the importance of the forest for provision of food, animals, firewood, water, croplands, and employment. However, the proportion of men citing timber as important was double that of women (75% and 37.5%). A difference persisted between women and men in responses about the forests' importance for sustaining life (25% and 6.3%). The responses in both 2001 and 2002 showed some differences between women's and men's views of the forest: women gave greater importance to food, and men were more likely to mention timber and employment.

Responses related to the provision of food changed between 2001 and 2002. For both women and men, there was a marked decrease in the percentage who mentioned the importance of the forest in providing food (women from 84.2% to 37.5% and men from 56% to 25%). Since most of the families participating in the forest project were involved in a new agricultural project focused on coffee planting, promoted by a group of politicians from the city of Santa Cruz, this new project could have influenced women's and men's perception of the forest as the main provider of food for the family. Families may have seen the sale of coffee as an alternative income source that could allow them to purchase food, reducing the importance of the forest for food provision.

While the responses that mentioned timber as an important aspect of the forest decreased slightly for women (from 42.1% to 37.5%), they increased slightly for men (from 62.5% to 75%). This divergence is not surprising since, when timber companies were located in the community, they mostly employed men as loggers or drivers. Extraction of timber was one of the main activities men had carried out in the past and through which they had gained income. Moreover, the focus of the forest project on the management of timber resources may have caused men to think that timber was the most important aspect of the forest.

Other forest products, such as fruits, medicinal plants, or animals, were not incorporated into the planning of the forest management project. Therefore, it was understandable that men prioritized timber over other uses of the forest, such as provision of food and animals.

Women's responses about the importance of animals as a feature of the forest remained fairly constant (42% and 31.3%). Despite the decrease in the percentage of the women's responses mentioning provision of food and animals as important aspects of the forest, these responses maintained their respective position in first and second place in the frequency with which they were mentioned (37.5% and 31.3%). During the interviews in the first and second year of the research, a group of women emphasized their concern about the extraction of timber, because some of the valuable timber species also provide food for animals. According to what they expressed in the interviews, the women believe animals are endangered by timber extraction: "Now animals do not have as much to eat. It should be forbidden to exploit *Coquino* (*Arbizia cubana*) because the *Jochi pintado* (*Agouti paca*) eat it and get fat; we also cook the *Coquino* and obtain a kind of honey. Animals have not been taken into account when people cut down trees; they only examine the value of the timber species."[1] Women also articulated the importance of the animals because "they are seed dispersers" and because they are a "food resource for the family." During the interviews, women mentioned the connection between different forest species and animals. Women referred to fruits such as *Lúcuma* (*Pouteria* sp.) and *Guapomó* (*Salacia elliptica*) and forest trees such as *Coquino* and *Bibosi* (*Ficus* sp.) as important food sources for animals.

By contrast, men's responses about the importance of animals decreased from 56.3% to 31.3%. Unlike women, men did not express concerns about the impacts of timber extraction on the maintenance of animal habitat. Men's interest in getting involved in cash-yielding activities, like timber extraction, may have made them consider the effects on animals as being relatively unimportant. In Santa María men were the main providers of cash to the family economy, given their direct involvement in wage labor and in the local market economy. Although wild animals provided some income, it was not significant in the cash economy of the family. Moreover, the focus of the forest project on timber extraction as the main potential means of increased income may have influenced men's perception about the importance of wild animals. It was apparent, however, that men also have valuable knowledge about animals and other nontimber forest products. Hunters of the community know not only the forest area, but also the places within the forest where hunting is more effective. These places are related to the location of trees that also provide food and shelter for animals.

While mention of firewood as an important aspect of the forest decreased in women's responses (from 42.1% to 18.8%), it rose slightly in men's responses (from 12.5% to 18.8%). Women had the primary responsibility for meeting the family's basic needs for firewood. Usually they collected firewood once a day, although in the winter this frequency diminished. Men supplied firewood for the family mainly during clearing of the *chaco* to plant. The changes in responses about firewood may have been caused by two particular developments during the second phase of the fieldwork.

In 2002, the research coincided not only with the preparation of the *chaco* for planting, but also with an intensification of agricultural labor in the community, motivated by the assessment of land requirement and land titles developed by the National Land Reform Institute, INRA.[2] Since the establishment of agricultural production was one of the main criteria in the assessment of land rights (Fundación Tierra 2002), people in the community were motivated to clear new areas within their *chacos* to demonstrate their need for the land and thus keep their property rights. Intensification of the slash-and-burn system to prepare the *chacos* provided an abundant supply of firewood for the family; men were the main members of the family in charge of this activity. Thus, the opportunity to collect and bring firewood to the household may have influenced men's perspectives about the importance of this forest product.

The villagers' ranking of water as an important resource of the forest showed a similar divergence between women's and men's views. Women's responses mentioning water decreased from 26.5% to 18.8%, but men's responses increased from 6.3% to 12.5%. This divergence could have been related to problems with the community's water supply, which had obliged the families to collect water from the polluted San Pablo River, because the generator used to distribute water to the houses was broken for four months (April–July). This problem was delegated to men, given their relation with municipality authorities. The task of finding a solution to repair or replace the generator that distributed the water to the houses may have made men more aware of the importance of this resource.

By 2002, women had slightly increased their evaluation of the importance of the forest for protecting the environment, providing cropland, sustaining life, and gaining employment. Men, in contrast, remained steady in their responses about cropland and lowered their responses concerning employment. Difficulties in the implementation of the forest project may have decreased men's expectations about job opportunities in forest operations.

In general, for both men and women, responses about the priority given to the different categories mentioned remained nearly the same in both years. The importance of the forest for the environment, for cropland, and for life rose in priority over firewood and water. The increasing ability to collect firewood from the *chaco* (avoiding traveling to other forest areas to collect it) could have modified perceptions of the importance of the forest as the main provider of firewood. An increasingly broad vision of the forest as important for the environment and for life may have subsumed the value of individual resources, such as water, which people continued to mention in the interviews as being essential to maintain animals, plants, and people.

Use of the Forest According to Gender

Table 14-2 presents the uses of the forest as reported by women and men. Although in both years men mentioned hunting as the second most frequent use (43.8 % and 56.3%, respectively), women mentioned hunting most frequently of all (73.7% and 100%, respectively). Because hunting was one of the main food resources for the family and women were in charge of most of the subsistence activities, they may have been inclined to value hunting above other forest uses. Women's views on hunting may also be connected with their responses in the interviews (as shown in Table 14-1), in which women mentioned animals as an important aspect of the forest and expressed their concern about animals being endangered by the extraction of timber and the implementation of the forest management project. There was also noticeable interrelation in responses by women about the importance and use of the forest for trees, animals, and food for the family: Trees provide food for animals, and the latter represent the main meat resource for families in Santa María.

Although hunting was a male activity, men's responses showed hunting was not as important for them as it was for women and was also not interrelated to any other use of the forest. In a conversation with a leader of the community, Bolaños was told that the men who specialized in this task hunted once a week. Although the family usually consumed the hunted animals, sometimes the meat was sold to other community members or exchanged for other food. The primary species hunted were *Agouti paca*, collared peccary, and white-lipped peccary.

Women's and men's responses about firewood differed markedly in 2001 (57.9% for women and 37.5% for men), but by 2002 both responses were similar (75% for women and 68.8% for men); both increased, while men's responses nearly doubled. These changes are consistent with those found in Table 14-1 and could again be related to men's increased collection of firewood during *chaco* preparation and to the intensification of land clearing as a requirement for land titles.

The percentage of responses that mentioned using the forest to get land was not greatly different for women and men (10.5% and 18.8%, respectively) in 2001; however, in 2002 men's responses increased (31.3%) and women's responses decreased (6.3%). The link between forest and cultivable land in Santa María is closely related to the history of settlement and the current conflicts about land tenure in the region, issues about which men were more likely to be well-informed than women. Most of the families settled in Santa María came from other areas of the country for two reasons: as employees of the timber companies or as colonists looking for land. The construction of roads for the transportation of timber facilitated immigration of families that were interested in getting land. Internal conflicts regarding distribution of land among members of local unions, identified in the first phase of the research, had increased in the second year. More than a simple

Table 14-2. *Percentages of Women and Men Who Cited Uses of the Forest, 2001 and 2002*

Use mentioned in responses	2001		2002	
	Women	Men	Women	Men
Extracting timber	31.6	87.0	31.3	56.3
Hunting	73.7	43.8	100.0	56.3
Gathering firewood	57.9	37.5	75.0	68.8
Expanding cropland	10.5	18.8	6.3	31.3
Gathering fruits and medicinal plants	5.3	0.0	25.0	0.0
Number of respondents	19	16	16	16

Source: Interviews with members of the Santa María forest group.

internal dispute for land distribution, the conflict has been part of the regional discussion about land rights in which the two major groups of the region—indigenous and colonist campesinos—are currently involved. Land tenure rights and access to forest resources have been discussed not only in relation to the rights of large landholders and local communities, but also between campesinos and indigenous groups.

There was also a change in the responses given by women about the uses of fruits and medicinal plants from the forest. During the second year, the percentage of women's responses increased from 5.3% to 25%. The responsibilities that women have with family subsistence, as well as their concerns regarding impacts of the forest project on trees and plants that provide food for people and animals, may explain the change in responses. In addition, internal community conditions, such as the scarcity of water and its consequences for people's health, may help to explain why the responses regarding fruits and medicinal plants increased. In the interviews, women mentioned that they collected diverse fruits and plants from the forest for different purposes, such as feeding the family and preparing medicines to treat illness.

Among the most widely used medicinal plants mentioned by women were Caré (*Chenopodium ambrosioides*) and Cedro (*Cedrela odorata*), described as useful for stomach ache. Palo Santo (*Triplaris Americana*) and leaves and husk of Guayabo (*Psidium guajava*) were used for diarrhea and vomiting; Paja Cedrón (*Cymbospogon utratus*) for the treatment of heart problems; Urucú (*Bixa orellana*) for kidney problems; and Piñon (*Jatropha curcas*) for foot swelling. Because women were the main collectors of medicinal plants, they were more interested in and better informed than men concerning the attributes of these plants. A group of women of the forest group asserted the need to learn more about medicinal plants, because most of the community's members did not know about these plants. They also insisted that the community should know about medicinal plants to guard against their disappearance as a result of the clearing and exploitation of the forest.

It is important to clarify that women who responded in this way were specifically those natives of the community descended from the last Sirionó Indian family who lived in the evangelical mission, and one of the first colonist families who came to the community during the time of the mission. Most of the trees, fruits, and medicinal plants and their uses explained by this group of women corresponded to the traditional uses by the indigenous Guarayos, described elsewhere (Bojanic and others 1986; VAIPO 1998).

The problems with the water supply also may have affected these responses. It was reported that many of the children and other members of the families suffered from stomach illness and that not many of them had the money to buy medicines. Using medicinal plants was an alternative to treat stomach problems. This situation made women more aware of the importance of medicinal plants for the family.

Tables 14-1 and 14-2 show some similarities, as well as significant differences, between men and women in the importance given to the forest and the resources most often used. In the men's responses, timber was the most important aspect of the forest, and it was also the resource most often used. Men's evaluation is likely to be related to their experience of working for timber companies. Men's association of forests with timber and employment makes them obvious participants in the BOLFOR forest management proposal, with its exclusive focus on timber. However, the forest management project could also become a new way to reinforce other values of the forest and its resources, especially if women became more involved. Women were more focused on food security, wild game, and medicinal plants than were men. These values could provide additional incentives for communities to manage forests sustainably.

Opportunities to Participate in the Forest Management Project

Table 14-3 shows how women and men would like to participate in the forest management project. In both years, women mentioned most frequently the provision of food for other forest workers (42.1% and 50%, respectively) as the main way they expected to participate in the forest project. Men changed their expectations: In the first year, men expected to participate in the forest project mainly through logging activities (56.3% of responses), but in the second year, this response decreased to 37.0%. In addition to logging, half of the men mentioned inventory and census as a way to participate in the forest project.

Conditions in the village help to explain the differences in men's responses. Logging was not only one of the most important men's activities in Santa María, but also represented a source of socioeconomic status within the community. According to the information collected in the interviews, informal conversations, and workshops, the regular income that men received while working in the timber companies gave them some economic status, and their previous logging experience made them more inclined to participate in this activity rather than forest inventory and census, which were new to them. However, with the planning of the forest project in the community, activities such as census and inventory were specifically emphasized by BOLFOR's team, not only as an important step in the sustainable management of the forest, but also as an activity that provided equal or

Table 14-3. *Percentages of Men and Women Who Anticipated Opportunities to Participate in the Forest Project, 2001 and 2002*

Opportunity mentioned in responses	2001		2002	
	Women	Men	Women	Men
Providing food	42.1	25.0	50.0	0.0
Administration	31.6	25.0	6.3	31.3
Inventory and census	36.8	12.5	6.3	50.0
Logging	10.5	56.3	0.0	37.0
Participating in meetings	10.5	25.0	6.3	12.5
Marketing	10.5	6.3	0.0	12.5
Motivating people	10.5	6.3	37.5	12.5
Preparing food	10.5	0.0	31.1	0.0
Carrying equipment	5.3	0.0	12.5	0.0
Number of respondents	19	16	16	16

Source: Interviews with members of the Santa María forest group.

higher income than logging. In a meeting held between the BOLFOR team and the forest group to plan the first exploration of the potential forest area, the group examined the budget, in which census and inventory had very good potential income because of the specific knowledge required.

Women, in contrast, because of their responsibility for family subsistence, considered providing food as a way to get involved in the project. Providing food is well respected within the community, because it represents and reinforces values of solidarity. Nonetheless, although in the first year men envisioned providing food (25% of responses) as a way to participate, it was not even mentioned in the second year.

It is noticeable in Table 14-3 how women's expectation to participate in the forest project in nontraditional ways, such as census and inventory, administration, logging, and marketing, declined considerably in 2002, while activities like motivating people and preparing food increased from 10.5% to 37.5% and from 10.5% to 31.3%, respectively. To explain the changes in women's perceptions about their opportunities to participate in the forest project, it is necessary to look at the internal situation of gender relations in Santa María and the strategy developed to promote women's participation in the project. These changes exposed how power relationships established different spheres of action for women and men in the community.

The gender division of labor in Santa María defined not only the tasks that men and women should perform, but also the spaces of participation. Women's tasks involved most subsistence activities, including domestic work and agricultural production. Most of those activities were related to the private sphere of the household. Women seldom participated in public activities like community meetings, and when they did they rarely participated in decisionmaking (although there have been some exceptions). Men were in charge of the agricultural activities (clearing, planting, and harvesting), and hunting. At the same time, men usually were the household members who represented the family in public activities, and their participation was expected in decisionmaking at the community level.[3]

When the forest management plan started in the community, BOLFOR also began addressing gender issues in community forestry activities. The main strategy was to invite women to participate in community meetings and encourage them to get involved in forest management training provided by BOLFOR. In general, the strategy had a good response from women, who expressed interest in the project and began participating in the meetings and training.

However, this strategy was undertaken without any knowledge of the structure of gender relationships in the community. The apparent willingness within the community, including men, to facilitate women's participation in the forest project was constrained by intrahousehold power relationships. During the first phase of the research, most of the women interested in participating in meetings and forestry training faced problems of balancing their responsibilities in the household with the time commitment of attendance at meetings. More than the time chosen to hold the meetings, in the evening and in weekends, it was the burden of women's responsibilities that constrained their participation.

By the second phase of the research, Bolaños was told about men's disapproval of women's participation in activities considered to be mainly for men. One of the women interested in participating in the forest project commented that her husband was worried that if she were paid for her work in the forest project she would abandon her family. For this reason, she decided that the best way to participate in the forest project would be through activities related to her traditional role at home, like providing food for those who were going to work in the forest and motivating others to participate in the project (including her own husband). Therefore, changes in women's responses showed gender and social constraints that women in Santa María faced in trying to participate in the forest management project. Promoting women's participation did not automatically guarantee equal opportunities to participate in forestry activities. In addition to these factors, delays in the implementation of the forest management activities discouraged women's participation. In the interviews, one woman expressed that "she preferred to wait until men found a good area for the forest project, before participating in any activity." Therefore, intrahousehold relationships and problems regarding access to forest timber resources may have prevented women from participating in the forestry project.

Men's and Women's Expectations for Each Other's Roles in the Project

Table 14-4 presents data that complement the analysis of women's and men's expectations about participation in the forest project: it shows the expectations of women and men about what each other's role in the project should be.

According to men, women should mainly participate by preparing food (62.5% and 68.8% of responses). These responses coincide not only with the responses given by women in the previous table (Table 14-3) about the way they expected to participate, but also with men's perception of what are considered appropriate roles for women. In the initial interviews in 2001, men also recognized that women could participate in administration (31.2%), washing clothes (25%), motivating people (18.8%), inventory and census (12.5%), and marketing (6.3%). By 2002, a smaller percentage favored these varied forms of participation; only inventory and census increased (to 18.8 from 12.5%), perhaps because these tasks had been the focus of BOLFOR training in the intervening year. The percentage of men mentioning women's participation in motivating people remained the same; the percentage mentioning that women could participate in administration dropped from 31.2% to only 6.3%; and no men mentioned marketing as an activity for women.

During the interviews in the second phase of the research, many men explained that "very little can be done by women in forest operations." Nevertheless, because the forest management plan was a communal project, men also stated "women should participate through activities proper for them and through those that did not require special skills, like preparing food." Two husbands of the women who participated in the first exploration of potential forest area said "they would not permit their wives to go to work in the forest again, because it is not a safe place for women. They would allow women to go to work in the forest only if they [the husbands] were part of the forest crew." In this community, it was considered that the forest was an inappropriate place for women to work. Therefore, the efforts made by BOLFOR to open new opportunities for women's involvement in forest

Table 14-4. *What Men Think Women Should Do in the Project and What Women Think Men Should Do in the Project, 2001 and 2002*

| | 2001 | | 2002 | |
| | Men about women (%) | Women about men (%) | Men about women (%) | Women about men (%) |
Responses				
Preparing food	62.5	0.0	68.8	0.0
Administration	31.2	26.3	6.3	12.5
Logging	0.0	63.2	0.0	43.8
Inventory and census	12.5	26.3	18.8	25.0
Washing	25.0	0.0	18.8	0.0
Marketing	6.3	10.5	0.0	12.5
Motivating people	18.8	0.0	18.8	18.8
Number of respondents	16	19	16	16

Source: Interviews with members of the Santa María forest group.

management would require additional analysis of the cultural factors that constrain their participation. These interviews showed the need for further understanding of gender power relations and the gender division of labor in developing strategies to motivate women's participation in forestry efforts. Although the promotion of women's participation in the project motivated both women and men to think about the possible involvement of other members of the family in forest activities, it also caused resistance from both genders.

Nonetheless, men's responses about food preparation still indicated that they saw a job for the women in forest management through a traditional role (cooking). Women might have a role in the forest project, but only through certain tasks that would not jeopardize the established local division of labor and structure of gender power relationships. In forestry operations, cooks, although necessary, do not have the same status as workers in forest extraction; they also have lower wages than the other forest workers. As cooks, women would maintain the power hierarchy between women and men. This made preparing food an acceptable and appropriate role for women in the implementation of the forest project. Moreover, the number of personnel needed to prepare food for forest workers was minimal; usually two people were enough. Therefore, women's opportunity to participate in the forest project would be reduced in every respect: status, range of opportunities, wages, and participation in decisions about the management of the forest.

By contrast, women considered logging as the main way for men to get involved in the project (63% and 43.8% in 2001 and 2002). Inventory and census were other ways women considered that men could participate (26% and 25%), as well as administration (26.3% and 12.5%). A final way men could participate was in marketing, mentioned by 10.5% of the women in 2001 and 12.5% in 2002. Motivating people emerged in 2002 as another role women envisioned for men (18.8%). No women expected men to cook.

Despite efforts by BOLFOR to integrate gender issues into forest management and to promote women's participation, there were still many cultural and economic factors that constrained their participation. Enhancing women's participation would require a focused and sustained effort, including an understanding of the social and power relationships between women and men, and mechanisms to address the many demands on women's time.

Impact of the Forest Management Project

In addition to understanding the diversity of peoples' views of the forest and its resources and their hopes and expectations for the forest management project, we also wanted to know how people expected the new forest management activities would affect other important aspects of the community's livelihood. Two gender analysis workshops were conducted to explore opinions about the effects of participation in the project. The first discussion generated information on women's and men's view of the division of labor in Santa María. Participants listed the activities normally carried out inside and outside the home and identified members of the family who were involved in each activity, differentiating their involvement according to the time they spent on each activity: "always" for consistent responsibilities and "sometimes" for occasional involvement.

Women's and Men's Views of the Division of Labor

Tables 14-5 and 14-6 show women's and men's views of the appropriate division of labor among members of the household. People generally agreed about which member of the family was in charge of most activities. Women participated in almost all of the activities listed, including those outside of the household. People also acknowledged that most of the women's responsibilities were constant rather than occasional. Men sometimes helped women with some activities, such as carrying wood and feeding animals.

Most of the men's activities were reported to be outside of the home. They were involved in income-earning activities, such as selling food, and public activities, such as community meetings and religious services. Women and men agreed that the whole family went to church. While men reported that they participated in childcare and animal feeding, women did not recognize this. Although men acknowledged that women participated only occasionally in crop field activities, women maintained that they engaged in this work consistently.

Men and women differed most in their views about boys' and girls' involvement in the daily labor division. While women recognized children's participation in many activities, men acknowledged their involvement only in a few. According to women, girls helped their mothers within and outside the home, and boys mainly helped fathers in outside activities. Although girls were involved in more than half of the mother's activities, boys were involved in less than half of the father's activities, according to the women. The men, however, thought that girls and boys participated only in feeding animals and in religious services.

Men saw themselves as being involved in activities that generated income for the family, such as marketing of the agricultural products planted in their *chacos*.

Table 14-5. *Women's View of the Appropriate Division of Labor*

Activities	Women	Men	Boys	Girls
Preparing food for family and animals	A	—	—	S
Carrying wood	A	S	S	S
Washing clothes	A	—	—	S
Caring for children	A	—	—	S
Sewing	A	—	—	S
Planting and harvesting crops[a]	A	A	S	S
Fishing and hunting[a]	S	A	S	S
Community meetings[a]	S	A	—	—
Buying food[a]	S	A	—	—
Selling food[a]	S	A	—	—
Going to church[a]	S	S	S	S

Note: A = always, S = sometimes.
[a]Activities outside of the household.
Source: Gender analysis workshop. Santa María, July 2001.

Table 14-6. *Men's View of the Appropriate Division of Labor*

Activities	Women	Men	Boys	Girls
Preparing food for family and animals	A	S	—	—
Carrying wood	A	S	S	S
Washing clothes	A	—	—	—
Caring for children	A	S	—	—
Sewing	A	—	—	—
Planting and harvesting crops[a]	S	A	—	—
Fishing and hunting[a]	S	A	—	—
Community meetings[a]	S	A	—	—
Buying food[a]	S	A	—	—
Selling food[a]	S	A	—	—
Going to church[a]	S	S	S	S

Note: A = always, S = sometimes.
[a]Activities outside of the household.
Source: Gender analysis workshop. Santa María, July 2001.

Men also were involved in outside activities that gave them a higher social status. It was more acceptable for men to participate in activities in the public sphere, such as community meetings, where they had the opportunity to discuss and to decide about community projects. Thus men were better informed about certain issues, such as the process of land titling or social and economic investments by the government, that were fundamental in the internal decisions about the participation of community members in development projects. Some men had become representatives in local and regional discussions related to the implementation of projects (as leaders of the *Organización Territorial de Base,* or OTB, the *Agente Cantonal,* and the *Corregidor*), which symbolized their higher status in the community and established differences between women's and men's opportunities to participate in decisionmaking. The wives of these leaders did not participate in any public meeting or community project. They stated in individual conversations that, because their husbands had important community responsibilities, they should stay at home attending to all the activities required for the family's subsistence. Therefore, women were excluded not only from public activities, but also from the opportunity to decide and to be informed about essential issues in the community (cf. the situation in Zimbabwean communities, Chapters 8 and 9).

Hunting also gave men social status. The fact that a man had a good hunting record made people recognize his knowledge of the area and of the forest. A good hunter would have the opportunity to become a guide for the foreigners who came to the area looking for land or those who came to hunt game. Although hunting was not the main source of income, it provided additional food and money for the family. The leader of the OTB was recognized for his abilities and knowledge about hunting strategies and the best areas to hunt. He was one of the main providers of wild animal meat within and outside the community. As he reported in an informal conversation, among the animals regularly hunted (*Agouti paca*, collared peccary, white-lipped peccary, deer), *Jochi pintado* (*Agouti paca*) was the most hunted animal because of the quality of meat. Although the family usually consumed the animal, sometimes it was sold in the community or in nearby towns like Ascención de Guarayos, where it was highly prized by owners of local restaurants. A kilogram of meat cost approximately US$1. Before restrictions on hunting game were imposed by the new forest law and the indigenous organization, COPNAG, foreign hunters came to the area and paid US$4–5 a night for a guide. This leader was the most requested and best known in the region by outsiders looking for guides.

These two activities alone—hunting and attending community meetings—established a large difference in the status of women and men within the family and within the community. Women in Santa María did not participate much in the public sphere, such as attending community meetings and participating in decisionmaking related to land tenure and community services, nor did they hunt. These status differentials must be considered in the forest project in order to promote women's participation. Although women were in charge of most of the activities in the family, their lack of involvement in public activities reduced their opportunities in the forest project.

Family Relationships and Responsibilities

In the second workshop, participants were asked to discuss and present their analysis of the likely impacts of the forest project on family relationships and responsibilities. Participants were divided into two groups of women and two groups of men. The analysis showed how family relations and division of labor within the household could affect, and be affected by, the implementation of the forest management project. During the second phase of the research (May–July 2002), members of the forest group were asked to review the previous analysis summarized in Table 14-7 and to examine changes in their perceptions. Through a group discussion, women and men expressed their views.

As we can see in Table 14-7, women and men generally agreed about the likely impacts of the forest project on the labor division and on the level of involvement of members of the family. Because of their previous experience

Table 14-7. *Likely Impacts of the Forest Project on Labor Division and Activities*

	Activities and Roles Affected	
Activities	Women's view	Men's view
Planting and harvesting crops (the *chaco*)	Only women will be in charge of this activity, transferring other activities to children. There will be a problem in families with young children. Some women will not be able to assume this responsibility permanently. Assuming responsibility for this activity will restrict women's participation in the forest project.	Families will pay less attention to the *chaco*. This activity will become women's responsibility. If the forest project provides enough income, and women can get involved in the forest project, they will pay another person to carry out this activity.
Carrying wood	Women and children will continue to be in charge of this activity.	Men will supply enough wood for the family before leaving for the forest.
Hunting	The number of animals will decrease because of the logging of trees that provide food for them. The project will decrease provision of meat from the forest. If the forest project provides money, people can pay other hunters.	The project will decrease provision of meat because women and children do not know how to hunt. The family requires other resources for meat. Women can raise chickens.
Fishing	Women and children will be involved in this activity.	Children and women can carry out this activity consistently.
Attending community meetings	If women are in charge of all activities, they will have no time to participate in community meetings.	Men cannot participate if they are working in the forest. Community meetings should be planned to take account of the new project.
Buying food	It will require planning to provide enough food for family.	It will require planning to provide enough food for family. Men will seek credit in local stores.
Selling food	If production is low, women will continue in charge of this activity.	Depending on the amount of *chaco* production, women will be in charge of selling food.
Preparing food Washing clothes Caring for children Helping children with homework Sewing	These activities will continue as women's responsibility. Children will consistently assume these responsibilities if women participate in the forest project.	Women will organize these activities according to their needs.
Going to church	Children will represent the family when women and men cannot participate.	Men will attend when staying in town. Women or children can represent the family.

Source: Gender analysis workshops, Santa María, 2001 and 2002.

working with timber companies, men expected to be the main participants in the forest project. With men's involvement in the forest project, families would lose important labor inputs in their agricultural plots, in hunting and fishing, and in buying and selling food.

The implementation of the forest project raised concerns about the maintenance of the *chaco,* the main source of food security for the community. Participants expressed their concern about the potential risk of concentrating their efforts on a new production system that could jeopardize their subsistence, despite the project's promises of better income levels. They feared that the participants in the forest project would be involved in new activities and therefore would "pay less attention to the *chaco.*"

Women and men differed, however, in their concerns about the provision of meat and game animals. Women established the interdependence between forests and animals, as the means of provision of meat for the family, and expressed concerns about the reduction of animals as a result of logging activities. Men predicted a likely decline of meat provision because women and children were not skilled hunters. They also proposed that women could add poultry-raising to their other responsibilities, as an alternative source of meat.

Despite general agreement about the expected impact of the project, each gender group resolved probable labor problems differently. Men, deploying the labor of other members of the family, increased the amount of work by women in the domestic sphere and in agricultural activities in order to facilitate their own participation in the project. Men's main duties, such as planting, harvesting, hunting, and fishing, would be transferred to women and children. While men said that this displacement of activities would not cause problems in family roles, women expressed their concern over the large list of permanent activities in which they would be involved.

This overburdening with tasks was one of the main impacts women feared would constrain their opportunity to participate in the forest project (cf. Chapter 12). In the interviews, women expressed their interest in participating in the project's forest census, inventory, and administration, but feared that assuming new duties in the household would restrict their chances to be directly involved in the project. Women's involvement in community meetings could also be restricted by new demands on their labor.

These impacts present a contradiction. Although the forest project offers an opportunity for women to expand their involvement in the public sphere, in which they are culturally restricted, the likely impacts of the project within the household suggest new limitations for women's participation in the project. As participants analyzed it, women's direct involvement in the project would require a redistribution of labor within the family that was contrary to established power relationships. Otherwise, the analysis showed, women's tasks within the household would increase, diminishing their opportunities to participate in the project.

Thus, the strategy to promote women's participation and integrate the forest project into the community requires the understanding not only of family dynamics, but also of the internal bargaining regarding redistribution of labor within the household. The household is not a homogeneous unit with common interests;

instead, family gender relationships are shaped by potentially conflicting interests and gender inequalities in the distribution of tasks, allocation of resources, and opportunities to participate in decision-making (Agarwal 1994; Moser 1993). In decisions made within the family in Santa María, men's interests prevail over women's desire to participate in the forest project. Women are aware of their disadvantaged position regarding the decision to participate in the project. BOLFOR's promotion of women's involvement in the project has lacked efforts to overcome these constraints.

Moreover, women who had participated directly in forest exploration did not obtain good outcomes from their experience. One of the women did not receive any payment for her work, because of problems created by another member; she subsequently left the village to move to Santa Cruz. The other woman who participated in the activity reported conflicts with her husband, who disapproved of her participation in an activity "exclusively for men." These unfavorable experiences discouraged women's participation in forest activities. Men continued to consider that the forest project was mainly for men and that women's cooperation was vital to facilitate their participation. As shown in Table 14-7, women's cooperation implied their permanent involvement in all productive tasks, in addition to their long list of daily activities within and outside of the household.

By documenting the complexity and importance of gender issues in community forestry efforts, this chapter demonstrates the challenge of integrating the forest management project into the community of Santa María. A strategy for integration must recognize the areas of difference and agreement between women's and men's perceptions of the forest and the forest project, because both men and women clearly expressed their interests in participating in the forest project through different activities (providing food, extraction of timber, forest inventory and census, motivating people, etc.), While these perceptions and interests may have offered some limits for women's and men's opportunities to participate, they also may have presented other possible strategies to promote women's involvement in the project. It may be possible to give value to the main ways through which women expected to participate, such as providing and preparing food or motivating people. Although these activities were seen in the community as secondary in forest operations, they are indispensable to maintain momentum and facilitate the development of the diverse stages of the forest project. Addressing these types of activities as important aspects of planning and implementing the forest project would give women better opportunities to participate and may represent a potential opportunity to broaden the focus of the current forest management project beyond its timber emphasis.

Conclusions and Recommendations

Community forest management is embedded inextricably within a social community and within a specific ecological and political setting; it is not simply a "forest enterprise." This embeddedness inserts the community forestry enterprise into the broader policy environment and markets, as well as the community's socioeconomic structure. Cultural divisions, gender roles and relations, diverse views about the value and uses of the forest, multiple and conflicting tenure forms, and unequal power and status could all affect the success of the community forest enterprise in Santa María.

As summarized in Table 14-8, gender differences in views of forest importance and use, potential participation in the timber management project, and the community division of labor could have implications for the project. The table suggests several ways in which an adaptive collaborative approach with the Santa María community could address these potential impacts.

One important finding is that the project's current exclusive focus on timber fails to address other interests, including environmental concerns that could be discussed as part of the community-level planning process. Differences between women's and men's views of the forest and the forest management project suggest that combining the needs and interests of the whole community has the potential to address a wider range of economic, ecological, and social issues that affect the forest management project. For example, the concern expressed by women about the relationship between trees and animals and the likely effects of the project on game animals represents an opportunity for community planning discussion to broaden the concept of sustainable forest management to include conservation of biological diversity. Incorporating their concerns in the planning process would allow women's interests to be represented and could broaden the basis for sustainability of the project. Adopting a broader view of community livelihood systems could help the community to develop goals and measures to monitor the aspects of forest management important to different groups.

The barriers to women's participation in project activities suggest the need to address cultural beliefs, power differences, and time constraints by opening opportunities for nontraditional participation, while emphasizing the value of supporting roles in food provision and motivation. Realizing this potential will require developing training programs appropriate for both men and women and providing opportunities for women to gain experience in public participation and leadership.

The findings also point to potentially disruptive impacts of the project on the community division of labor, jeopardizing food security and overburdening women and children with subsistence activities. The way in which the forest project was proposed to the community failed to consider the cycle of agricultural production in the preliminary planning of the forest project. As a result, men who are involved in the forest project are worried about maintenance of the *chaco;* they are also seeking to reduce subsistence risks. Given the existing division of labor and power relations, women's labor (and children's) is likely to be deployed to the tasks required for the maintenance of the family. Issues such as the timing and distribution of labor demands for essential activities should be considered in planning forest activities, as a way to protect food security and encourage broad community participation in the project.

Women are interested in participating in the project. Yet, according to the analysis of the division of labor, women's multiple responsibilities and their lack of involvement in public activities reduce their opportunities to participate. This was

Table 14-8. Gender Differences and Recommendations for the Forest Management Project

Research findings	Gender differences		Implications for the project	Recommendations
	Women	Men		
Forest importance and uses	Food most important; also concerned with animals and life	Timber and employment of primary importance	Exclusive focus on timber neglects other interests, including the environment	Adopt a systems view and work with the community to analyze linkages, potential conflicts, and complementarities for different groups.
Potential participation in the project	Food provision, motivation, and some nontraditional activities	Logging, census and inventory, administration, and other activities	Opportunity to recruit some women in nontraditional tasks and to enhance the value of women's traditional roles	Recruit and train women for nontraditional tasks, and emphasize the value of women's supporting tasks.
Existing division of labor in the community	Activities inside the home	Outside activities with more income and higher status	Women lack experience and confidence in public meetings and decisionmaking	Work with women and men to facilitate women's leadership and public participation.
Impact of the project on the division of labor	Increased burden of women's and children's tasks in subsistence activities	Men's participation in logging displaces other activities to women and children	Potential overburdening of women, inhibiting participation, and risking food security	Work with the community to plan project activities, protect food security, and facilitate women's participation.

clearly observable during the implementation of administrative training offered by BOLFOR. Only one woman participated consistently; the rest of the women interested in participating decided to involve their sons or husband instead, because they had "previous duties" at home. If women do not overcome the limitations to their participation in meetings, in which training is usually carried out, they will not be able to get involved in the administrative and other tasks of the project. A strategy of working with women and men to address the labor constraints of both might help to address this problem.

Because men dominate the public sphere of participation, they have more control over the decisions about the forest project than do women. The promotion of women's participation in the forest project must go beyond the invitation to participate. It requires exploring and opening up conditions that allow their introduction to new spheres of participation. The knowledge and skills that men have acquired working with timber companies have given them an advantage over women to participate directly in the forest project. Although women have general knowledge and specific concerns that could help to guide the planning of the project, they do not have the necessary weight to influence the decisions that the forest group is taking. The previous involvement of a small group of women in forest activities (such as the training on Forest Management in TCOs) and in the forest group directorate, as well as the leadership that a few women have in the local union, indicate that opportunities do arise to gain space in the public sphere and at the level of decisionmaking. It is important to encourage and maintain the participation of these women in the project, because they may serve as motivation for others to get involved.

Divergent interests within the community may negatively affect the forest project by limiting the opportunity of women and men to participate in it. Understanding and integrating the diverse interests of the groups of the community will be required to plan an adequate forest management project for the Santa María community. In keeping with the approach of CIFOR's Adaptive Collaborative Management program, facilitation by BOLFOR, CIFOR, and UF researchers could help the community members address the diverse issues that concern them, so that they can become directly involved in planning, monitoring, and adapting the forest management project.

Notes

The authors wish to acknowledge the generous support of the BOLFOR project, the Tinker Foundation, the Compton Foundation, and the University of Florida's Tropical Conservation and Development program for funds to support the research reported here.

1. Bolaños's translation of the interviews (July 2001–2002) in Spanish.

2. The INRA law was approved in 1996 to clarify property land rights in order to redistribute land.

3. See the next section on gender analysis for further discussion of the gender division of labor.

Implications of Adaptive Collaborative Management for More Equitable Forest Management

Carol J. Pierce Colfer

T HE CHAPTERS IN THIS BOOK lead the reader through forests in Asia, Africa, and South America, describing both persistent inequities in forest management and efforts to make beneficial changes therein. We have tried to provide a flavor of life in tropical forests, with all the disruption, dilemmas, and change that characterize forest peoples' lives. And we have tried to provide some practical guidance for others with similar interests.

This brief concluding chapter begins with some straightforward, practical tips for forest managers. It then addresses the policy context and related implications of our work, including suggestions for how the adaptive collaborative management (ACM) approach can be integrated into existing bureaucracies. This is followed by a brief somewhat philosophical discussion of two issues pertaining to the work described in this book: the meaning of power and the place of cultural relativity. The book closes with final, brief conclusions about enhancing equity.

Tips for Forest Managers

The chapters of this book, of course, provide new information about the importance of women and other marginalized groups in forest management. There is evidence of their active involvement in informal management, and there are ideas on how to involve them more meaningfully in formal management. Most central to our purpose here, there are tips on how formal forest managers may be able to integrate women and marginalized groups more effectively (also addressed in the introduction to this volume). Some of these tips are recounted below:

- People's roles in forests vary greatly from one place to another and by gender, ethnicity, caste, and wealth group. Forest management planning must take those existing roles into account and be mindful of the division of labor, the costs and benefits of planned actions, and the distribution of benefits within households and within communities. Greater understanding of these local realities can be obtained by directly involving rural women and men of all locally relevant social categories in the planning process (Chapters 1–4, 6, 9, and 11–14) or by formal study (Chapters 5, 7, 8, and 10).

- Participatory Action Research is one broad and comparatively long-term mechanism whereby the voices of members of marginalized groups can be made more audible. Members of such groups can, through this mechanism, both benefit from and contribute to improved forest management.
- Rapid Rural Appraisal techniques can be used, in the form of games, to interest and raise the consciousness of various stakeholders (including marginalized peoples living in forested areas) about their environment.
- Workshops within and between stakeholder groups can be potent tools for arriving at shared goals, determining priorities, dividing up responsibilities, and clarifying benefit distribution. Where great power differences exist between stakeholders, it will probably be necessary to hold separate workshops initially and bring strong facilitation skills to the table, so that the less powerful can feel free to speak.
- Skill in facilitation is a key both to equitable presentation of views and to resolving the conflicts that inevitably arise in attempts to manage resources more equitably.

Policy and Its Implementation

We have focused on the micro level throughout much of this book. But we can and should shift scale, to look briefly at the policy level. As Chitiga and Nemarundwe (2002) have pointed out for southern Africa, many policies turn out, inadvertently, to affect women adversely. Oyono's chapter provides a comparable analysis for another marginalized group, the Pygmies (Chapter 5). One important issue is how we (including communities) can influence policy-makers usefully. Another is how we can inform them of the needs of communities and their subgroups.

Gender specialists have been analyzing the roles of women in agriculture and forestry for decades. Much information is also available from the anthropological literature on community differentiation in the use of forests. Yet, this information seems not to reach policy-makers, and policy inequities persist.

A primary hindrance to better policy interestingly involves the absence of action as much as the presence of inappropriate action. Marchbank (2000) uses "nondecisionmaking theory" in an analysis of how childcare has continued to be ignored in Great Britain, despite considerable interest in and demonstrated need for it among the populace. She documents how widely desired policy changes can be derailed by lack of action at various points in the policymaking hierarchy. Many of the issues that affect natural resource policies as they relate to disadvantaged groups are similar. Compare, for example, Kusumanto and Sari's analysis of the lengthy and circuitous path taken in an Indonesian, district-level decision to undertake a resettlement project supposedly requested by a village (2001). Only the involvement of individuals with special connections within the bureaucracy resulted in action.

Human beings who fail to learn about and fail to act on issues that might improve conditions for the marginalized thereby also make policy. Although communities can make many important changes to improve their own situations, policymakers also will probably need to take action. An easy solution is improbable, but some potentially useful actions may provide checks and balances and smooth the way for more equitable forest management:

- Training of officials about the needs, interests, and desires of the various segments of communities.
- Similar training among marginalized groups to strengthen their self-confidence in interaction with outsiders.
- Collective action among such marginalized groups.
- Networking between them and stakeholders with more wealth and prestige.
- Creation of safe platforms for dialogue between the marginalized and other stakeholders, through use of facilitators and shared rules of behavior.
- Involvement of third parties, such as nongovernmental organizations (NGOs), as advocates and links between stakeholder groups.

Basically, we need to develop a new policy narrative (Roe 1994)—in the sense of a simplified "story" about a policy's purpose—that mandates more equitable involvement of marginalized groups in forests. A common existing one (Policy Narrative No. 1)—held by many nonmarginalized stakeholders—goes something like this:

Forests are dangerous places that are most valuable for their trees. The physical strength of men is necessary in cutting down trees, so forests are a masculine affair. Furthermore, forests are valuable national assets, maintained by the nation to benefit the nation. They therefore need to be managed using scientific methods by specially trained experts. There are few or no people in forests, and the few who are there are primitives, practicing destructive slash-and-burn agriculture. They have little or nothing to contribute to sustaining forests.

The kinds of information provided in this book suggest that Policy Narrative No. 1 has some major flaws. In thinking about a possible substitute, we might consider Policy Narrative No. 2:

Forests are home to many different plants and animals, including people with complex and valuable cultural systems that interact closely with those forests. Forests provide a wealth of products, and they have aesthetic, spiritual, and existence value for many people. Both local men and women use the forest as part of their livelihoods, have valuable knowledge about it, and have strong emotional ties to it. Women have an additional importance, in their reproductive roles, in maintaining the balance between people and their environment. Forests, being important to many (men and women, wealthy and poor, old and young, divergent castes, ethnic groups, and classes), must be managed in a pluralistic fashion, taking these diverse interests into account. Those closest to the forest have the strongest claims on it, bringing unique knowledge, motivation, and opportunity to manage sustainably—in collaboration with others with lesser, but also important, claims.

Some readers may, of course, find this also an unappealing policy narrative. Even for those who consider it a congenial image, there may still be questions about how to realize such an approach to forests and their management, given what we currently have to work with (centralized policies; inflexible, top-down bureaucracies; disempowered field personnel; and dependent, impoverished communities).

In this book, we have talked a lot about how facilitators can work successfully with people in communities, as well as about how communities can work together and with other stakeholders. I think we have been able to show some exciting things that a facilitator—with various qualifications and liberated from an oppressive bureaucracy—can do in a typical community. There is actually a wealth of useful knowledge "out there" about how to work with communities and with marginalized groups. Our findings also confirm the importance of tailoring "solutions" to local conditions (compare the different conditions described in Chapters 13 and 14, which discuss two nearby communities).

How can our experiences be replicated on a broader scale, when many, perhaps most, bureaucracies are characterized by top-down and inflexible "cultures" wherein field personnel have virtually no decision-making rights or authority? Policies are made at the highest level, often designed for vast geographic areas, with people in the lower levels of the bureaucracy simply told to implement them. Can existing field personnel serve in a role that replicates the ACM facilitators' roles?

Here I argue that they can. I believe successful replication will require some high-level, sympathetic bureaucrats, a fair amount of training and sensitization, and on-going support. I think a secret to successful realization of Policy Narrative No. 2 on a broader scale is in the structure—of ideas, of incentives, of sanctions—in which the facilitators work. I think CIFOR's experience along these lines is pertinent.

The ACM team was given considerable freedom within CIFOR's bureaucracy to plan and implement our research, within very broad guidelines. We were able to create for ourselves a congenial, adaptive, and collaborative culture (though very different from other teams at CIFOR). The team members based in Bogor were able to work together to fashion an initial vision for our work in a formal conceptual framework. We agreed from the outset to maintain a flat structure in which input from each team member was as valued as that from any other. Mutual respect was important and generally observed, despite occasional disagreements. We agreed that our diversity of ethnicity, discipline, experience, and gender was a valuable resource on which we should draw, that we could all learn from each other in an on-going way. We agreed that our most important outcomes would happen in the field and that those of us in the office should try our best to serve those on that "front line." We also recognized that those in the field would be confronting both great human and natural diversity and dramatic change; it would be necessary for them to have considerable autonomy and decisionmaking authority. We thought it was important that we be adaptive and collaborative in our work at all levels, in the field and in the office. We also recognized the importance of risk-taking and the necessity to move into the unknown. There was no way that we could predict what specific communities with specific facilitators would be willing or able to plan and implement.

We used the need to be accountable to motivate ourselves to succeed. We recognized that the level of autonomy and flexibility that we enjoyed would quickly dissipate if we were not learning useful things in the process. Team members valued their integral role in program decision-making and worked hard to accomplish our goals. We were also personally committed to the goals, since we had been intimately involved in developing them. We committed ourselves to on-going self-monitoring, through various means (routine and special assessment and planning meetings, CIFOR's performance appraisals, an International Steering Committee, National Steering Committees in each country, an internal impact assessment, etc.). Rarely have I seen such enthusiasm and personal commitment as developed within the ACM program. Similarly, ACM facilitators report outstanding effects in communities where many of the same processes were at work.

This is not to suggest that we had no problems. Some team members, used to a more top-down approach, were initially uncomfortable with the responsibility implied by the approach we took. Some still longed for more conventional, hierarchical leadership, rather than the diffuse and multicentered leadership that characterized our program. Diversity also brought with it differences in perspective and occasional resulting disagreements, which we coped with through discussion and internal facilitation. The great distances among our sites meant we had more difficulty integrating team members in the field than those in accessible offices. To overcome this difficulty, we had (expensive) meetings, worked on ensuring e-mail accessibility, developed a shared virtual space on the web (which we occasionally copied onto a CD and distributed for those with the worst internet connections), used the web page, developed *ACM News* (a quarterly electronic program newsletter), and e-mailed each other a lot. Still this problem plagues us.

But my point is that the kinds of processes we advocate *are* possible to implement. We've done it. And, I am convinced, so could a government or company bureaucracy where some key actors were interested in bringing about a more pluralistic, democratic approach to forest management.

Two Problematic Concepts

Throughout this book two thorny issues have popped up recurrently: the question of power and the question of cultural relativity. Although I have no final answers, I do want to discuss briefly some of the questions and dilemmas that I think anyone dealing with forest communities and their environments will be forced to confront.

On Power

Power and empowerment are controversial terms, with vague meanings in popular discourse and divergent meanings in various bodies of literature. I think of power simply as the ability to act. In tropical forest contexts, the abilities of some people ("the powerless") to act are constrained by others ("the powerful"). The characterizations—powerless and powerful—are not entirely accurate, of course, since the "powerless" can seize power by force, they can achieve their objectives in indirect ways, they can play their parts in a shared cultural drama, etc. (cf. Foucault 1980, Scott 1985). There are structural, cognitive, and interactional elements to the relationships between what we loosely term the powerless and the powerful.

The structural elements are the organizations (government bureaucracies, private industries, village elders, clans, etc.) and institutions (rules, norms) that define the roles individuals hold in relation to others. In these resides the formal authority to use power. The cognitive elements pertain to what Ardener (1975) and her contributors describe as the "muting" of women's voices. The perceptions of women and other "powerless" people are more difficult to articulate in the terms of dominant groups; their perceptions and experience do not fit neatly into the conceptual categories of the "powerful," who are often unaware of the realities of marginalized groups. The voices of the marginalized are less clearly heard (cf. Colfer 1983; Jordan et al. 1993; Davis-Floyd and Sargent 1997; Jordan 1997).

I wonder though if it might not be time for a shift from the structural and cognitive focus we have had, to paying more attention to power as it relates to human interactions. Stacey et al. (2000), management specialists, have argued against our common emphasis on systems (though I remain reluctant to abandon this concept myself), in favor of more attention to connections among individuals within those systems. The authors begin with the observation—one now widely shared—that it is impossible for human beings to make objective observations of human systems in the same way that we can stand aside and look at physical or other biological systems (and some argue against this possibility as well). Stacey et al. argue that we are part of such systems, and thereby affected by our involvement in them. They also emphasize the vital role of human agency in human systems. Such views are consistent with the many authors who argue for pluralistic approaches to forest management, but Stacey et al. make a persuasive case for shifting our emphasis to the connections among people within systems (in their case, in management structures). This suggestion is supported by their observations of how things really get done in organizations (regardless of the formal management systems designed by management specialists and administrators).

Our experience of ACM suggests to me some parallels. In my study designed to identify the conditions that support or interfere with the ACM process, I concluded that the enthusiasm and motivation of the facilitator was far more central than the various characteristics we had anticipated might affect our success. The emotional and cognitive support these individuals received from other team members and colleagues was also critical. The links between the facilitator and other CIFOR staff (or surrogates, like the University of Florida faculty members) were key. One of the facilitators' primary tasks, in turn, was to develop or strengthen equitable links between individuals in their respective sites—in their efforts to catalyze collective action. Such links were between men and women, between ethnic groups, between villagers and government or NGOs, and even between villagers and researchers or facilitators.

Krishna (2002), in a study of social capital, undertook an unusually systematic and careful examination of 69 Indian villages with more intensive case studies in

16 communities. His intention was to evaluate the role of social capital (as a community characteristic) in communities' "development" success. He found social capital—a concept that also depends on the nature of links among people—to be an important element. But its significance was magnified when effective agency was available, usually in the form of a comparatively well-educated villager with good contacts to the outside world—another example of the importance of connections or links. I would like to see us looking at these connections more carefully in future work. It is possible that the concept of power as something that flows like electricity may prove more illuminating than our more typical, and more static, views of power. I would not, however, abandon our interest in the systems within and between which these connections exist; these are also important.

On Cultural Relativity

Another important dilemma to which we have found no real answer is implied in this quote from Porro and Stone (Chapter 12) who say:

> Rather than focusing on how to transform local gender relations so that they *fit into* preestablished development frameworks, development practitioners should build on locally defined gender relations that reflect the realities and complexities of women and men in those communities without (and this is the challenge) reproducing local gender inequities. [Italics in original.]

Sithole, in Chapter 8, makes similar points.

We—and many others—have agreed that equity is an important value and that it has implications for the management of forests. We have also accepted that it is both impractical and unjust to unilaterally deprive the people who depend on a forest of access to it and a say in its future.

We have also accepted the value and importance of cultural systems, of people's familiar and cherished ways of doing things. People depend to varying degrees on their cultural systems for their preferred subsistence strategies, their desires and expectations for the future, their ways of raising their young, their security in old age, the meaning in their lives. In many instances, these two concerns or commitments—to equity and to cultural relativity—clash. Cultural systems (without which we would not be human, as widely defined) have tended to be inequitable. This is equally true of the cultures that are linked to forests.

Although only two of the contributions have highlighted this issue dramatically, it has been a dilemma on all sites. It calls for great care in the conduct of processes like ACM. Facilitators can expect to have to make difficult decisions in the process of working with community members to strengthen the positions of the weak, while respecting the very systems that define those people as weak. But it is worth the effort.

Fundamental Conclusions about Enhancing Equity

Our primary conclusions from this experience are:

- Women and other marginalized peoples residing in and around forests have something significant to offer to, and gain from, improved equity in forest management.
- The effort to involve both communities and relevant segments of communities in the forest management process is important for maintaining and enhancing forest health and human well-being.
- Involving these previously ignored groups is feasible, using the kinds of methods and approach outlined in this book.

The approaches we have identified in this collection complement each other; no one is sufficient alone or relevant everywhere. Indeed, the most fundamental conclusion from our work is that the diversity, complexity, and dynamism that characterize human and forest systems *require* first that we learn to respect, rely on, and strengthen the knowledge and creativity of communities living in forests; and, second, that we train, mentor, and learn to trust those who work in forests and with forest communities jointly to devise locally appropriate solutions to equity and forest management problems. Otherwise, we are doomed to prolong the standardized, top-down, and demonstrably ineffective techniques we have been wedded to for so long—which we know contribute neither to more sustainable forest management nor to more equitable human systems.

References

Abbott, J., and I. Guijt. 1998. *Changing Views on Change: Participatory Approaches to Monitoring the Environment*. London: International Institute for Environment and Development.

Abilogo, E. 2001. Un Peuple en Péril. *Bubinga* 50: 4.

ACM (Adaptive Collaborative Management) Team of CIFOR (Center for International Forestry Research). 1999. Report of the Workshop on Building an Agenda Together *(Bangun Rencana Bersama)* and Mapping Training *(Pelatihan Pengenalan Pemetaan)*. Long Loreh, East Kalimantan, November 1999. Bogor, Indonesia: CIFOR.

———. 2001. Report to International ACM Steering Committee, October 2001, Manila, Philippines. Bogor, Indonesia: CIFOR.

Adhikary, G. 1987. Institutional Practices and the Nepalese Poor. Rural Poverty Research Paper Series No. 7. Kathmandu, Nepal: Winrock International Institute for Agricultural Development.

Agarwal, B. 1994. *A Field of One's Own: Gender and Land Rights in South Asia*. Cambridge, UK: Cambridge University Press.

———. 2000. Conceptualising Environmental Collective Action: Why Gender Matters. *Cambridge Journal of Economics* 24: 283–331.

Agrawal, A. 1995. Dismantling the Divide between Indigenous and Scientific Knowledge. *Development and Change* 26: 413–439.

AGRITEX (Department of Agricultural, Technical and Extension Services). 1998. *Learning Together through Participatory Extension: A Guide to an Approach Developed in Zimbabwe*. Harare, Zimbabwe: AGRITEX.

Akono, J.-M. 1995. Le Braconnage dans la Réserve du Dja: Une Grave Menace. *Moabi* 2(5).

Allen, W. 2001. Working Together for Environmental Management: The Role of Information Sharing and Collaborative Learning. Ph. D. thesis in Development Studies, Massey University, New Zealand, 12–29.

Allen, W. J., K. Brown, T. Gloag, J. Morris, K. Simpson, J. Thomas, and R. Young. 1998. Building Partnerships for Conservation in the Waitaki/Mackenzie Basins. Landcare Research Contract Report LC9899/033. Lincoln, New Zealand: Manaaki Whenua Landcare Research.

Almeida, A. 1995. *Quebradeiras de Côco Babaçu: Identidade e Mobilização. Legislação Específica e Fontes Documentais e Arquivísticas (1915–1995)*. III Encontro Interestadual das Quebradeiras de Côco Babaçu. São Luís, Brazil: Estação.

———. 2001. Preços e Possibilidades: A Organização das Quebradeiras de Côco Babaçu Face à Segmentação dos Mercados. In *Economia do Babaçu: Levantamento Preliminar de Dados,* edited by

A. Almeida, J. Shiraishi Neto, and B. Mesquita. Movimento Interestadual das Quebradeiras de Côco Babaçu. São Luís, Brazil: Balaios Typographia, 27–46.

Alrøe, H. F., and E. S. Kristensen. 2002. Towards a Systemic Methodology in Agriculture: Rethinking the Role of Values in Science. *Agriculture and Human Values* 19: 3–23.

Althabe, G. 1965. Changements Sociaux chez les Pygmées Baka de l'Est-Cameroun. *Cahiers d'Etudes Africaines* 5/4(20): 561–592.

Amaral, P., and M. Amaral Neto. 2000. *Manejo Florestal Comunitário na Amazônia Brasileira: Situaçao Atual, Desafios e Perspectivas.* Brasília, Brazil: Instituto Internacional de Educaçao do Brasil (IIEB).

Anau, N. 1999. Laporan Studi Pengamatan Partisipatif di Desa Langap, Paya Seturan, Tanjung Nanga, Long Jalan, Nunuk Tanah Kibang, Long Rat, Setarap dan Setulang, Kecamatan Malinau, Kabupaten Bulungan, Mei, Juni dan Agustus 1999. Report to CIFOR. Bogor, Indonesia: CIFOR.

Anderson, A., P. May, and M. J. Balick. 1991. *The Subsidy from Nature: Palm Forests, Peasantry, and Development on an Amazon Frontier.* New York: Columbia University Press.

Anderson, D., and R. Grove (eds.). 1987. *Conservation in Africa: People, Policies and Practices.* New York: Cambridge University Press.

Anderson, J. (ed.). 1998. Reconciling Multiple Interests in Forestry. *Unasylva* 49(194).

Anderson, J., J. Clément, and L. V. Crowder. 1999. Pluralism in Sustainable Forestry and Rural Development: An Overview of Concepts, Approaches and Future Steps. In *Pluralism and Sustainable Forestry and Rural Development: Proceedings of the International Workshop on Pluralism and Sustainable Forestry and Rural Development.* December 1997, Rome, Italy. Rome: FAO (Food and Agriculture Organization of the United Nations).

Annaud, M., and S. Carrière. 2000. *Les Communautés des Arrondissements de Campo et Ma'an: Etat des Connaissances.* Projet Campo-Ma'an. Paris: Groupe d'Etude des Populations Forestières Equatoriales.

Anonymous. 1958. Pygmeeën Worden Landbouwers (Pygmies Become Farmers). *Africa Christo* 6: 23–24.

———. 1999. *Kecamatan Batu Sopang dalam Angka Tahun 1999* (Batu Sopang Sub-district in Figures, 1999). Batu Kajang, Indonesia: Pemerintah Kabupaten Pasir, Kecamatan Batu Sopang.

Antona, M., and D. Babin. 2001. Multiple Interest Accommodation in African Forest Management Projects: Between Pragmatism and Theoretical Coherence. *International Journal of Agriculture, Resources, Governance and Ecology (IJARGE)* 1(3/4): 286–305. Special issue on accommodating multiple interests in local forest management.

Arce, A., M. Villarreal, and P. de Vries. 1994. The Social Construction of Rural Development: Discourses, Practices and Power. In *Rethinking Social Development: Theory, Research, and Practice,* edited by David Booth. Harlow, UK: Addison Wesley Longman, 152–171.

Ardener, S. 1975. *Perceiving Women.* New York: Wiley.

Arda-Minas, L. 2002. Cut Flowers in Lantapan. *ACM News* 3(2): 5–6.

Arnold, J. 1991. *Community Forestry: Ten Years in Review.* Community Forestry Note 7. Rome, Italy: FAO (Food and Agriculture Organization of the United Nations).

Arnstein, S. R. 1969. Ladder of Citizen Participation. *Journal of American Institute of Planners* 35: 216–224.

Asanga, C. A. 2001. Facilitating Viable Partnerships in Community Forest Management in Cameroon: The Case of the Kilum-Ijim Mountain Forest Area. In *Social Learning in Community Forestry,* edited by E. Wollenberg, D. Edmunds, L. Buck, J. Fox, and S. Brodt. Bogor, Indonesia: CIFOR; Honolulu, HI: East-West Center, 21–44.

Assembe, S. 2001. Participation des Minorités à la Cogestion d'un Massif Forestier: Cas des Pygmées dans la Forêt Communale de Dimako. Yaoundé, Cameroon: CIFOR.

Atsiga Essala, L. 1998. Le Projet 'Intégration Socio-Economique des Pygmées du Cameroun.' Document de travail. Yaoundé, Cameroon: Planet-Survey.

———. 1999. L'Exploitation des Populations Marginales: Le Cas des Pygmées du Cameroun. *Cahier Africain des Droits de l'Homme* 2: 12–16.

Bahuchet, S. 1988. Food Sharing among the Pygmies of Central Africa: A Comparison between the Aka and the Baka, Gyeli (Kola) and Mbuti. *African Study Monographs* 11(1): 27–53.

——— (ed.). 1992. *La Situation des Populations Indigènes des Forêts Denses Humides.* Brussels: ULB.

———. 1993. Afrique Centrale. In *Situation des Populations Indigènes des Forêts Denses Humides,* edited by S. Bahuchet and P. de Maret. Brussels: European Commission, 389–441.

Bahuchet, S., and H. Guillaume. 1982. Aka–Farmers' Relations in the Northwest Congo Basin. In *Politics and History of Band Societies,* edited by E. P. Leacock and R. B. Lee. Cambridge, UK: Cambridge University Press, 189–211.

Bailey, R. C. 1988. The Significance of Hypergyny for Understanding Subsistence Behaviour among Contemporary Hunters and Gatherers. In *Diet and Subsistence: Archaeological Perspectives,* edited by B. Kennedy and V. LeMoine. Calgary, Canada: Calgary University Press.

Bailey, R. C., S. Bahuchet, and B. Hewlet. 1992. Development in the Central Africa Rainforests: Concern for Forest Peoples. In *Conservation of West and Central African Rainforests,* edited by K. Cleaver, M. Munasinghe, M. Dyson, N. Egli, A. Peuker, F. Wencélius. Washington, DC: World Bank, 202–212.

Bailey, R. C., G. Head, M. Jenike, B. Owen, R. Rechtman, and E. Zechenter. 1989. Hunting and Gathering in Tropical Rainforest: Is It Possible? *American Anthropologist* 91(1): 59–82.

Bajracharya, B. 1994. *Gender Issues in Nepali Agriculture: A Review.* Kathmandu, Nepal: HMG Ministry of Agriculture and Winrock International.

Barbier, J.-C. 1978. *Les Pygmées de la Plaine de Tikar au Cameroun.* Yaoundé, Cameroon: ONAREST (Office National de la Recherche Scientifique et Technique).

Barr, C., E. Wollenberg, G. Limberg, N. Anau, R. Iwan, I. M. Sudana, M. Moeliono, and T. Djogo. 2001. *The Impacts of Decentralization on Forests and Forest-Dependent Communities in Kabupaten Malinau, East Kalimantan.* CIFOR Monograph. Bogor, Indonesia: CIFOR.

Bassett, T. 1993. Introduction: The Land Question and Agricultural Transformation in Sub-Saharan Africa. In *Land in African Agrarian Systems,* edited by T. Bassett and D. Crummey. Madison, WI: University of Wisconsin, 3–34.

Bateson, G. 1972. *Steps to the Ecology of Mind.* New York: Ballantine Books.

Baviskar, A. 2001. Forest Management as Political Practice: Indian Experiences with the Accommodation of Multiple Interests. *International Journal of Agriculture, Resources, Governance and Ecology (IJARGE)* 1(3/4): 243–263. Special issue on accommodating multiple interests in local forest management.

Bawden, R. 1991. Systems Thinking and Practice in Agriculture. *Journal of Dairy Science* 74: 2362–2373.

Bayart, J.-F. 1993. *The State in Africa: The Politics of the Belly.* London: Longman.

Bebbington, A. 1994. Theory and relevance in indigenous agriculture: Knowledge, agency, and organization. In *Rethinking Social Development: Theory, Research, and Practice,* edited by David Booth. Harlow, UK: Addison Wesley, Longman, Ltd., 202–205.

Berezin, M. 1999. Democracy and Its Others in a Global Polity. *International Sociology* 14(3): 227–243.

Berkes, F. 1997. New and Not-So-New Directions in the Use of the Commons: Co-Management. In *Proceedings of the Workshop on Future Directions for Common Property Theory and Research, 28 February 1997,* edited by B. J. McCay and B. Jones. New Brunswick, NJ: Ecopolicy Center for Agricultural, Environmental and Resource Issues. http://www.indiana.edu/~iascp/webdoc.html (accessed 01/03/2002).

Berkes, F., P. George, and R. J. Preston. 1991. Co-Management. *Alternatives* 18(2): 12–18.

Berry, S. 1993. *No Condition Is Permanent: The Social Dynamics of Agrarian Change in Sub-Saharan Africa.* Madison, WI: University of Wisconsin Press.

Bickford, S. 1999. Reconfiguring Pluralism: Identity and Institutions in the Egalitarian Polity. *American Journal of Political Science* (43)1: 86–108.

Biesbrouck, K. 1999. Agriculture among Equatorial 'Hunters-Gatherers' and the Process of Sedentarization: The Case of the Bagyeli in Cameroon. In *Central African 'Hunters-Gatherers' in a Multidisciplinary Perspective: Challenging Elusiveness,* edited by K. Biesbrouk, S. Elders, and G. Rossel. Leiden, Netherlands: Research School CNWS, 189–206.

Bigombé, P. 1999. Pipeline Tchad-Cameroun: Le Point de Vue d'une ONG Camerounaise. *Le Journal d'ICRA* 34: 6–8.

———. 2001. Trajectoires du Déclassement Social des Pygmées: Les Logiques d'Une Citoyenneté Banalisée. *Bubinga* 50: 4–6.

———. 2002. Les Pygmées Entre la Forêt et le Village Global: Quelles Chances de Survie? *Enjeux* 13: 5–6.

———. 2003. The Decentralized Forestry Taxation System in Cameroon: Local Management and State Logic. Working Paper. Washington, DC: World Resources Institute.

Bigombé, P., and J.-P. Bell. 1998. Les Modes de Vie Traditionnels et Actuels des Pygmées. Document de travail. Yaoundé, Cameroon: Planet-Survey.

Bikié, H., J.-G. Collomb, L. Djomo, S. Minnemeyer, R. Ngoufo, and S. Nguiffo. 2000. *An Overview of Logging in Cameroon.* Research Report. Washington, DC: World Resources Institute.

Bisilliat, J. 1992. Introduction. In *Relation de Genre et Développement: Femmes et Sociétés,* edited by J. Bisilliat, F. Pinton, and M. Lecarme. Paris: ORSTOM (French Institute for Scientific Research in Overseas Development and Cooperation).

Bojanic-Helbingen, A., et al. 1986. La Agricultura Guaraya: Informe de Visita a la Zona de Guarayos. Documento de Trabajo No. 56. Santa Cruz, Bolivia: CIAT (International Center for Tropical Agriculture).

Booth, D. 1994. *Rethinking Social Development: Theory, Research, and Practice.* Harlow, UK: Addison Wesley Longman.

Bopda, A. 1993. Le Secteur Vivrier Sud-Camerounais Face à la Crise de l'Économie Cacaoyère. *Travaux de l'Institut de Géographie de Reims* 83–84: 109–122.

Borrini-Feyerabend, G. 1996. *Collaborative Management of Protected Areas: Tailoring the Approach to the Context in Issues in Social Policy.* Gland, Switzerland: IUCN (International Union for the Conservation of Nature). http://www. iucn. org/themes/spgeng/Tailor/Tailor. html (accessed 07/02/2002).

———. 1997. *Beyond Fences: Seeking Social Sustainability in Conservation.* Volume 2: A Resource Book. Gland, Switzerland: IUCN.

Borrini-Feyerabend, G., M.T. Farvar, J. C. Nguinguiri, and V. A. Ndangang. 2000. *Co-Management of Natural Resources: Organising, Negotiating and Learning-by-Doing.* Heidelberg, Germany: Kasparek Verlag for GTZ (German Agency for Technical Cooperation) and IUCN.

Bradley, P. 1991. *Woodfuel, Women and Woodlots.* Volume 1. London: Macmillan.

Braidotti, R., E. Charkiewicz, et al. 1994. *Women, the Environment, and Sustainable Development: Towards a Theoretical Synthesis.* London: Zed Books/INSTRAW.

Bratton, M. 1994. Peasant and State Relations in Postcolonial Africa: Patterns of Engagement and Disengagement. In *State Power and Social Force: Domination and Transformation in the Third World,* edited by J. S. Midgal, A. Kohli, and V. Shue. New York: Cambridge University Press, 231–255.

Bratton, M., and N. van de Walle. 1997. *Democratic Experiments in Africa. Regime Transitions in Comparative Perspective.* Cambridge, UK: Cambridge University Press.

Brito, M. 1999. O Papel do Técnico Como Facilitador Nos Processos de Capacitação. Série Cadernos Temáticos 4. Recife-PE, Brazil: Projeto Banco do Nordeste/PNUD (UNDP).

Bromley, D.W., and M. Cernea. 1989. *The Management of Common Property Natural Resources.* World Bank Discussion Paper 27. Washington, DC: World Bank.

Brown, D. 1999. *Principle and Practice of Forest Co-Management: Evidence from West-Central Africa.* European Union Tropical Forestry Paper 2. London: Overseas Development Institute.

Brown, K., and S. Lapuyade. 2001. Changing Gender Relationships and Forest Use. A Case Study from Komassi, Cameroon. In *People Managing Forests: The Link between Human Well-Being and Sustainability,* edited by C. J. P. Colfer and Y. Byron. Washington, DC: Resources for the Future; Bogor, Indonesia: CIFOR, 90–110.

Brubaker, R. 1992. *Citizenship and Nationhood in France and Germany.* Cambridge, MA: Harvard University Press.

Bruce, J. W. 1990. Legal Issues in Land Use and Resettlement. Background paper prepared for the World Bank Zimbabwe Agriculture Sector Memorandum. Washington, DC: World Bank.

Buck, L., E. Wollenberg, and D. Edmunds. 2001. Social Learning in the Collaborative Management of Community Forests: Lessons from the Field. In *Social Learning in Community Forestry,* edited by E. Wollenberg, D. Edmunds, L. Buck, J. Fox, and S. Brodt. Bogor, Indonesia: CIFOR; Honolulu, HI: East-West Center, 1–20.

Buck, S. J. 1999. Multiple-Use Commons, Collective Action, and Platforms for Resource Use Negotiation. *Agriculture and Human Values* 16: 237–239.

Bunker, S. G. 1985. *Underdeveloping the Amazon: Extraction, Unequal Exchange, and the Failure of the Modern State.* Urbana, IL: University of Illinois Press.

Buttel, F. H., and P. McMichael. 1994. Reconsidering the Explanandum and Scope of Development Studies: Toward a Comparative Sociology of State-Economy Relations. In *Rethinking*

Social Development: Theory, Research, and Practice, edited by David Booth. Harlow, UK: Addison Wesley Longman, 42–61.

Buvinic, M. 1983. Women's Issues in Third World Poverty: A Policy Analysis. In *Women and Poverty in the Third World,* edited by M. Buvinic, M. Lycette, and W. P. McGreevey. Baltimore, MD: John Hopkins University Press, 14–33.

Calhoun, C. 1994. Social Theory and the Politics of Identity. In *Social Theory and the Politics of Identity,* edited by C. Calhoun. Cambridge, UK: Blackwell, 9–36.

Campell, C. 1996a. Forest, Field and Factory: Changing Livelihood Strategies in Two Extractive Reserves in the Brazilian Amazon. Ph. D. thesis, University of Florida, Gainesville.

———. 1996b. Out on the Front Lines but Still Struggling for a Voice: Women in the Rubber Tappers' Defense of the Forest in Xapuri, Acre, Brazil. In *Toward a Feminist Political Ecology: Global Perspectives and Local Experiences,* edited by D. E. Rocheleau, B. Thomas-Slayter and E. Wangari. New York: Routledge, 27–61.

Carret, J.-C. 2000. La Réforme de la Fiscalité Forestière au Cameroun: Débat Politique et Analyse Economique. *Bois et Forêts des Tropiques* 264: 37–51.

Carroll, R. W. 1986. Status, Distribution, and Density of Lowland Gorilla, Forest Elephant, and Associated Fauna. Study Report. New Haven, CT: Yale University, School of Forestry and Environmental Studies.

Carter, J. 1996. *Recent Approaches to Participatory Forest Resource Assessment.* Rural Development Forestry Study Guide 2 in association with the Forestry Research Programme. London: Overseas Development Institute.

Castillo-Fiel, C. de. 1949. Los Bayele, un Grupo de Pigmoide de la Guinea Espanola. *Antropologica y Etnologica* 4: 17–28.

Castro, M. G., and M. Abromovay. 1997. *Gênero e meio ambiente.* São Paulo, Brazil: Cortez Editora; Brasilia, Brazil: UNESCO and UNICEF.

Cavalli-Sforza, L. L. (ed.). 1986. *African Pygmies.* New York: Academic Press.

Cayres, G. 1999. Nazarenos e Marias do Rio Capim: Análise de Gênero em Uma Comunidade Amazônica. M. Sc. thesis, Núcleo de Altos Estudos Amazônicos (NAEA), Universidade Federal do Pará (UFPA), Belém-PA, Brazil.

Cayres, G., and R. Costa. 2000. Análise de Mão-de-Obra No Sistema de Produção Familiar de uma Comunidade Amazônica. In *X Congresso Internacional de Sociologia Rural.* Rio de Janeiro, Brazil: Sociedade Brasileira de Economia e Sociologia Rural.

Cernea, M. M. (ed.). 1991. *Putting People First: Sociological Variables in Rural Development.* New York: Oxford University Press for the World Bank.

———. 1995. *Putting People First. Sociological Variables in Rural Development.* Second edition, revised and expanded. New York: Oxford University Press for the World Bank.

Cerqueira, R. R. 1997. Técnicas de Dinâmica de Grupo para uma Capacitação Ativa. *Série Cadernos Metodológicos* 3. Recife, Brazil: Projeto Banco do Nordeste/PNUD (UN Development Programme).

Chambers, R. 1983. *Rural Development: Putting the Last First.* London: Longmans.

———. 1997. *Whose Reality Counts? Putting the First Last.* London: ITGD (Intermediate Technology Development Group) Publishing.

———. 1998. Beyond "Whose Reality Counts?" New Methods We Now Need. In *People's Participation: Challenges Ahead,* edited by O. Fals Borda. New York: APEX Press; London: ITGD (Intermediate Technology Development Group) Publishing, 105–130.

Chauveau, J. P. 1992. Le "Modèle Participatif" de Développement Rural Est-Il Alternatif? Éléments pour une Anthropologie de la Culture des "Développeurs." *Bulletin APAD* (Euro-African Association for the Anthropology of Social Change and Development) 3: 20–30.

Chayanov, A. V. 1986. *A. V. Chayanov on the Theory of Peasant Economy.* Madison, WI: University of Wisconsin Press.

Checkland, P. B. 1981. *Systems Thinking, Systems Practice.* Chichester, UK: John Wiley.

———. 1989. Soft Systems Methodology. *Human Systems Management* 8: 273–289.

Chein, I., S. W. Cook, and J. Harding. 1948. The Field of Action Research. *American Psychologist* 3: 43–50.

Cheneaux-Repond, M. 1992. The Economic and Social Situation of Widows and Divorced Women on a Resettlement Scheme in Zimbabwe. In *Working Papers of Inheritance,* edited by J. Stewart.

Women and Law in Southern Africa Research Project Working Paper No. 5. Harare, Zimbabwe: Women and Law in Southern Africa Research Project, 22–52.

Chitiga, M., and N. Nemarundwe. 2002. Policies and Gender Relationships and Roles in the Miombo Woodland Region of Southern Africa. Harare, Zimbabwe: CIFOR.

CIFOR (Center for International Forestry Research). 2001. *Working to Sustain Tropical Forests and the Communities They Serve*. Annual Report 2000. Bogor, Indonesia: CIFOR.

———. Undated. *Local People, Devolution and Adaptive Collaborative Management of Forests: Researching Conditions, Processes and Impacts*. Bogor, Indonesia: CIFOR.

Cleaver, F. 2002. Institutions, agency, and the limitations of participatory approaches to development. In *Participation: The New Tyranny?* edited by B. Cooke and U. Kothari. London: Zed Books, 36–55.

Cleavey, F. 2000. Analyzing Gender Roles in Community Natural Resources Management: Negotiations, Life Courses and Social Inclusion. *IDS Bulletin* 31(2): 60–68.

Cohen, J. L. 1999. Changing Paradigms of Citizenship and the Exclusiveness of the Demos. *International Sociology* 14(3): 245–268.

Colfer, C. J. P. 1983. On Communication among "Unequals." *International Journal of Intercultural Communication* 7: 263–283.

———. Forthcoming. *The Complex Forest: Communities, Uncertainty, and Adaptive Collaborative Management*. Washington, DC: Resources for the Future.

Colfer, C. J. P., M. A. Brocklesby, C. Diaw, P. Etuge, M. Günter, E. Harwell, C. McDougall, N. M. Porro, R. Porro, R. Prabhu, A. Salim, M. A. Sardjono, B. Tchikangwa, A. M. Tiani, R. L. Wadley, J. Woelfel, and E. Wollenberg. 1999a. *The BAG (Basic Assessment Guide for Human Well-Being)*. C&I Toolbox Series No. 5. Bogor, Indonesia: CIFOR.

———. 1999b. *The Grab Bag: Supplementary Methods for Assessing Human Well-Being*. C&I Toolbox Series No. 6. Bogor, Indonesia: CIFOR.

Colfer, C. J. P., and Y. Byron. 2001. *People Managing Forests: The Link between Human Well-Being and Sustainability*. Washington, DC: Resources for the Future; Bogor, Indonesia: CIFOR.

Colfer, C. J. P., with R. G. Dudley, in cooperation with H. Hadikusumah, Rusyidi, N. Sakuntaladewi and Amblani. 1997. *Peladang Berpindah di Indonesia: Perusak atau Pengelola Hutan?* (Shifting Cultivators of Indonesia: Marauders or Managers of the Forest?) FAO Community Forestry Case Study Series, No. 6. Rome: FAO (Food and Agriculture Organization of the United Nations).

Colfer, C. J. P., R. Prabhu, et al. 1999. *Who Counts Most? Assessing Human Well-Being in Sustainable Forest Management*. Bogor, Indonesia: CIFOR.

Commons, J. R. 1968. *Legal Foundations of Capitalism*. Madison, WI: University of Wisconsin Press.

Congels, S., and P. Pasquet. 2000. Vivre à Mvi'ilimengalé: Activités Quotidiennes et Gestion du Temps chez les Ntumu du Sud Cameroun. In *L'Homme et la Forêt Tropicale,* edited by S. Bahuchet, D. Bley, H. Palezy, and N. Vernazza-Licht. Marseille, France: SEH/APFT, Editions de Bergier, 175–189.

Contreras, A., L. Dachang, D. Edmunds, G. Kelkar, D. Nathan, M. Sarin, N. Singh, and E. Wollenberg. 2001. Creating Space for Local Forest Management. Draft Report. Bogor, Indonesia: CIFOR.

Cooke, B. 2002. The social psychological limits of participation. In *Participation: The New Tyranny?* edited by B. Cooke and U. Kothari. London: Zed Books: 103–121.

Cooke, B., and U. Kothari (eds.). 2002. *Participation: the New Tyranny?* London: Zed Books.

Cornwall, A., and S. Fleming. 1995. Context and Complexity: Anthropological Reflections on PRA. *PLA Notes* 24: 8–12.

Cornwall, A., I. Guijt, and A. Welbourne. 1994. Acknowledging Process: Methodological Challenges for Agricultural Research and Extension. In *Beyond Farmer First: Rural People's Knowledge, Agricultural Research and Extension Practice,* edited by I. Scoones and J. Thompson. London: IIED (International Institute for Environment and Development); London: ITGD (Intermediate Technology Development Group) Publishing, 98–117.

Croll, E., and D. Parkin (eds.). 1992. *Bush Base, Forest Farm: Culture, Environment and Development*. London: Routledge.

CTA (Center of Amazonian Workers). 1995. Manejo Florestal de Uso Múltiplo na Reserva Extrativista de Porto Dias. Proposal submitted to the Pilot Program to Conserve the Brazilian Rain Forest (PPG-7). Unpublished.

Cunha dos Santos, M., and I. Salgado. Undated. Manejo Florestal: Potencialidades e Limites de Sua Aplicação na Amazônia Brasileira. Unpublished.

Dangbégnon, C. 1998. Collective Action for Regenerative Natural Resource Management: Case Studies for Benin. Ph. D. thesis, Wageningen Agricultural University, Wageningen, Netherlands.

Davies, S. 1994. Introduction: Information, Knowledge and Power. *IDS Bulletin* 2: 1–13.

Davis-Floyd, R., and C. Sargent (eds.). 1997. *Childbirth and Authoritative Knowledge: Cross-Cultural Perspectives.* Berkeley, CA: University of California Press.

De Boo, H. L., and K. F. Wiersum. 2002. *Adaptive Management of Forest Resources: Principles and Process.* Forest and Nature Conservation Policy Group Discussion Paper 2002–04. Wageningen, Netherlands: Forest and Nature Conservation Policy Group, Wageningen University.

De Foy, G. P. 1984. *Les Pygmées d'Afrique Centrale.* Paris: Éditions Parenthèses.

Denholm, J. 1990. Reaching Out to Forest Users: Strategies for Involving Women. In *Perspectives on the Role of Women in Mountain Development: Selected Papers,* edited by D. Bachracharya. Kathmandu, Nepal: ICIMOD (International Center for Integrated Mountain Development).

Diamond, I., and G. Orenstein (eds.). 1990. *Reweaving the World: The Emergence of Ecofeminism.* San Francisco, CA: Sierra Club Books.

Diaw, M. C. 1997. *Si, Nda Bot and Ayong: Shifting Cultivation, Land Use and Property Rights in Southern Cameroon.* Rural Development Forestry Network, Paper 21e. London: Overseas Development Institute.

———. 1998. *From Sea to Forest: An Epistemology of Otherness and Institutional Resilience in Nonconventional Economic Systems.* Vancouver: International Association for the Study of Common Property.

Diaw, M. C., and J.-C. S. Njomkap. 1998. La Terre et le Droit: Une Anthropologie Institutionnelle de la Tenure Coutumière, de la Jurisprudence et du Droit Foncier chez les Peuples Bantou et Pygmée du Cameroun Méridional Forestier. Document de travail. Yaoundé, Cameroon: INADES.

Diaw, M. C., and P. R. Oyono. 1998. Déficits et Instrumentalité des Itinéraires de Décentralisation de la Gestion des Forêts Camerounaises. *Bulletin Arbres, Forêts et Communautés Rurales* 15/16: 20–25.

Diaw, M. C., P. R. Oyono, F. Sangkwa, C. Bidja, J. Nguiébouri, and S. Efoua. 1998. *Social Science Methods Assessing Criteria and Indicators of Sustainable Forest Management: Tests Conducted in the Cameroon Humid Forest Benchmark and in the Lobé and Ntem River Basins.* Yaoundé, Cameroon: IITA/CIFOR.

Dinas Perhutanan dan Konservasi Tanah in cooperation with PPLH (the Environmental Studies Center), Mulawarman University. 1999. *Identifikasi Potensi dan Data Keanekaragaman Hayati pada Kawasan Lindung dan Daerah Penyangga: Kawasan Lindung Gunung Lumut Desa Rantau Buta.* (Identification of Potential and Biodiversity Data for the Protected Area and its Buffer Zone: Gunung Lumut Protected Area, Rantau Buta, Pasir District). Tanah Grogot, Indonesia: Dinas Kehutanan Kabupaten Pasir (Pasir District Forestry Service).

Dissnayake, W. 1986. Communication Models in Knowledge Generation, Dissemination and Utilization Activities. In *Knowledge Generation, Exchange and Utilization,* edited by G. M. Beal, W. Dissanayake, and S. Konoshima. Boulder, CO: Westview Press.

DiZerega, G. 2000. *Persuasion, Power and Polity: A Theory of Democratic Self-Organization.* Oakland, CA: Institute for Contemporary Studies.

Dkamla, G. P. 2003. Associations, Argent et Pouvoir chez les Pygmées Baka de l'Est-Cameroun. Document de travail. Yaoundé, Cameroon: INADES.

Dounias, E. 1993. Dynamique et Gestion Différentielle du Système de Production à Dominante Agricole des Mvae du Sud Cameroun Forestier. Ph. D. thesis, Université de Sciences et Techniques du Languedoc, Montpellier, France.

Dourojeanni, D. 1997. Informe de Trabajo de Campo (Fieldwork Report). Investigación sobre la Economía Familiar en la Carretera Puerto Maldonado-Río Mañuripe (Research on Household Economy along the Puerto Maldonado-Río Mañuripe Highway). Unpublished report for Conservation International-Peru Program and FADEMAD (Madre de Dios Agrarian Federation).

Dove, M. R. 1993. A Revisionist View of Tropical Deforestation and Development. *Environmental Conservation* 20(1): 17–24.

Droulers, M., and P. Maury. 1980. Colonização da Amazônia Maranhense. *Ciência e Cultura* 33(8): 1033–1050.

Dupré, W. 1962. Die Babinga-Pygmäen. *Ann. Lateranensi* 26: 102–172.

Dyson, M. 1992. Concern for Africa's Forest People: A Touchstone of a Sustainable Development Policy. In *Conservation of West and Central African Rainforests,* edited by K. Cleaver, M. Munasinghe, M. Dyson, N. Egli, A. Peuker, F. Wencélius. Washington, DC: World Bank, 212–221.

Edmunds, D., and E. Wollenberg. 2000. Historical Perspectives on Forest Policy Change in Asia: An Introduction. *Journal of Environmental History* 6(2): 190–212.

———. 2001. A Strategic Approach to Multistakeholder Negotiations. *Development and Change* 32: 231–253.

Edwards, M. 1989. The Irrelevance of Development Studies. *Third World Quarterly* 11(1).

———. 1994. Rethinking Social Development: The Search for Relevance. In *Rethinking Social Development: Theory, Research, and Practice,* edited by David Booth. Harlow, UK: Addison Wesley Longman, 279–297.

Efoua, S. 2002a. *Situation des Forêts Communautaires de la Région de Lomié.* Yaoundé, Cameroon: CIFOR.

———. 2002b. *Les Baka Face aux Changements des Politiques Forestières au Cameroun: L'Expérience du Village Moangue-Le-Bosquet dans la Région Lomié-Messok.* Yaoundé, Cameroon: CIFOR.

Engel, P. G. H. 1997. *The Social Organization of Innovation: A Focus on Stakeholder Interaction.* Amsterdam: KIT Publishers.

Engel, P. G. H., A. Hoeberichts, and L. Umans. 2001. Accommodating Multiple Interests in Local Forest Management: A Focus on Facilitation, Actors, and Practices. *International Journal of Agriculture, Resources, Governance and Ecology (IJARGE)* 1(3/4): 306–326. Special issue on accommodating multiple interests in local forest management.

Engel, P. G. H., and M. Salomon. 1997. *Facilitating Innovation for Development: A RAAKS Resource Box.* Amsterdam: KIT Publishers.

Eoné, M. E. 2003. Stratégies de Gestion Durable des Forêts Communautaires de Lomié sur la Base des Expériences Acquises par les Communautés: Aspects Socio-Economiques. M. Sc. thesis, University of Dschang, Dschang, Cameroon.

ERE Développement. 2001. Etude Socio-économique dans l'UTO de Campo-Ma'an. Rapport principal de la phase I: Analyse et synthèse des données générales et des enquêtes. Document provisoire. Yaoundé, Cameroon: Campo-Ma'an Project.

Escobar, A. 1995a. *Encountering Development: The Making and Unmaking of the Third World.* Princeton, NJ: Princeton University Press.

———. 1995b. Imagining a Post-Development Era. In *Power of Development,* edited by Jonathan Crush. New York: Routledge.

Etoungou, P. 2003. Decentralization Viewed from Inside: The Implementation of Community Forests in East Cameroon. Working Paper. Washington, DC: World Resources Institute.

Fals-Borda, O. (ed.). 1998. *People's Participation: Challenges Ahead.* New York: APEX Press; London: ITGD (Intermediate Technology Development Group) Publishing.

Fals-Borda, O., and M. A. Rahman. 1985. The Theory and Practice of Participatory Action Research. In *The Challenge of Social Change,* edited by O. Fals-Borda. London: Sage Publications, 107–132.

———(eds.). 1991. *Action and Knowledge: Breaking the Monopoly with Participatory Action Research.* New York: Apex Press.

FIMAC (Fonds d'Intervention pour les Micro-Réalisations Agricoles et Communautaires). 1999. Situation des Groupes d'Initiative Commune dans la Province de l'Est. Bertoua: FIMAC/MINAGRI.

Finger, M. Undated. Heinz Moser's Concept of Action Research. Unpublished paper. Faculty of Psychology and Educational Science, University of Geneva, Geneva, Switzerland.

Fisher, R. 1995. Collaborative Management of Forests for Conservation and Development. Gland, Switzerland: IUCN and World Wildlife Fund for Nature.

Fisher, R. J., and W. J. Jackson. 1998. Action Research for Collaborative Management of Protected Areas. In *Collaborative Management of Protected Areas in the Asian Region: Proceedings of a Workshop Held at Royal Chitwan National Park, Sauraha, Nepal,* edited by K. Prasad Oli. Kathmandu, Nepal: IUCN Nepal, 235–243.

Forestry Planning Agency. 2001. *Forestry Statistics of Indonesia 2000*. Jakarta: Ministry of Forestry.

Fortmann, L. 1995. Talking Claims: Discursive Strategies in Contesting Property. *World Development* 23: 1053–1064.

Fortmann, L., and J. Bruce. 1988. *Whose Trees? Proprietary Dimensions of Forestry*. Boulder, CO: Westview Press.

Fortmann, L., and N. Nabane. 1992. *The Fruits of Their Labours: Gender, Property and Trees in Mhondoro District*. Occasional Paper No. 7. Centre for Applied Social Sciences. Harare, Zimbabwe: University of Zimbabwe.

Foucault, M. 1980. *Power/Knowledge: Selected Interviews and Other Writings, 1972–1977*. Edited by Colin Gordon. New York: Pantheon Books.

———. 1990. *The History of Sexuality: An Introduction*. Volume 1. New York: Vintage Books.

Fox, J. 1998. Mapping the Commons: The Social Context of Spatial Information Technologies. *Common Property Resource Digest* 45: 1–4.

———. Forthcoming. Siam Mapped and Mapping in Cambodia: Boundaries, Sovereignty, and Indigenous Conceptions of Space. *Society and Natural Resources*.

Freire, P. 1970. *Pedagogy of the Oppressed*. New York: Herder and Herder.

———. 1972. *Pedagogy of the Oppressed*. London: Penguin.

———. 1980. *Conscientização: Teoria e Prática da Libertação: Uma Introdução ao Pensamento de Paulo Freire*. Third edition. São Paulo, Brazil: MORAES.

———. 1981. Criando Métodos de Pesquisa Alternativa: Aprendendo a Fazê-la Melhor Através da Ação. In *Pesquisa Participante*, edited by C. R. Brandão. São Paulo, Brazil: Brasiliense, 34–41.

Freire, P., and A. M. Araújo Freire. 1994. *Pedagogy of Hope: Reliving Pedagogy of the Oppressed*. New York: Continuum.

Fultan, B. 1992. The Efficiency of the Forestry Taxation System in Cameroon. In *Conservation of West and Central African Rainforests*, edited by K. Cleaver, M. Munasinghe, M. Dyson, N. Egli, A. Peuker, and F. Wencélius. Washington, DC: World Bank.

Fundación Tierra. 2002. *Cinco Propuestas de Modificación o Sustitución de la Ley INRA*. La Paz, Bolivia: Fundación Tierra.

Gaidzanwa, R. B. 1988. Women's Land Rights in Zimbabwe: An Overview. Rural and Urban Planning Occasional Paper No. 13. Harare, Zimbabwe: University of Zimbabwe.

———. 1995. Land and the Economic Empowerment of Women: A Gendered Analysis. *Southern African Feminist Review* 1(1): 1–12.

GEOVIC-Republic of Cameroon. 2002. *Mining Convention between the Republic of Cameroon and GEOVIC*. Yaoundé, Cameroon: MINMEE.

Giddens, A. 1987. *Social Theory and Modern Society*. Oxford: Blackwell.

Gittinger, J. P. 1982. *Economic Analysis of Agricultural Projects*. Baltimore, MD: Johns Hopkins University Press.

Goebel, A. 1998. Process, Perception and Power: Notes from 'Participatory' Research in a Zimbabwean Resettlement Area. *Development and Change* 29: 277–305.

———. 1999. Then It's Clear Who Owns the Trees: Common Property and Private Control in Social Forest in a Zimbabwean Resettlement Area. *Rural Sociology* 64(4): 625–641.

Government of Acre. 2000. *Zoneamento Ecológico-Econômico do Estado do Acre*. Documento Final-la Fase. Rio Branco, Acre, Brazil: Secretaria de Estado de Ciência, Tecnologia e Meio Ambiente-SECTMA. www. ac. gov. br.

Greenwood, D. J., and M. Levin. 1998. *Introduction to Action Research: Social Research for Social Change*. London: Sage Publications.

Grimble, R. J., and M. K. Chan. 1995. Stakeholder Analysis for Natural Resource Management in Developing Countries. *Natural Resources Forum* 19: 113–124.

Groot, A., and M. Maarleveld. 2000. *Demystifying Facilitation in Participatory Development*. Gatekeeper Series No. 89. London: International Institute for Environment and Development.

Groot, A., N. van Dijk, and J. Jiggins. 2002. Three Challenges in the Facilitation of System-wide Change. In *Wheelbarrows Full of Frogs: Social Learning in Rural Resources Management. International Research and Reflections*, edited by C. Leeuwis and R. Pyburn. The Netherlands: Koninklijke van Gorcum, 199-213.

Guba, E. G. (ed.). 1990. *The Paradigm Dialog*. Newbury Park, CA: Sage Publications.

Guijt, I., J. A. Berdegué, and M. Loevinsohn (eds.). 2000. *Deepening the Basis of Rural Resource Management: Proceedings of a Workshop.* The Hague: International Service for National Agricultural Research (ISNAR) with support from Red Internacional de Metodología de Investigación de Sistemas de Producción (Chile), International Institute for Environment and Development (UK), International Support Group (Netherlands), Centre de Coopération Internationale en Rechereche Agronomique pour le Développement (CIRAD-TERA, France), Instituto Nacional de Tecnología (Argentina), Pesquisa e Desenvolvimento (ECOFORÇA, Brasil), European Union, and International Development Research Centre.

Guijt, I., and A. Cornwall. 1995. Critical Reflections on the Practice of PRA. *PLA Notes* 24: 2–7.

Guijt, I., and M. K. Shah. 1998. *The Myth of Community: Gender Issues in Participatory Development.* London: ITDG (Intermediate Technology Development Group) Publishing.

Guillaume, H. 1989. "L'Etat Sauvage …": Pygmées et Forêts d'Afrique Centrale. *Politique Africaine* 34: 74–82.

Gupta, A. K., and IDS Workshop. 1989. Maps Drawn by Farmers and Extensionists. In *Farmer First: Farmer Innovation and Agricultural Research,* edited by R. Chambers, A. Pacey, and L. A. Thrupp. London: ITDG (Intermediate Technology Development Group) Publishing, 86–92.

Gurung, J. D. 1995. *Participatory Approaches to Agricultural Technology Promotion with Women in the Hills of Nepal.* Kathmandu, Nepal: ICIMOD.

Habermas, J. 1995. Citizenship and National Identity: Some Reflections on the Future of Europe. In *Theorizing Citizenship,* edited by R. Beiner. Albany, NY: State University of New York Press.

Hailey, J. 2002. Beyond the formulaic: Process and practice in South Asian NGOs. In *Participation: The New Tyranny?* edited by B. Cooke and U. Kothari. London: Zed Books, 88–101.

Haraway, D. J. 1991. *Simians, Cyborgs, and Women: The Reinvention of Nature.* New York: Routledge.

Harding, S. 1986. *The Science Question in Feminism.* Ithaca, NY: Cornell University Press.

Hartanto, H., with L. Arda-Minas, L. Burton, A. Estanol, M. C. Lorenzo, and C. Valmores. 2002. *Planning for Sustainability of Forests through Adaptive Comanagement: Philippines Country Report.* Bogor, Indonesia: CIFOR.

Hartanto, H., M. C. Lorenzo, C. Valmores, L. Arda-Minas, L. Burton, and A. Frio. 2003. *Learning Together: Responding to Change and Complexity to Improve Community Forests in the Philippines.* Bogor, Indonesia: CIFOR.

Havnevik, K. J., and M. Harsmar. 1999. *The Diversified Future: An Institutional Approach to Rural Development in Tanzania.* Stockholm: EGDI (Expert Group on Development Issues).

Hecht, S., and A. Cockburn. 1990. *The Fate of the Forest: Developers, Destroyers and Defenders of the Amazon.* New York: Harper Perennial.

Hermosa, W. V. 1972. Los Pueblos Guarayos: Una Tribu del Oriente Boliviano. *Academia Nacional de Ciencias de Bolivia,* Publicación No. 27. La Paz, Bolivia: Academia Nacional de Ciencias de Bolivia.

Hildebrand, P. E., E. Bastidas, and A. Araújo. 1998. *Introduction to Linear Programming: A Training Manual.* Gainesville, FL: University of Florida.

Hildyard, N., P. Hegde, et al. 2002. Pluralism, participation, and power: Joint forest management in India. In *Participation: The New Tyranny?* edited by B. Cooke and U. Kothari. London: Zed Books, 56–71.

Hobley, M. 1996. *Participatory Forestry: The Process of Change in India and Nepal.* London: Overseas Development Institute.

INE (Instituto Nacional de Estadística). 2002. Bolivia: Mapa de Pobreza 2001 Necesidades Básicas Insatisfechas, Serie 1. *Resultados Censo Nacional de Población y Vivienda 2001,* Series 1, Volume 2. La Paz, Bolivia: INE.

INEI (Instituto Nacional de Estadística e Informática). 1999. *Estado de la Población Peruana.* Lima, Peru: INEI.

Ingles, A. W., A. Musch, and H. Qwist-Hoffman. 1999. *The Participatory Process for Supporting Collaborative Management of Natural Resources: An Overview.* Community Forestry Unit, FAO (Food and Agriculture Organization of the United Nations). Rome: FAO.

Jackson, W. J. 1993. Action Research for Community Forestry: The Case of the Nepal Australia Community Forestry Project. Paper presented at the ICIMOD (International Center for Integrated Mountain Development) Methodology Workshop on Rehabilitation of Degraded Mountain Ecosystems of the Hindu Kush-Himalaya Region. 29 May–3 June 1993, Kathmandu, Nepal.

Jacobs, S. 1991. Changing Gender Relations in Zimbabwe: The Case of Individual Family Re-settlement Areas. In *Male Bias in the Development Process*, edited by D. Elson. Manchester, UK: Manchester University Press, 51–82.

Jiggins, J. 1997. Women and the Re-making of Civil Society. In *From the Bottom Up: Participation in the Rise of Civil Society*, edited by J. Burbidge. New York: PACT Publications (published for the Institute of Cultural Affairs International).

Jiggins, J., and N. Röling. 2000. Inertia and Inspiration: Three Dimensions of the New Profession-alism. In *Deepening the Basis of Rural Resource Management: Proceedings of a Workshop*, edited by I. Guijt, J. A. Berdegué, and M. Loevinsohn. The Hague, Netherlands: ISNAR (International Service for National Agricultural Research), 212–221.

Joiris, D. V. 1986. *Techno-Economic Changes among the Sedentarised BaGyeli Pygmies*. London: London School of Economics.

———. 1992. Entre le Village et la Forêt. Place des Femmes Baka et Bakola dans les Sociétés en Voie de Sédentarisation. In *Relation de Genre et Développement: Femmes et Sociétés*, edited by J. Bisilliat, F. Pinton, and M. Lecarme. Paris: ORSTOM.

———. 1997. La Nature des Uns et la Nature des Autres: Mythe et Réalité du Monde Rural Face aux Aires Protégées d'Afrique Centrale. *Civilisations* 44 (1–2 Special): 94–103.

Jordan, B. 1997. Authoritative Knowledge and Its Construction. In *Childbirth and Authoritative Knowledge: Cross-Cultural Perspectives*, edited by R. Davis-Floyd and C. Sargent. Berkeley, CA: University of California Press, 55–79.

Jordan, B., et al. 1993. *Birth in Four Cultures: A Cross-Cultural Investigation of Childbirth in Yucatan, Holland, Sweden, and the United States*. Prospect Heights, IL: Waveland Press.

Kabeer, N. 1988. *Reversed Realities: Gender Hierarchies in Development Thought*. London: Verso.

———. 1994. *Reversed Realities: Gender Hierarchies in Development Thought*. New York: Verso.

Kaimowitz, D., A. Faune, and R. Mendoza. 1999. Your Biosphere Is My Backyard: The Story of Bosawas in Nicaragua. Unpublished manuscript.

Kainer, K. A., and M. L. Duryea. 1992. Tapping Women's Knowledge: Plant Resource Use in Extractive Reserves. *Economic Botany* 46(6): 408–425.

Kanji, N., and K. Menon-Sen. 2001. What Does the Feminisation of Labour Mean for Sustainable Livelihoods? *Opinion: World Summit on Sustainable Development*. London: International Institute for Environment and Development (IIED) in cooperation with the Regional and Interna-tional Networking Group (RING). http://www.iied.org/pdf/gender13.pdf.

Kaski Team. 2001. *Bamdibhirkhoria Community Forest User Groups, Kaski District, Nepal*. Socio-Economic C & I Background Study Report. Adaptive and Collaborative Management Re-search Project. Bogor, Indonesia: CIFOR.

———. 2002. *Bamdibhir Community Forest User Groups, Kaski District, Nepal*. Final Report. Adap-tive and Collaborative Management Research Project. Bogor, Indonesia: CIFOR.

Kaskija, L. 2000. *Punan Malinau and the Bulungan Research Forest*. Research Report. Bogor, Indone-sia: CIFOR.

Katerere, Y., S. Moyo, and L. Mujakachi. 1993. The National Context: Land, Agriculture and Struc-tural Adjustment, and the Forestry Commission. In *Living with Trees: Policies for Forest Manage-ment in Zimbabwe*, edited by P. N. Bradley and K. McNamara. World Bank Technical Paper No. 210. Washington, DC: World Bank.

Kaya, B., P. E. Hildebrand, and P. K. R. Nair. 2000. Modeling Changes in Farming Systems with the Adoption of Improved Fallows in Southern Mali. *Agricultural Systems*.

Keeley, J., and I. Scoones. 1999. *Understanding Environmental Policy Processes: A Review*. IDS Working Paper 89. Brighton, UK: Institute of Development Studies, University of Sussex.

Kemmis, S., and R. McTaggart (eds.). 1988. *The Action Research Planner*. Geelong, Australia: Deakin University Press.

Koch, C. W. H. 1912. Das Zwergvolk der Bagielli. *Dt. Kolonialzeitung* 29: 23–28.

———. 1968. *Magie et Chasse dans la Forêt Camerounaise*. Paris: Berger-Lavrault.

Kolb, D. A. 1984. *Experiential Learning: Experience as a Source of Learning and Development*. Englewood Cliffs, NJ: Prentice Hall.

Kouna, C. 2001. *Décentralisation de la Gestion Forestière et Développement Local au Cameroun: Economie Politique de la Performance et de la Reddition de Comptes dans la Gestion des Revenus Forestiers à Dimako et à Lomié (Est-Cameroun)*. Yaoundé, Cameroun: CIFOR.

Krishna, A. 2002. *Active Social Capital*. New York: Columbia University Press.

Kuhn, T. S. 1996. *The Structure of Scientific Revolutions*. Third edition. Chicago, IL: University of Chicago Press.

Kurian, P. 2000. *Engendering the Environment? Gender in the World Bank's Environmental Policy*. Burlington, VT: Ashgate Publishing Company.

Kusumanto, T., and E. P. Sari. 2001. *Up-and-Down the Ladder: A Jambi Case Study of Policy-Related Information Flows*. Bogor, Indonesia: CIFOR.

Laburthe-Tolra, P. 1985. *Les Seigneurs de la Forêt: Essai sur le Passé Historique, l'Organisation Sociale et les Normes Ethiques des Anciens Béti du Cameroun*. Paris: Publications de la Sorbonne.

Lammerts van Bueren, E.M., and E. Blom. 1997. *Hierarchical Framework for the Formulation of Sustainable Forest Management Standards*. Netherlands: Veeman Drukkers.

Le Roy, A. 1929. *Les Pygmées, Négrilles d'Afrique et Negritos d'Asie*. Paris: Procure Générale des Pères du St-Esprit.

Leach, M. 1994. *Rain Forest Relations: Gender and Resource Use among the Mende of Gola, Sierra Leone*. Washington, DC: Smithsonian Institution Press; London: Edinburgh University Press for the International African Institute.

Leach, M., and J. Fairhead. 2001. Plural Perspectives and Institutional Dynamics: Challenges for Local Forest Management. *International Journal of Agriculture, Resources, Governance and Ecology (IJARGE)* 1(3/4): 223–242. Special issue on accommodating multiple interests in local forest management.

Leach, M., R. Mearns, and I. Scoones. 1997. Challenges to Community-Based Sustainable Development: Dynamics, Entitlements, and Institutions. *IDS Bulletin* 28(4): 4–14.

Lee, K. N. 1993. *Compass and Gyroscope: Integrating Science and Politics for the Environment*. Washington, DC: Island Press.

Leeuwis, C., and R. Pyburn (eds.). 2002. *Wheelbarrows Full of Frogs: Social Learning in Rural Resources Management: International Research and Reflections*. Amsterdam, Netherlands: Koninklijke van Gorcum.

Letouzey, R. 1967. Note sur les Pygmées de la région de Tikar au Cameroun. *Journal d'Agriculture Tropicale et de Botanique Appliquée* 14.

———. 1975. Noms d'Arbres des Pygmées Bagyeli dans le Sud-Ouest du Cameroun. *Journal d'Agriculture Tropicale et de Botanique Appliquée* 22(1–2–3): 23–45.

Lewin, K. 1946. Action Research and Minority Problems. *Journal of Social Issues* 2 (4): 34–46.

Lincoln, Y. S. 1990. The Making of a Constructivist. A Remembrance of Transformations Past. In *The Paradigm Dialog*, edited by E. G. Guba. Newbury Park, CA: Sage Publications.

Lindgren, B. 2002. The Politics of Ndebele Ethnicity: Origins, Nationality, and Gender in Southern Zimbabwe. Ph. D. thesis, Department of Cultural Anthropology and Ethnology, Uppsala University, Uppsala, Sweden.

Litow, P. 2000. Food Security and Household Livelihood Strategies in the Maya Biosphere Reserve: The Importance of Milpa in the Community of Uaxactun, Petén, Guatemala. Master's thesis, University of Florida, Gainesville.

Long, N. (ed.). 1989. *Encounters at the Interface: A Perspective on Social Discontinuities in Rural Development*. Wageningen Studies in Sociology No. 27. Wageningen, Netherlands: Wageningen Agricultural University.

Long, N. 1992. From Paradigm Lost to Paradigm Regained? The Case for an Actor-Oriented Sociology of Development. In *Battlefields of Knowledge: The Interlocking of Theory and Practice in Social Research and Development*, edited by N. Long and A. Long. London: Routledge, 17–43.

Long, N., and J. D. van der Ploeg. 1994. Heterogeneity, Actor and Structure: Towards a Reconstitution of the Concept of Structure. In *Rethinking Social Development: Theory, Research, and Practice*, edited by David Booth. Harlow, UK: Addison Wesley Longman, 62–89.

Long, N., and M. Villareal. 1994. The Interweaving of Knowledge and Power in Development Interfaces. In *Beyond Farmer First: Rural People's Knowledge, Agricultural Research and Extension Practice*, edited by I. Scoones and J. Thompson. London: IIED (International Institute for Environment and Development); London: ITGD (Intermediate Technology Development Group) Publishing, 41–56.

Loung, J.-F. 1959. Les Pygmées de la Forêt de Mill. Un Groupe de Pygmées Camerounais en Voie de Sédentarisation. *Les Cahiers d'Outre-Mer* 12(48): 1–20.

————. 1981. *La Population Pygmée de la Région Côtière Camerounaise.* Yaoundé, Cameroon: Institut des Sciences Humaines.

————. 1992. *Prise en Compte des Populations Pygmées dans le Cadre des Projets' Réserves de Faune,' 'Parcs Nationaux' et 'Forêts.'* Yaoundé, Cameroon: Institut des Sciences Humaines.

————. 1998. *Le Peuplement Pygmée au Cameroun.* Yaoundé, Cameroon: Planet-Survey.

Luckert, M., N. Nemarundwe, and A. Mandondo. 2000. *Tenure Survey in Romwe and Mutangi Questionnaire.* Institute of Environmental Studies. Harare, Zimbabwe: University of Zimbabwe.

Lukes, S. 1986. *Power.* New York: New York University Press.

Maarleveld, M., and C. Dangbégnon. 1998. Managing Natural Resources in Face of Evolving Conditions: A Social Learning Perspective. Paper presented at Crossing Boundaries, the seventh conference of the International Association for the Study of Common Property, June 1998, Vancouver, Canada.

————. 1999. Managing Natural Resources: A Social Learning Perspective. *Agriculture and Human Values* 16: 267–280.

Maarleveld, M., N. Röling, and S. Seegers. 1997. FASOLERAN (Facilitation of Social Learning). Technical Proposal for Second Phase of the Environment and Climate RTD Program (1997–1998). Unpublished. Wageningen Agricultural University, Wageningen, Netherlands.

Maclure, R., and M. Bassey. 1991. Participatory Action Research in Togo: An Inquiry into Maize Storage Systems. In *Participatory Action Research,* edited by W. F. Whyte. London: Sage Publications, 191–209.

MacPherson, C. B. 1978. *Property: Mainstream and Critical Positions.* Toronto: University of Toronto Press.

Malla, Y. B. 2001. Changing policies and the persistence of patron-client relations in Nepal. *Environmental History* 8 (2):287–307.

Malla, Y. B., P. Branney, H. Neupane, and P. Tamrakar. 2001. *Participatory Action and Learning: A Field Worker's Guidebook for Supporting Community Forest Management.* Kathmandu, Nepal/Reading, UK: Livelihoods and Forestry Programme (HMGN/DFID)/Department of International and Rural Development.

Mancilla, R., and A. Andaluz. 1996. *Cambios Sustanciales en la Legislación Forestal Nacional.* Boletín BOLFOR 7. La Paz, Bolivia: BOLFOR.

Mandondo, A. 2000. *Forging (Un)Democratic Governance Systems from the Relic of Zimbabwe's Colonial Past.* Institute of Environmental Studies, University of Zimbabwe and the CIFOR (Center for International Forestry Research). Harare, Zimbabwe: Institute of Environmental Studies; Bogor, Indonesia: CIFOR.

————. 2001. Allocation of Governmental Authority and Responsibility in Tiered Governance Regimes: The Case of Chivi Rural District Council Land-Use Planning and Conservation By-Laws. *African Studies Quarterly* 5 (3) http://web. africa. ufl. edu/asq/v5/v5i3a3. htm.

Marchbank, J. 2000. *Women, Power and Policy: Comparative Studies of Childcare.* London: Routledge.

Marquet, J. 1971. *Power and Society in Africa.* London: World University Library.

Masters, J. 1995. The History of Action Research. In *Action Research Electronic Reader,* edited by I. Hughes. Sydney, Australia: University of Sydney. http://www. behs. cchs. usyd. edu. au/arow/ Reader/rmasters. htm (accessed on 07/02/2002).

Matondi, P. B. 2001. The Struggle for Access to Land and Water Resources in Zimbabwe: The Case of Shamva District. Ph. D. thesis, Department of Rural Development Studies, Swedish University of Agricultural Sciences, Uppsala, Sweden.

Matose, F. 1994. Local People and Forestry Reserves in Zimbabwe. M. Sc. thesis submitted in Rural Economy, University of Alberta, Alberta, Canada.

————. 1997. Conflicts around Forest Reserves in Zimbabwe: What Prospects for Community Management? *IDS Bulletin* 28(4): 69–79.

Maués, M. 1993. *Trabalhadeiras e Camarados: Relações de Gênero, Simbolismo e Ritualização Numa Comunidade Amazônica.* Belém, Brazil: UFPA/Centro de Filosofia e Ciências Humanas.

May, J. 1979. Social Aspects of the Legal Position of Women in Zimbabwe-Rhodesia. M. Phil. thesis submitted to the Centre for Inter-Racial Studies, University of Rhodesia, Salisbury.

May, P. 1990. *Palmeiras em Chamas: Transformação Agrária e Justiça Social na Zona do Babaçu.* São Luís, Brazil: EMAPA/FINEP/Ford Foundation.

Mazarire, G. C. 1999. Women, Politics, and the Environment in Pre-Colonial Chivi circa 1840 to 1900. Paper presented to the Women and the Environment Conference, August 1999, Harare, Zimbabwe.

Mbembé, A. 1986. Pouvoir des Morts et Langage des Vivants. *Politique Africaine* 22: 37–72.

———. 1989. La Bouche, le Ventre et le Pénis: Notes sur le Pouvoir et l'Obscénité en Postcolonie. Paper presented at the Annual Congress of the African Study Association, September 1989, Atlanta, GA.

McCay, B. J., and J. M. Acheson. 1987. Human Ecology of the Commons. In *The Question of the Commons: The Culture and Ecology of Communal Resources,* edited by B. J. McCay and J. M. Acheson. Tucson, AZ: University of Arizona Press, 1–34.

McDougall, C. 1998. *Final Methods Test, Bulungan, East Kalimantan.* Bogor, Indonesia: CIFOR.

———. 2000. *Adaptive and Collaborative Management of Community Forests Research Project Update* (August). Bogor, Indonesia: CIFOR.

———. 2001. *Adaptive and Collaborative Management of Community Forests: Research Project Update* (August). Bogor, Indonesia: CIFOR.

McNamara, K. 1993. Key Policy Issues. In *Living with Trees: Policies for Forest Management in Zimbabwe,* edited by P. N. Bradley and K. McNamara. World Bank Technical Report No. 210. Washington, DC: World Bank.

McNeely, J. A. (ed.). 1995. *Expanding Partnerships in Conservation.* Washington, DC: Island Press.

MDSMA (Ministerio de Desarrollo Sostenible y Medio Ambiente). 1997. *Ley Foresta.* Edited by BOLFOR. Santa Cruz, Bolivia: Ministerio de Desarrollo Sostenible y Medio Ambiente.

Meggers, B. J. 1977. *Amazônia: A Ilusão de um Paraíso.* Rio de Janeiro, Brazil: Civilização Brasileira.

Meinzen-Dick, R. S. 1997. Gendered Participation in Water Management: Issues and Illustrations from Water Users Association of South-East Asia. Paper presented at the Workshop on Cooperative Management of Water Resources at the Center for India and South Asian Research, Institute of Asian Research, University of British Columbia. December 1997, Vancouver, Canada.

Meinzen-Dick, R. S., L. R. Brown, H. S. Feldstein, and A. R. Quisumbing. 1997. Gender, Property Rights, and Natural Resources. *World Development* 25(8): 1303–1315.

Meinzen-Dick, R. S., and M. Zwarteveen. 2001. Gender Dimensions of Community Resource Management: A Case of Water Users' Associations in South Asia. In *Communities and the Environment: Ethnicity, Gender, and the State in Community-based Conservation,* edited by A. Agrawal and C. C. Gibson. New Brunswick, NJ: Rutgers University Press, 63–88.

Meka, P. 2001. Oeuvrer pour une Meilleure Prise en Compte des Minorités. *Bubinga* 50: 3.

Milol, A., and J.-M. Pierre. 2000. *Volet Additionnel de l'Audit Economique et Financier du Secteur Forestier: Impact de la Fiscalité Décentralisée sur le Développement Local et les Pratiques d'Utilisation des Ressources Forestières au Cameroun.* Yaoundé, Cameroun: World Bank.

Mimboh, P.-F. 1998. *Les Pygmées Bakola-Bagyéli du Sud-Ouest Cameroun et le Phénomène de la Déforestation: Vivre dans Deux Modes de Vie.* Document de travail. Yaoundé, Cameroun: Planet-Survey.

MINPAT (Ministère du Plan et de l'Aménagement du Territoire du Cameroun). 1985. *Rapport de la Réunion Provinciale de Concertation sur l'Intégration Socio-Economique des Pygmées de l'Est.* Bertoua, Cameroun: MINPAT.

MIQCB (Movimento Interestadual das Quebradeiras de Côco Babaçu). 1993. Second Interstate Meeting of the Movimento Interestadual das Quebradeiras de Côco Babaçu Report 1993. São Luís: MIQCB.

MMA (Ministry of Environment, Water Resources and the Amazon) and SCA (Secretariat for Coordination of Legal Amazon Affairs). 1996. *Pilot Program to Conserve the Brazilian Rain Forest.* Brasília, Brazil: MMA/SCA.

Mohan, G. 2002. Beyond participation: Strategies for deeper empowerment. In *Participation: The New Tyranny?* edited by B. Cooke and U. Kothari. London: Zed Books, 153–167.

Molano, A. 1998. Cartagena Revisited: Twenty Years On. In *People's Participation: Challenges Ahead,* edited by O. Fals-Borda. New York: APEX Press; London: ITGD (Intermediate Technology Development Group) Publishing, 3–10.

Momberg, F., K. Atok, and M. Sirait. 1996. *Drawing on Local Knowledge: A Community Mapping Training Manual. Case Studies from Indonesia.* Jakarta, Indonesia: Ford Foundation with Yayasan Karya Sosial Pancur Kasih and WWF Indonesia Programme.

Momsen, J. H., and J. Townsend (eds.). 1987. *The Geography of Gender in the Third World.* Albany, NY: State University of New York Press.

Monceaux, P. 1891. La Légende des Pygmées et des Nains de l'Afrique Equatoriale. *Revue Historique* 17: 2–7.

Moser, C. 1993. *Gender Planning and Development: Theory, Practice and Training.* London and New York: Routledge.

Moser, C., A. Tornqvist, and B. van Bronkhorst. 1999. *Mainstreaming Gender and Development in the World Bank: Progress and Recommendations.* Washington, DC: World Bank.

Moyo, S. 1995. A Gendered Perspective of the Land Question. *Southern African Feminist Review* 1(1): 13–31.

Mpol, F., M. Mendouka, A. Bikoi, and L. A. Ndongo. 1994. *Evaluation Participative de la Pauvreté au Cameroun.* Yaoundé, Cameroon: World Bank and CARE International.

Mukamuri, B. B. 2000. Local Institutions and Management of Indigenous Woodland Resources in Zimbabwe. In *Forests, Chiefs and Peasants in Africa: Local Management of Natural Resources in Tanzania, Zimbabwe and Mozambique,* edited by P. Virtanen and M. Nummelin. Silva Carelica 34. Joensuu, Finland: University of Joensuu, 15–33.

Mukonyora, I. 1999. Women and Ecology in Shona Religion. *Word and World* 19(3): 276–284.

Municipio de Ascención. 2000. Capacitación Para la Producción Forestal en TCOs del Municipio de Ascensión: Memorias Bases Institucionales Para la Producción Forestal. Santa Cruz, Bolivia: Municipio de Ascensión.

Murombedzi, J. C. 1994. The Dynamics of Conflict in Environmental Management Policy in the Context of the Communal Areas Management Programme for Indigenous Resources (CAMP-FIRE). Ph. D. thesis, Centre for Applied Social Sciences, University of Zimbabwe, Harare, Zimbabwe.

Murphree, M. W. 1991. Communities as Institutions for Resource Management. Centre for Applied Social Sciences Occasional Paper. Harare, Zimbabwe: University of Zimbabwe Centre for Applied Social Sciences.

Mveng, E. 1984. *Histoire du Cameroun.* Volume 2. Yaoundé, Cameroon: CEPER.

Nabane, N. 1997. A Gender Sensitive Analysis of a Community-Based Wildlife Management Utilisation Initiative in Zimbabwe's Zambezi Valley. M. Phil. thesis, Centre for Applied Social Sciences, University of Zimbabwe, Harare.

Nabane, N., and G. Matzke. 1997. A Gender Sensitive Analysis of a Community-Based Wildlife Management Utilisation Initiative in Zimbabwe's Zambezi Valley. *Society and Natural Resources Journal* 10(6): 519–535.

Narayan, D. 2000. *Voices of the Poor: Can Anyone Hear Us?* Oxford, UK: Oxford University Press for the World Bank.

Nasi, R., A. M. Tiani, and J. Nguiébouri. 2001. Tournée dans l'UTO Campo Ma'an: Village de Nkoelon. Internal report. Bogor, Indonesia: CIFOR.

Ndoye, O., M. Ruiz-Perez, and A. Eyebé. 1998. Non-Timber Forest Products, Markets, and Potential Forest Resource Degradation in Central Africa: The Role of Research for a Balance between Welfare Improvement and Forest Conservation. Paper presented at the International Experts Workshop on Nontimber Forest Products, August 1998, Limbé, Cameroon.

Nelson, R. 1995. *Public Lands and Private Rights: The Failure of Scientific Management.* Lanham, MD: Rowman and Littlefield.

Nemarundwe, N. 2001. Institutional Collaboration and Shared Learning for Forest Management in Chivi District, Zimbabwe. In *Social Learning in Community Forestry,* edited by E. Wollenberg, D. Edmunds, L. Buck, J. Fox, and S. Brodt. Bogor, Indonesia: CIFOR; Honolulu, HI: East-West Center, 85–108.

———. Forthcoming. *Rules, Organizations and Social Instruments for Access to Common Pool Resources in Chivi District, Zimbabwe.* Harare, Zimbabwe: Institute of Environmental Studies, University of Zimbabwe.

Nemarundwe, N., M. Mutamba, and W. Kozanayi. 1998. An Overview of Woodland Utilization and Management in Three Communal Areas in Zimbabwe: Results of PRA Research in Chivi, Mangwende, and Gokwe South. IES (Institute of Environmental Studies) Special Report No. 16. Harare, Zimbabwe: Institute of Environmental Studies, University of Zimbabwe.

Ngueguim, J., with R. Ohanda I. 2001. *Etude de la Chasse Villageoise dans la Périphérie Nord du Parc National de Campo-Ma'an: Cas du Village Bifa.* Final Report. Kribi, Cameroon: Campo-Ma'an Project.

Nguiébouri, J. 2001. *Adaptativité des 'Minorités Forestières' aux Enjeux et Contraintes de la Gestion Durable des Forêts: Cas des Bagielli de Campo-Ma'an.* Yaoundé, Cameroon: CIFOR.

———. 2002. Personal communication between J. Nguiébouri and C. J. P. Colfer, November 13.

Nguiébouri, J., A. Tiani, G. A. Neba, and M. C. Diaw. 2001. *Les Critères et Indicateurs et la Gestion Durable des Forets.* Yaoundé, Cameroon: CIFOR.

Nhira, C., and L. Fortmann. 1993. Local Woodland Management: Realities at the Grassroots. In *Living with Trees: Policies for Forestry Management in Zimbabwe,* edited by P. N. Bradley and K. McNamara. World Bank Technical Paper No. 210. Washington, DC: World Bank, 139–155.

Niekisch, M. 1992. Nontimber Forest Products from the Tropics: The European Perspective. In *Sustainable Harvest and Marketing of Rain Forest Products,* edited by M. Plitkin and L. Famolare. Washington, DC: Island Press, 280–288.

Nkoumbélé, F. 1997. Connaissances Locales et Mutations Socio-Economiques: Perception et Exploitation des Forestières par les Bagyéli de l'Arrondissement d'Akom2. Paper presented at Séminaire sur La Contribution des Sciences Sociales à l'Élaboration d'un Schéma Directeur d'Aménagement Forestier dans la Zone Tropenbos, June 1997, Kribi, Cameroon.

North, D. 1990. *Institutions, Institutional Change, and Economic Performance.* Cambridge, UK: Cambridge University Press.

Nugent, D. 1993. Property Relations, Production Relations, and Inequality: Anthropology, Political Economy, and the Blackfeet. *American Ethnologist* 20(2): 336–362.

Nuñes, W. 2001. Potencial Florestal e Suas Restrições de Uso Comercial: O Caso das Comunidades Ribeirinhas do Alto Rio Atuá/Muaná/Pará. M. Sc. thesis, Faculdade Ciências Agrárias do Pará, Belém, Brazil.

Nyirenda, R. 2001. Zimbabwe Adaptive Collaborative Management Research Project—Mafungautsi Forest: Report on Local Level Context Studies, Based on Participatory Rural Appraisals. Unpublished manuscript.

Olivier De Sardan, J.-S. 1990. Populisme Développementiste et Populisme en Sciences Sociales: Idéologies, Action, Connaissance. *Cahiers d'Etudes Africaines* 120(30): 475–492.

Ollagnon, H. 1991. Vers une Gestion Patrimoniale de la Protection de la Qualité Biologique des Forêts. *Forests, Trees and People Newsletter* 3: 2–35.

Ondo, S. C., and N. J. Mbarga. 2001. *Etude de la Chasse Villageoise dans la Périphérie Nord du Parc National de Campo-Ma'an: Cas du Village Bifa.* Final Report. Kribi, Cameroon: Campo-Ma'an Project.

Ostrom, E. 1990. *Governing the Commons: The Evolution of Institutions for Collective Action.* New York: Cambridge University Press.

———. 1999. *Self-Governance and Forest Resources.* CIFOR Occasional Paper No. 20. Bogor, Indonesia: CIFOR.

Oyono, P. R. 1998a. *Les Pygmées Baka de la Région de Lomié et le Flux des Projets de Développement et de Conservation.* Yaoundé, Cameroon: Planet-Survey.

———. 1998b. Cameroon Rainforest: Economic Crisis, Rural Poverty, Biodiversity. Local communities in the Lomié Region. *Ambio* 27(7): 557–559.

———. 2002. Usages Culturels de la Forêt au Sud-Cameroun: Rudiments d'Ecologie Sociale et Matériau pour le Pluralisme. *Africa* 57(3): 334–355.

Oyono, P. R., and M. C. Diaw. 1998. Méthodes des Sciences Sociales sur la Gestion Durable des Forêts. Tests Conduits dans la Forêt Humide Camerounaise. Yaoundé, Cameroon: CIFOR.

Oyono, P. R., M. C. Diaw, and S. Efoua. 2000. Structure et Contenu Anthropologique du *Bilik:* Le Potentiel de la 'Maison Naturelle' et de la 'Maison Culturelle' pour la Foresterie Communautaire au Sud-Cameroun. *Environnement Africain* 41/42: 12–17.

Oyono, P. R., and S. Efoua. 2001. Rapport d'Atelier d'Échanges sur les Critères et Indicateurs de Gestion Durable des Forets, March 2001, Lomié, Cameroon. Yaoundé, Cameroon: CIFOR.

Oyono, P. R., W. Mala, and J. Tonyé. 2003. Adaptation Versus Rigidity: Contribution to the Debate on Agricultural Viability and Forest Sustainability in Central Africa. *Culture & Agriculture* 25(2): 32–40.

PACBCM (Projet d'Aménagement et de Conservation de la Biodiversité de Campo-Ma'an). 2002. Schéma Directeur pour le Développement de l'Unité Opérationnelle de Campo-Ma'an. Preliminary report. Kribi, Cameroon: MINEF (Ministry of Environment and Forestry), Tropenbos International, and SNV (Netherlands Development Organisation).

Pacheco, P. B., and D. Kaimowitz. 1998. *Municipios y Gestión Forestal en el Trópico Boliviano.* La Paz, Bolivia: CIFOR, CEDLA (Centro de Estudios para el Desarrollo), TIERRA (Taller de Iniciativos en Estudios Rurales y Reforma Agraria), and BOLFOR.

Pandey, R. K. 2002. Self-Monitoring as an Effective Tool to Enhance Adaptive and Collaborative Management (ACM) in Community Forestry in Nepal. *ACM News* 3(3): 9–10.

Pandey, R. K., and C. McDougall. 2002. *Stakeholder Analysis, Participation and Institutional Change.* CIFOR draft report. Bogor, Indonesia: CIFOR.

Parajuli, P. 1998. Beyond Capitalized Nature: Ecological Ethnicity as an Arena of Conflict in the Regime of Globalization. *Ecumene* 5(2): 186–217.

Paulson, S. 2001. Plan de acción para incorporar la Dimensión de Género en el Proyecto BOLFOR. Documento Administrativo 56/2001. Santa Cruz, Bolivia: BOLFOR.

Pavez, I. L., and A. Bojanic-Helbingen. 1998. *El Proceso Social de Formulación de la Ley Forestal de Bolivia de 1996.* La Paz, Bolivia: CIFOR, CEDLA, TIERRA, and PROMAB (Programa de Manejo de Bosques de la Amazonia Boliviana).

Peluso, N. L. 1995. Whose Woods Are These? Counter-Mapping Forest Territories in Kalimantan, Indonesia. *Antipode* 27: 383–406.

Perz, S. 2001. Household Demographic Factors as Life Cycle Determinants of Land Use in the Amazon. *Population Research and Policy Review* 20: 159–186.

Pinton, F. 1992. Les Stratégies de Genre Favorisent-Elles le Développement ? Des Femmes en Forêt Colombienne. In *Relations de Genre et Développement: Femmes et Sociétés,* edited by J. Bisilliat, F. Pinton, and M. Lecarme. Paris: ORSTOM.

PLA (Participatory Learning and Action). 1995. Critical Reflections from Practice. *PLA Notes* 24. London: IIED (International Institute for Environment and Development).

Poffenberger, M. (ed.). 1990. *Keepers of the Forest: Land Management Alternatives in Southeast Asia.* West Hartford, CT: Kumarian Press.

Pokam, J., and W. D. Sunderlin. 1999. L'Impact de la Crise Économique sur les Populations, les Migrations, et le Couvert Forestier du Sud-Cameroun. Unpublished paper. Bogor, Indonesia: CIFOR.

Pokorny, B., G. Cayres, W. Nuñes, D. Segebart, and R. Drude. 2003a. First Experiences with Adaptive Co-Management in Pará, Brazilian Amazon. In *Integrated Management of Neotropical Rain Forests by Industries and Communities,* edited by C. Sabogal and N. Silva. Belém, Brazil: Embrapa, 258–280.

Pokorny B., G. Cayres, W. Nuñes, D. Segebart, R. Drude, and M. Steinbrenner. 2003b. *Adaptive Collaborative Management: Criteria and Indicators to Assess Sustainability.* Bogor, Indonesia: CIFOR.

Pokorny, B., G. Cayres, and W. Nuñes. 2003c. Participatory analysis of heterogeneity, an approach to consolidate collaborative initiatives at community level. *Forests, Trees and Livelihoods* 13(2): 161–175.

Porro, N. M. 2002. Rupture and Resistance: Gender Relations and Life Trajectories in Babaçu Palm Forests of Brazil. Ph. D. thesis, University of Florida, Gainsville.

Porro, R., and N. Miyasaka Porro. 1998. Methods for Assessing Social Science Criteria and Indicators for the Sustainable Management of Forests: Brazil Test. Bogor, Indonesia: CIFOR.

Porter, D. R., and D. A. Salvesen (eds.). 1995. *Collaborative Planning for Wetlands and Wildlife: Issues and Examples.* Washington, DC: Island Press.

Prabhu, R., C. J. P. Colfer, and C. Diaw. 2001. *Sharing Benefits from Forest Utilization: Trojan Horses, Copy Cats, Blind Mice and Busy Bees.* Georgetown, Guyana: Cropper Foundation, Iwokrama Centre, Woods Hole Research Center.

Prabhu, R., C. J. P. Colfer, and R. G. Dudley. 1999. *Guidelines for Developing, Testing and Selecting Criteria and Indicators for Sustainable Forest Management.* C&I Tool 1. Bogor, Indonesia: CIFOR.

Prabhu, R., H. J. Ruitenbeek, T. J. B. Boyle, and C. J. P. Colfer. 1998. Between Voodoo Science and Adaptive Management: Towards a Better Understanding of the Role and Research Needs for Indicators for Sustainable Forest Management. Overview paper presented at IUFRO/FAO/CIFOR International Conference on Indicators for Sustainable Forest Management, August 1998, Melbourne, Australia. London: CABI-IUFRO.

Prabhu, R., with ACM Team. 2000. ACM Handbook (draft). Bogor, Indonesia: CIFOR.

Pretty, J. 1994. Alternative Systems of Inquiry for a Sustainable Agriculture. *IDS Bulletin* 25(2): 37–48.

———. 1999. Capital Assets and Natural Resources Improvements: Linkages and New Challenges. In *Issues and Options in the Design of Soil and Water Conservation Projects,* edited by M. McDonald and K. Brown. Bangor School of Agricultural and Forest Sciences Publications No. 17. Llandudno, UK: University of Wales, Bangor School of Agricultural and Forest Sciences, 11–21.

Rahman, M. A. (ed.). 1984. *Grass-Roots Participation and Self-Reliance: Experiences in South and South East Asia.* New Delhi: Oxford and IBH Publishing Company.

Ramírez, R. 2001. Understanding the Approaches for Accommodating Multiple Stakeholders' Interests. *International Journal of Agriculture, Resources, Governance and Ecology (IJARGE)* 1(3/4): 264–285. Special issue on accommodating multiple interests in local forest management.

Ranger, T. 1993. The Invention of Tradition Revisited: The Case of Colonial Africa. In *Legitimacy and the State in Twentieth Century Africa: Essays in Honour of A. H. M. Kirk Greene,* edited by T. Ranger and O. Vaughan. Oxford, UK: Macmillan Press, 62–111.

———. 1999. *Women and Environment in African Religion: The Case of Zimbabwe.* Paper presented to the Women and the Environment Conference, August 1999, Harare, Zimbabwe.

de Ravignan, A. 2000. Tchad-Cameroun: Pour Qui le Pétrole Coulera-t-Il? *Lettre Mensuelle de la FIDH* (Fédération Internationale des Ligues des Droits de l'Homme) 295: July.

Reardon, G. 1995. *Power and Process: A Report from the Women Linking for Change Conference, Thailand, 1994.* Oxford, UK: Oxfam Publications.

Rees, F. 1998. *The Facilitator Excellence Handbook: Helping People Work Creatively and Productively Together.* San Francisco: Jossey-Bass/Pfeiffer.

Rescher, N. 1993. *Pluralism: Against the Demand for Consensus.* Oxford, UK: Oxford University Press.

Resosudarmo, I. A. P., and A. Dermawan. 2002. Forests and Regional Autonomy: The Challenge of Sharing the Profits and Pains. In *Which Way Forward? People, Forests and Policymaking in Indonesia,* edited by C. J. P. Colfer and I. A. P. Resosudarmo. Washington, DC: Resources for the Future, 325–357.

Rhee, S. Forthcoming. De Facto Decentralization and Community Conflicts in East Kalimantan: Indonesia: Explanations from Local History and Implications for Community Forestry. In *Decentralization and Communities,* edited by K. Abe, W. de Jong, and L. Tuck-Po. Osaka, Japan: Japanese Center for Area Studies.

Ribot, J. C. 1998. Integral Rural Development: Authority and Accountability in Decentralised Natural Resource Management. Africa Technical Division Working Paper. Washington, DC: World Bank.

———. 1999. Decentralisation, Participation and Accountability in Sahelian Forestry: Legal Instruments of Political-Administrative Control. *Africa* 69(1): 23–65.

———. 2001. Integral Local Development: "Accommodating Multiple Interests" Through Entrustment and Accountable Representation. *International Journal of Agriculture, Resources, Governance and Ecology (IJARGE)* 1(3/4): 327–350. Special issue on accommodating multiple interests in local forest management.

Ribot, J. C., and N. Peluso. 2002. A Theory of Access. *Rural Sociology* 68(2): 153–181.

Richards, P. 1995. Participatory Rural Appraisal: A Quick-and-Dirty Critique. *PLA Notes* 24. London: IIED (International Institute for Environment and Development).

Rocheleau, D. 1987. Women, Trees, and Tenure: Implications for Agroforestry Research and Development. In *Land, Trees, and Tenure: Proceedings of an International Workshop on Tenure Issues in Agroforestry,* edited by J. B. Raintree. Madison, WI: Land Tenure Center, University of Wisconsin.

———. 1990. *Gender Complementarity and Conflict in Sustainable Forest Development: A Multiple User Approach.* Paper presented to IVRO World Congress Quinquennial, August 1990, Montreal, Canada.

———. 1991. Gender, Ecology, and the Science of Survival: Stories and Lessons from Kenya. *Agriculture and Human Values* 8: 9–16.

Rocheleau, D., et al. 1995. Gendered Resource Mapping: Focusing on Women's Spaces in the Landscape. *Cultural Survival Quarterly* Winter: 62–68.

Rocheleau, D., and D. Edmunds. 1997. Women, Men, and Trees: Gender, Power, and Property in Forest and Agrarian Landscapes. *World Development* 25(8): 1351–1371.

Rocheleau, D., and R. Slocum. 1995. Participation in context: Key questions. In *Power, Process and Participation: Tools for Change,* edited by R. Slocum, L. Wichhart, D. Rocheleau, and B. Thomas-Slayter. London: Intermediate Technology Development Group Publishing, 17–30.

Rocheleau, D. E., B. Thomas-Slayter, and E. Wangari. 1996. Gender and Environment: A Feminist Political Ecology Perspective. In *Toward a Feminist Political Ecology: Global Perspectives from Local Experience,* edited by D. E. Rocheleau, B. Thomas-Slayter, and E. Wangari. New York: Routledge, 3–23.

Roe, E. 1994. Narrative Policy Analysis: Theory and Practice. Durham, NC: Duke University.

Röling, N. 1992. The Emergence of Knowledge Systems Thinking: A Changing Perception of Relationships among Innovation, Knowledge Process and Configuration. *Knowledge and Policy: The International Journal of Knowledge Transfer and Utilization* 5(Spring/1): 42–64.

———. 2002. Beyond the Aggregation of Individual Preferences: Moving from Multiple to Distributed Cognition in Resource Dilemmas. In *Wheelbarrows Full of Frogs: Social Learning in Rural Resources Management. International Research and Reflections,* edited by C. Leeuwis and R. Pyburn. Amsterdam, Netherlands: Koninklijke van Gorcum, 25–47.

Röling, N., and J. Jiggins. 1993. Policy Paradigm and Sustainable Farming. Unpublished paper, Wageningen Agricultural University, Netherlands.

———. 1998. The Ecological Knowledge System. In *Facilitating Sustainable Agriculture: Participatory Learning and Adaptive Management in Times of Environmental Uncertainty,* edited by N. Röling and M. A. E. Wagemakers. Cambridge, UK: Cambridge University Press.

Röling, N. G., and M. Maarleveld. 1999. Facing Strategic Narratives: An Argument for Interactive Effectiveness. *Agriculture and Human Values* 16: 295–308.

Röling, N. G., and M. A. E. Wagemakers (eds.). 1998. *Facilitating Sustainable Agriculture: Participatory Learning and Adaptive Management in Times of Environmental Uncertainty.* Cambridge, UK: Cambridge University Press.

Röling, N., and J. Woodhill. 2001. From Paradigm to Practice: Foundations, Principles and Elements for Dialogue on Water, Food, and Environment. Background Document for the National and Basin Dialogue Design Workshop, December 2001, Bonn, Germany.

Rossi, J. 1997. Participation Run Amok: The Costs of Mass Participation for Deliberative Agency Decision Making. *Northwestern University Law Review* 92(1): 173–247.

Rudel, T. K., and J. M. Gerson. 1999. Postmodernism, institutional change, and academic workers: A sociology of knowledge. *Social Science Quarterly* 80(2): 213–228.

Sachs, C. E. 1997. Introduction: Connecting women and the environment. In *Women Working in the Environment,* edited by C. E. Sachs. Bristol, PA: Taylor and Francis, 1–11.

Santos, A. M. 1982. Aritapera: Uma Comunidade de Pequenos Produtores na Várzea Amazônica (Santarém-PA). *Boletim do Múseo Paraense Emílio Goeldi, Antropologia* 83.

Sarin, M. Forthcoming. De-democratization in the Name of Devolution? Findings from Three States in India. Draft report. Bogor, Indonesia: CIFOR.

Saywell, D., and A. Cotton. 1999. *Spreading the Word: Practical Guidelines for Research Dissemination Strategies.* Interim findings. Water, Engineering and Development Centre. Loughborough, UK: Loughborough University.

Schebesta, P. 1936. Die Grundlinie der Wittschaftskultur der Kongo-Pygmäen. *Forschunge und Fortschritte* 12: 303–304.

Schkopp, Von E. 1903. Zwergvölker in Kamerun. *Globus* 83(18).

Schlichter, H. 1892. The Pygmy Tribes of Africa. *Scottish Geographical Magazine* 8: 289–301.

Schmidt, E. 1992. *Peasants, Traders, and Wives: Shona Women in the History of Zimbabwe, 1870–1939.* Portsmouth, NH: Heinemann; Harare, Zimbabwe: Boabab; London: James Currey.

Schmink, M., and C. H. Wood. 1992. *Contested Frontiers in Amazonia.* New York: Columbia University Press.

Schroeder, R. 1993. Shady Practices: Gender and the Political Ecology of Resource Stabilisation in Gambian Garden/Orchards. *Economic Geography* 69(4): 349–365.

———. 1997. Reclaiming Land in the Gambia: Gendered Property Rights and Environmental Intervention. *Annals of the Association of American Geographers* 87(3): 487–508.

Schumacher, P. 1947. Les Batwa sont-ils des Pygmées authentiques? *Aequatoria* 4.

Scoones, I., and J. Thompson. 1994. Knowledge, Power and Agriculture: Towards a Theoretical Understanding. In *Beyond Farmer First: Rural People's Knowledge, Agricultural Research, and Extension Practice,* edited by I. Scoones and J. Thompson. London: IIED (International Institute for Environment and Development); London: ITGD (Intermediate Technology Development Group) Publishing.

———(eds.). 1994. *Beyond Farmer First: Rural People's Knowledge, Agricultural Research and Extension Practice.* London: IIED (International Institute for Environment and Development); London: ITGD (Intermediate Technology Development Group) Publishing.

Scott, J. C. 1985. *Weapons of the Weak: Everyday Forms of Peasant Resistance.* New Haven, CT: Yale University Press.

Scott, R. A., and A. R. Shore. 1979. *Why Sociology Does Not Apply: A Study of The Uses of Sociology in Public Policy.* Oxford, UK: Elsevier.

Scott, W. R. 1995. *Institutions and Organisations.* London: Sage Publications.

Selaya, N. 2002. Forestería Comunitaria con Perspectiva de Género: Diagnóstico a TCOs de Guarayos. Documento de Trabajo 3/2002. Santa Cruz, Bolivia: BOLFOR.

Selener, D. 1997. *Participatory Action Research and Social Change.* Cornell Action Research Network, Cornell University. Ithaca, NY: Cornell Action Research Network.

Sellato, B. 2001. *Forest, Resources, and People in Bulunga: Elements for a History of Settlement, Trade, and Social Dynamics in Borneo, 1880–2000.* Bogor, Indonesia: CIFOR.

Seltz, S. 1993. *Pygmées d'Afrique Centrale.* Paris: Editions Peeters/SELAF.

Shipton, P., and M. Goheen. 1992. Understanding African Land Holding: Power, Wealth and Meaning. *Africa* 62: 307–326.

Shiraishi, J. 2001. Babaçu Livre: Conflito entre Legislação Extrativa e Práticas Camponesas. In *Economia do Babaçu: Levantamento Preliminar de Dados,* edited by A. Almeida, J. Shiraishi Neto, and B. Mesquita. Movimento Interestadual das Quebradeiras de Côco Babaçu. São Luís, Brazil: Balaios Typographia.

Shiva, V. 1989. *Staying Alive: Women, Ecology and Development.* London: Zed Books.

Shreshtha, M. L., S. P. Joshi, U. R. Bhuju, D. B. Joshi, and M. Gautam. 1995. *Community Forestry Manual.* Community and Private Forest Division, Department of Forest. Kathmandu, Nepal: HMG Department of Forest.

Sigot, A., L. A. Thrupp, et al. (eds.). 1995. *Towards Common Ground: Gender And Natural Resource Management in Africa.* Nairobi, Kenya: African Centre for Technology Studies; Washington, DC: World Resources Institute.

Sihlonyane, M. F. 2001. The Rhetoric of the Community in Project Management: The Case of Mohlakeng Township. *Development in Practice* 11(1): 34–44.

Simonian, L. T. 1995. Mulheres Seringueiras na Amazônia Brasileira—Uma Vida de Trabalho Silenciado. In *A Mulher Existe? Uma Contribuição ao Estudo da Mulher e Gênero Na Amazônia,* edited by M. L. Álvares and M. D'incao. Belém, Brazil: Múseo Paraense Emílio Goeldi, 97–115.

Sithole, B. 2002. *Where the Power Lies: Multiple Stakeholder Politics over Natural Resources: A Participatory Methods Guide.* Bogor, Indonesia: CIFOR.

———. 2004. New Configurations of Power in Mafungautsi State Forest, Zimbabwe. In *The Myths and Realities of Community Based Natural Resource Management in Southern Africa,* edited by C. Fabricius and E. Koch with H. Magome and S. Turner. London: Earthscan.

Sithole, B., and W. Kozanayi. 2001. Uncovering Shrouded Identities! Institutional Considerations and Differentiation in an Adaptive Co-Management Project in Zimbabwe. Consultants' report submitted to CIFOR ACM Program. Bogor, Indonesia: CIFOR.

Skarwan, D. 2002. *Análisis del Los Factores que Aceleran o Frenan Procesos Locales Comunitarios y el Uso Sostenible de Recursos Naturales.* Red Internacional sobre las Metodologías Participativas y el Desarrollo Rural en Áreas Forestales. Red Ecoregional para América Latina Tropical (REDECO). Cali, Colombia: CIAT (International Center for Tropical Agriculture).

Slocum, R., L. Wicchart, D. Rocheleau, and B. Thomas-Slayter. 1995. *Power, Process and Participation: Tools for Change.* London: ITDG (Intermediate Technology Development Group) Publishing.

Somers, M. R. 1994. The Narrative Constitution of Identity: A Relational and Network Approach. *Theory and Society* 23: 605–649.

Souza, M. 1997. A Participação Política de Homens e Mulheres na Implantação de uma Reserva Ecológica na Amazônia. *Papéis Sociais, Divergências e Convergências.* Monografia (Especialização em Teoria Antropológica). Belém, Brazil: CFCH (Centro de Filosofia e Ciências Humanas), Universidade Federal do Pará.

Stacey, R. D., D. Griffin, et al. 2000. *Complexity and Management: Fad or Radical Challenge to Systems Thinking?* London: Routledge.

Stamp, P. 1989. *Technology, Gender, and Power in Africa.* Ottawa, Canada: International Development Research Centre (IDRC).

Steins, N. A., and V. Edwards. 1999. Platforms for Collective Action in Multiple Use Common-Pool Resources. *Agriculture and Human Values* 16(3): 241–315.

Stocks, A. 1999. Iniciativas Forestales Indígenas en el Trópico Boliviano: Realidades y Opciones. Documento Técnico 78/1999. Santa Cruz, Bolivia: BOLFOR.

Stone, S. 2003. From Tapping to Cutting Trees: Participation and Agency in Two Community-Based Timber Management Projects in Acre, Brazil. Ph. D. thesis, University of Florida, Gainesville.

Sullivan, A. 2000. Decoding Diversity: Strategies to Mitigate Household Stress. Master's thesis, University of Florida, Gainesville.

Sundar, N. 2001. Is Devolution Democratisation? *World Development* 29(12): 2007–2023.

Superintendencia Forestal. 2002a. *De la Accion Transparente a la Reflexión Responsable: Una Invitación al Control Social Sobre el Aprovechamiento Forestal en las TCOs de Bolivia.* Unidad de Coordinación con Pueblos Origenarios Indígenas (UCIPOI). Santa Cruz, Bolivia: Superintendencia Forestal.

———. 2002b. *Informe Anual 2001 Superintendencia Forestal.* Santa Cruz, Bolivia: Superintendencia Forestal.

Tamburini, L. 2000. Nuevo Régimen Forestal y Territorios Indígenas en Bolivia. In *Atlas Territorios Indígenas en Bolivia: Situación de las Tierras Comunitarias de Origen (TCOs) y Proceso de Titulación,* edited by J. A. Martínez. Santa Cruz, Bolivia: CPTI (Planning Centre for Indigenous Territories), 213–228.

Tax, S. 1960. Action Anthropology. In *Documentary History of the Fox Project 1948–1959: A Program in Action Anthropology,* edited by F. Gearing, R. McNetting, and L. R. Peattie. Chicago, IL: University of Chicago Press, 167–171.

Taylor, H. 2002. Insights into participation from critical management and labour process perspectives. In *Participation: The New Tyranny?* edited by B. Cooke and U. Kothari. London: Zed Books, 122–138.

Team, New Era. 2002. *Collaborative Monitoring Process in Sankhuwasabha Site, Nepal.* Bogor, Indonesia: CIFOR.

Thomas-Slayter, B., D. Rocheleau, et al. (eds.). 1995. *Gender, Environment, and Development in Kenya.* Boulder, CO: Lynne Rienner Publishers.

Thorbecke, F. 1913. Eine neue Zwergenrasse. *Dt. Kolonialzeitung* 30 (11): 176–178.

Tiani, A. M., G. Akwah, and J. Nguiébouri. 2001. Les Communautés Riveraines Face au Parc National de Campo-Ma'an: Perceptions, Vécu Quotidien et Participation. Bogor, Indonesia: CIFOR.

Tim Pendampingan Pemetaan Partisipatif. 2000. Pemetaan Partisipatif di Sungai Malinau: Kegiatan dalam Rangka "Pengelolaan Hutan Bersama." Bogor, Indonesia: CIFOR.

Townsend, J. G., with U. Arrevillaga, et al. (eds.). 1995. *Women's Voices from the Rainforest.* London: Routledge.

Trilles, P. 1932. *Les Pygmées de la Forêt Equatoriale.* Münster, Germany: Anthropos Bibliotek.

Tsafack, E. M., and O. P. Ngah Obama. 2001. Etude de la Chasse Villageoise dans la Périphérie Nord du Parc National de Campo-Ma'an: Cas des Villages de Messama I, Messama II et Bindem. Final report. Kribi, Cameroon: Campo-Ma'an Project.

Tumbahangphe, N. 2002. Conceptual Note for Case Study of Bamdibhir CFUG and Daurali Bagadada CFUG. In Participatory Action Research Program on Equity and Poverty in the Management of Common Property Resource in the Hindu-Kush Himalayas. Report submitted to ICIMOD.

Turnbull, C. 1966. Wayward Servants: The Two Worlds of the African Pygmies. Unpublished research report. London.

Uphoff, N. 1986. *Local Institutional Development: An Analytical Source Book with Cases*. West Hartford, CT: Kumarian Press.

Vabi, M., with J. B. Endeley, C. Njebet-Ntamag, and E. P. Menong Memo. 2001. *Intégration Genre dans la Mise en Œuvre des Projets de Développement et de Conservation de la Diversité Biologique. Etat Actuel: Contraintes et Opportunités au Cameroun*. Yaoundé, Cameroon: World Wildlife Fund Cameroon.

Vabi, M., C. Ngwasiri, P. Galega, and P. R. Oyono. 2000. The Devolution of Management Responsibilities to Local Communities: Context and Implementation Hurdles in Cameroon. Working Paper. Yaoundé, Cameroon: World Wildlife Fund Cameroon.

VAIPO (Viceministerio de Asuntos Indígenas y Pueblos Originarios). 1998. *Identificación y Consolidación de Tierras Comunitarias de Origen y Areas Territoriales Indígenas de Bolivia: Informe de Necesidades Espaciales del Peublo Indígena Guarayo*. Santa Cruz, Bolivia: Ministerio de Desarrollo Sostenible.

Vallejos, C. 1998. Ascensión de Guarayos: Indígenas y Madereros. In *Municipios y Gestión Forestal en el Trópico Boliviano*, edited by P. Pacheco and D. Kaimowitz. La Paz, Bolivia: CIFOR, CEDLA, TIERRA, and BOLFOR.

Vallois, H.V. 1947. Missions Anthropologiques en Afrique Française. *L'Anthropologie* 51: 568–572.

———. 1948. Chez les Pygmées du Cameroun. *La Nature* 76: 17–20.

Vallois, H. V., and P. Marquer. 1974. *Les Pygmées Baka du Cameroun: Anthropologie et Ethnographie, avec une Annexe Démographique*. Paris: Museum National d'Histoire Naturelle.

van den Berg, Y., and K. Biesbrouck. 2000. *The Social Dimension of Rainforest Management in Cameroon: Issues for Co-Management*. Kribi, Cameroon: Tropenbos-Cameroon Programme.

van Heist, M. 2000. Participatory Mapping of Village Territories, Malinau, East Kalimantan, January–December 2000: Some Lessons in "Adaptive Use and Management of Geographic Data." CIFOR Report. Bogor, Indonesia: CIFOR.

Van Soest, D. P. 1995. *Tropical Rainforest Degradation in Cameroon*. Groningen, Netherlands: University of Groningen.

Vandergeest, P. 2003. Racialization and Citizenship in Thai Forest Politics. *Society and Natural Resources* 16: 19–37.

Vansina, J. 1990. *Paths in the Rainforests: Toward a History of Political Tradition in Equatorial Africa*. Madison, WI: University of Wisconsin Press.

Velho, O. 1972. *Frentes de Expansão e Estrutura Agrária: Estudo do Processo de Penetração numa Área da Transzamazônica*. Rio de Janeiro, Brazil: Zahar Editores.

Venkateswaran, S. 1995. *Environment, Development and the Gender Gap*. New Delhi: Sage Publications.

Vermuelen, S. 1994. Consumption, Harvesting, and Abundance of Wood along the Boundary between Mafungabusi State Forest and Gokwe Communal Area, Zimbabwe. M. Sc. thesis, Biological Sciences, University of Zimbabwe, Harare.

Wadsworth, Y. 1998. *What Is Participatory Action Research?* Action Research International Paper 2. http://www.scu.edu.au/schools/gcm/ar/ari/p-ywadsworth98.html.

Warner, M. W., R. M. Al-Hassan, and J. C. Kydd. 1996. Beyond Gender Roles? Conceptualizing the Social and Economic Lives of Rural Peoples in Sub-Saharan Africa: Some Evidence from Northern Ghana. Unpublished paper.

Weber, J. 1977. Structures Agraires et Évolution des Milieux Ruraux: Le Cas de la Région Cacaoyère du Centre-Sud Cameroun. *Cahiers ORSTOM* 16(2).

———. 1998. Perspective de Gestion Patrimoniale des Ressources Renouvelables. In *Quelle Politique Foncière en Afrique Rurale?* edited by P. Lavigne Delville. Paris: Karthala-Coopération Française.

Weber, M. 1957. *The Theory of Social and Economic Organisation*. New York: Free Press.

———. 1962. *Basic Concepts in Sociology*. New York: Free Press.

Wenger, E. 1998. *Communities of Practice: Learning, Meaning and Identity. (Learning in Doing: Social, Cognitive and Computational Perspectives)*. Cambridge, UK: Cambridge University Press.

White, S. C. 1996. Depoliticising Development: The Use and Abuse of Participation. *Development and Practice* 6(1).

Wilkie, D. S. 1988. Hunters and Farmers of the African Forest. In *People of the Tropical Rainforest*, edited by J. S. Denslow and C. Padoch. Berkeley and Los Angeles, CA: University of California Press, 111–126.

Wilson, K., and J. E. B. Morren, Jr. 1990. *Systems Approaches for Improvement in Agriculture and Resource Management.* New York: Macmillan.

Winichakul, T. 1994. *Siam Mapped: A History of the GeoBody of a Nation.* Honolulu: University of Hawaii Press.

Winterbottom, R. 1992. Tropical Forestry Action Plans and Indigenous People: The Case of Cameroon. In *Conservation of West and Central African Rainforests,* edited by K. Cleaver, M. Munasinghe, M. Dyson, N. Egli, A. Peuker, F. Wencélius. Washington, DC: World Bank, 222–228.

Wolff, C. S. 1998. Marias, Franciscas e Raimundas: Uma história das Mulheres da Floresta Alto Juruá, Acre, 1870–1945. Ph. D. thesis, University of São Paulo, Brazil.

Wollenberg, E., with D. Edmunds, and L. Buck. 2000. *Anticipating Change: Scenarios as a Tool for Adaptive Forest Management. A Guide.* Bogor, Indonesia: CIFOR.

———. 2001a. *Mengantisipasi Perubahan: Skenario sebagai Sarana Pengelolaan Hutan Secara Adaptif: Suatu Panduan.* (Anticipating Change: Scenarios as a Tool for Adaptive Forest Management: A Guide). Bogor, Indonesia: CIFOR.

Wollenberg, E., D. Edmunds, and J. Anderson. 2001b. Editorial. *International Journal of Agriculture, Resources, Governance and Ecology (IJARGE)* 1(3/4): 193–198. Special issue on accommodating multiple interests in local forest management.

Wollenberg, E., J. Anderson, and D. Edmunds. 2001c. Pluralism and the Less Powerful: Accommodating Multiple Interests in Local Forest Management. *International Journal of Agriculture, Resources, Governance and Ecology (IJARGE)* 1(3/4): 199–222. Special issue on accommodating multiple interests in local forest management.

Wollenberg, E., D. Edmunds, L. Buck, J. Fox, and S. Broch (eds.). 2001d. *Social Learning in Community Forests.* Bogor, Indonesia: CIFOR.

Wood, C. H. 2002. Introduction: Land Use and Deforestation in the Amazon. In *Deforestation and Land Use in the Amazon,* edited by C. H. Wood and R. Porro. Gainesville, FL: University Press of Florida, 1–38.

Woodhill, J. 2002. Sustainability, Social Learning, and the Democratic Imperative. In *Wheelbarrows Full of Frogs: Social Learning in Rural Resources Management. International Research and Reflections,* edited by C. Leeuwis and R. Pyburn. Amsterdam, Netherlands: Koninklijke van Gorcum, 317–331.

Woodhill, J., and N. Röling. 1998. The Second Wing of the Eagle: The Human Dimension in Learning Our Way to More Sustainable Futures. In *Facilitating Sustainable Agriculture,* edited by N. Röling and M. Wagemakers. Cambridge, UK: Cambridge University Press.

World Bank. 1997. Rural Poverty Alleviation Project, Maranhão. Northeast Rural Poverty Alleviation Program. Staff Appraisal Report, Brazil. Report No. 16758-BR. Washington, DC: World Bank.

———. 2002. http://lnweb18.worldbank.org/ESSD/essdext.nsf/22DocByUnid/.

World Commission on Environment and Development. 1987. *Our Common Future.* London: Oxford University Press.

Wyckoff-Baird, B. 1994. *Women and Water Lilies.* Windhoek, Namibia: WWF (World Wide Fund for Nature).

Index

Books from RFF Press *and*
the Center for International Forestry Research

- *China's Forests: Global Lessons from Market Reforms*
  WILLIAM F. HYDE, BRIAN BELCHER, JINTAO XU, EDITORS
  CLOTH, ISBN 1-891853-67-8 PAPER, ISBN 1-891853-66-X

- *People Managing Forests: The Links between Human Well-Being and Sustainability*
  CAROL J. PIERCE COLFER AND YVONNE BYRON, EDITORS
  CLOTH, ISBN 1-891853-05-8 PAPER, ISBN 1-891853-06-6

- *Which Way Forward? People, Forests, and Policymaking in Indonesia*
  CAROL J. PIERCE COLFER AND IDA AJU PRADNJA RESOSUDARMO, EDITORS
  CLOTH, ISBN 1-891853-44-9 PAPER, ISBN 1-891853-45-7

Other Books *by* RFF Press

- *Choosing Environmental Policy: Comparing Instruments and Outcomes in the United States and Europe*
  WINSTON HARRINGTON, RICHARD D. MORGENSTERN, AND THOMAS STERNER, EDITORS
  CLOTH, ISBN 1-891853-87-2 PAPER, ISBN 1-891853-88-0

- *Climate Change Economics and Policy: An RFF Anthology*
  MICHAEL A. TOMAN, EDITOR
  PAPER, ISBN 1-891853-04-X

- *India and Global Climate Change: Perspectives on Economics and Policy from a Developing Country*
  MICHAEL A. TOMAN, UJJAYANT CHAKRAVORTY, AND SHREEKANT GUPTA, EDITORS
  CLOTH, ISBN 1-891853-61-9

- *The Measurement of Environmental and Resource Values: Theory and Methods, Second Edition*
  MYRICK FREEMAN III
  CLOTH, ISBN 1-891853-63-5 PAPER, ISBN 1-891853-62-7

- *National Environmental Accounting: Bridging the Gap between Ecology and Economy*
  JOY E. HECHT
  CLOTH, ISBN 1-891853-93-7 PAPER, ISBN 1-891853-94-5

- *Policy Instruments for Environmental and Natural Resource Management*
  THOMAS STERNER
  CLOTH, ISBN 1-891853-13-9 PAPER, ISBN 1-891853-12-0

- *Private Rights in Public Resources: Equity and Property Allocation in Market-Based Environmental Policy*
  LEIGH RAYMOND
  CLOTH, ISBN 1-891853-69-4 PAPER, ISBN 1-891853-68-6

- *The RFF Reader in Environmental and Resource Management*
  WALLACE E. OATES, EDITOR
  PAPER, ISBN 0-915707-96-9

- *Technological Change and the Environment*
  ARNULF GRÜBLER, NEBOJSA NAKICENOVIC, AND WILLIAM D. NORDHAUS, EDITORS
  CLOTH, ISBN 1-891853-46-5

For more information, visit www.rffpress.org